Microscopic Dynamics of Plasmas and Chaos

Series in Plasma Physics

Series Editors:

Steve Cowley, Imperial College, UK
Peter Stott, CEA Cadarache, France
Hans Wilhelmsson, Chalmers University of Technology, Sweden

Other books in the series

Forthcoming titles

Series in Plasma Physics

Microscopic Dynamics of Plasmas and Chaos

Yves Elskens

CNRS—Université de Provence (Marseilles)

Dominique Escande

CNRS—Université de Provence (Marseilles)

Institute of Physics Publishing
Bristol and Philadelphia

British Library Cataloguing-in-Publication Data

A catalogue record for this book is available from the British Library.

ISBN 0 7503 0612 2

Library of Congress Cataloging-in-Publication Data are available

Commissioning Editor: John Navas
Production Editor: Simon Laurenson
Production Control: Sarah Plenty
Cover Design: Victoria Le Billon
Marketing: Nicola Newey and Verity Cooke

Published by Institute of Physics Publishing, wholly owned by The Institute of Physics, London

Institute of Physics Publishing, Dirac House, Temple Back, Bristol BS1 6BE, UK

US Office: Institute of Physics Publishing, The Public Ledger Building, Suite 929, 150 South Independence Mall West, Philadelphia, PA 19106, USA

Typeset in LaTeX 2_ε by Text 2 Text, Torquay, Devon
Printed in the UK by MPG Books Ltd, Bodmin, Cornwall

I know the tendency of the human mind to do anything rather than think. None of us expect to succeed without labour, and we all know that to learn any science requires mental labour, and I am sure we would all give a great deal of mental labour to get up our subjects. But mental labour is not thought, and those who have with great labour acquired the habit of application, often find it much easier to get up a formula than to master a principle. I shall endeavour to show you here, what you will find to be the case afterwards, that principles are fertile in results, but the mere results are barren, and that the man who has got up a formula is at the mercy of his memory, while the man who has thought out a principle may keep his mind clear of formulæ, knowing that he could make any number of them when required.

(James Clerk Maxwell,
Inaugural lecture at King's College,
London, October 1860)

Permissions

Figures reprinted from

- *Phys. Plasmas* **4** Bénisti and Escande ©1997 with permission from the American Institute of Physics
- *Phys. Lett.* A **284** Doveil *et al* ©2001 with permission from Elsevier Science
- *Physica* D **62** Elskens and Escande ©1993 with permission from Elsevier Science
- *Nonlinearity* **4** Elskens and Escande ©1991 with permission from the Institute of Physics
- *Phys. Rev.* E **64** Firpo *et al* ©2001 with permission from the American Physical Society
- *Phys. Rev. Lett.* **84** Firpo and Elskens ©2000 with permission from the American Physical Society
- *Transport, Chaos and Plasma Physics* vol 2, Guyomarc'h *et al* ©1996 with permission from World Scientific Publishing Co

Contents

Preface

Etymologically, 'plasma' means 'shapeless' but plasma physics has many aspects. This book aims at providing both students and experts with a description of one of these aspects, the resonant wave–particle interaction. This is done by using classical mechanics only and by focusing on Langmuir waves which correspond to the collective vibration of electrons with respect to the ions of a plasma. These simple waves often have an intricate interaction with electrons moving close to their phase velocity. This interaction then involves strong nonlinear effects like trapping or chaos for the electrons and fluctuating growth or damping for the waves. In order to understand this interaction, fundamental concepts and methods must be introduced. These are concepts of nonlinear dynamics and Hamiltonian chaos, relevant to all of classical mechanics, celestial mechanics, non-dissipative hydrodynamics, condensed matter and other fields of physics. These concepts underlie breakthroughs in physical sciences as well as key results in applied mathematics and engineering.

This book provides an innovative description of collective phenomena in plasmas which has been developed in the last 13 years. Some of these phenomena are made accessible with undergraduate tools though until now they required graduate level tools to be understood. More globally, the new description at last realizes a grand dream of the 19th century: the non-trivial evolution of a macroscopic many-body system is described by taking into account the true character of its chaotic motion. It is striking to note that this occurs in plasma physics, often considered as an application-oriented field of physics and more complicated than other fields due to the long range of the particle interaction. As a result, this book is of interest for scientists far beyond the restriction of plasma physics, while providing stronger foundations to this physics.

Difficulties in the traditional approach

This section reveals the motivations for the approach used in this book for experts, by recalling the difficulties in the classical Vlasovian approach. In the 19th century, describing the evolution of a gas or a fluid as a mechanical N-body problem was so formidable a task that it was put aside, and statistical approaches were elaborated. When plasma physics appeared, these approaches were already well developed in other fields of physics and their plasma counterpart was naturally constructed without attempting a classical mechanics description. Starting from the Liouville equation for the probability distribution of N-body plasmas, the Bogoliubov–Born–Greene–Kirkwood–Yuon (BBGKY) hierarchy enabled the Vlasov(–Poisson) equation which describes the one-particle distribution function of a collisionless plasma to be derived. Most of the microscopic description of such plasmas has been derived from this equation which has proved to be extremely fruitful.

Wave–particle interactions are fundamental processes characteristic of plasmas. Their prototypical signature is the evolution of Langmuir waves which is correctly described by a one-species fluid model, in one space dimension. In the linear theory of these waves, the Vlasov kinetic equation enables one to derive the celebrated Landau damping of a Langmuir wave by resonant particles. The rigorous derivation of the Landau effect involves a search for poles and analytic continuation over complex half-planes for the Fourier–Laplace transformed, linearized Vlasov–Poisson equations. Growth and decay rates have the same formal expression though instability (Landau growth) corresponds to exponentially growing eigenmodes while, mysteriously, decay appears only as a time-asymptotic effect obtained through an integral representation and analytical continuation. A proper mathematical description (not just a rate calculation) of damping actually involves a continuum of singular (beam-type) velocity distributions as discussed by van Kampen (1955, 1957) and Case (1959), and shows this damping to be the result of the phase mixing among a continuum of modes with the same wavenumber and a broad frequency spectrum—which questions the meaning of a dispersion relation for Langmuir waves.

However, even in one space dimension, both Landau and van Kampen–Case approaches fail to unveil the intuitive physical nature of the Landau effect (what happens to the particles?) and, in particular, a simple reason why damping is not described by an eigenmode. This is somewhat disappointing to the student who has already raided through the new continents of Liouville, BBGKY and Vlasov equations, Fourier–Laplace transforms, pole tracking and analytical continuation! To fill in this gap, textbooks introduce heuristic models with test particles and one wave or make the analogy with a surfer's motion. The picture of energy balance between particles slower and faster than the wave may first satisfy the student who sees an analogy with the absorption and stimulated emission of light by atoms but thinking it over she/he may wonder how just a single wave may account for the van Kampen–Case broad spectra. She/He may also wonder how the trapping feature of the surfer's dynamics may be compatible with the linear character of Landau's theory. It is worth recalling that even for specialists, the reality of collisionless Landau damping was fully recognized only after its experimental observation (Malmberg and Wharton 1964), almost two decades after its prediction (Landau 1946). Finally, when trying to make a complete analogy with the interaction of atoms and light, the student finds that the computation of the spontaneous emission of Langmuir waves by particles still requires another tool, adding an individual test particle to a (continuous) Vlasovian plasma.

Turning to the description of nonlinear wave–particle interactions, the Vlasovian approach has until now often been deceptive for Langmuir waves. The simplest problem which a student encounters in this frame is the one-dimensional cold beam–plasma interaction. As yet it has not been dealt with by the Vlasov equation but by a specific approach in which the velocity distribution of electrons is cut into two parts: bulk particles for which one keeps a Vlasovian description and whose collective oscillations are Langmuir waves; and N_t tail (beam) particles which keep their status of individual particles. The interaction of the most unstable Langmuir mode with tail particles is described by a finite set of so-called self-consistent equations coupling N_t particles to one wave. The student may wonder why we first go through Liouville, BBGKY and Vlasov before dealing with this problem when a fluid description of the bulk gives the same result. She/He may also wonder why this finite set of equations is not derived directly from a mechanical N-body description of the plasma, since real plasmas are sets of finite numbers of particles.

A more involved issue is the one-dimensional nonlinear interaction of a warm beam with a plasma. For this problem, the simplest one in kinetic plasma turbulence, the Vlasovian approach may be used to derive the so-called quasilinear equations, but their derivation (which until now rested on the Vlasov equation) has been debated for the past two decades for the strongly nonlinear regime.

Benefits of modern tools

The approach presented in this book is a direct consequence of this long-lasting controversy about quasilinear equations. Indeed this controversy appeared in the 1980s in parallel with the development of nonlinear dynamics and chaos which involved, in particular, plasma physics. An intuitive description of chaos and new tools were introduced especially for finite-dimensional dynamics. This suggested that a finite-number-of-degrees-of-freedom approach of the one-dimensional warm beam–plasma problem could benefit from these description and tools. The basic mechanical model for such an approach was available as a generalization to M resonant waves of the self-consistent equations introduced for the cold beam–plasma interaction. A first step was to recover the traditional Landau–van Kampen–Case theory from this dynamical system, with the additional benefit of unveiling their intuitive mechanical content. Then it was natural to avoid the diversion through the statistical description of plasmas to obtain this system coupling $N_t + M$ Hamiltonian degrees of freedom, and it was derived from an N-body model of the plasma. On this basis the classical problem of the interaction of a warm beam with a unique Langmuir wave was revisited through a Gibbsian description of the plasma. The construction of a theory for the chaotic diffusion in a prescribed wave field was the prelude to that of the warm beam–plasma instability, which proved the quasilinear equations to be relevant in the strongly nonlinear regime as well.

The mechanical approach provides a new paradigm for the foundation of microscopic plasma physics. It makes the physics intuitive and brings in tools to deal with the chaotic regime of self-consistent wave–particle interaction. In particular, it provides an instance where the ubiquitous quasilinear approximation can be justified in a strongly nonlinear regime. Furthermore, the higher Fourier components of the Coulomb interaction turn out to be inessential in the theory, which shows its applicability to a wider class of dynamics than just plasmas. The present approach has indeed been generalized to other wave–particle systems like travelling-wave tubes and free electron lasers. It also captures the essential features of the interaction of vortices with finite-velocity flow in hydrodynamics.

With additional terms for particle sources, particle diffusion and mode relaxation, a numerical code based on the self-consistent equations was used to study the bump-on-tail instability and the interaction of α-particles with Alfvén waves in conditions typical of energetic particles in tokamaks.

Structure of the book

This book aims at a pedagogical presentation of these theoretical developments with the double concern of understanding and of computing. It uses classical mechanics and introduces statistical averages only when they are necessary to obtain analytical results. The outcome for the neophyte reader should be both to master the physical aspects of plasma dynamics and to put its modern tools on strong intuitive foundations.

The first four chapters involve almost no nonlinear dynamics. Chapter 1 presents the basic N-body model and wave–particle resonance. Chapter 2 shows how this model leads to considering the self-consistent dynamics of N_t tail particles with M Langmuir modes or collective vibrations of the plasma bulk, with $N_t + M \ll N$. This dynamics is described by the so-called self-consistent Hamiltonian where the waves appear as harmonic oscillators. Going from the N-body description to field–particle interaction is of fundamental importance for the remainder of the book. An intuitive derivation of the self-consistent Hamiltonian is provided to the reader who would like to avoid entering the full derivation.

Then chapter 3 derives those results of the Landau–van Kampen–Case theory that can be recovered by considering a single mechanical realization of the plasma: the Landau instability, the van Kampen–Case modes and Landau damping in the infinite-N limit. In chapter 4, a statistical average of the mechanical equations shows that a stable Langmuir wave relaxes exponentially to a finite thermal level with the Landau damping rate. The Vlasovian limit turns out to be singular, as Landau damping is recovered in the limit of an infinite number of particles, where the thermal level vanishes. These two chapters show why the

Hamiltonian character of the dynamics forbids Landau damping to correspond to an eigenmode and that even the Landau instability involves, backstage, analogues of van Kampen–Case eigenmodes.

The mechanical approach unifies the Landau theory, the van Kampen–Case theory, their simple physical description, the theory of spontaneous emission and the theory of cold beam–plasma interaction. This approach provides, in particular, a classical description resembling that of the atom–light interaction with a quantized electromagnetic field, which altogether yields absorption and both spontaneous and stimulated emissions. The mathematical tools we use are no more elaborate than Fourier series and the model is explicitly solvable in spite of its high dimensionality. As the derivation stays close to the mechanical intuition at each step, the fate of particles can be monitored. In particular, wave–particle resonance occurs in the weak sense of a frequency broadening due to exponential growth or damping and not in the strong sense of a trapping in the wave potential. Particles are shown to be acted upon by a force which tends to synchronize them with the wave. At this point, the previous mysteries of the Vlasovian approach are explained by unveiling the mechanical roots of the kinetic theory of Langmuir waves. However, for practical matters, our approach does not supersede Landau's efficient pole computation and the Vlasovian approach stands unchallenged for computational purposes in the linear regime. The dialectics between both approaches may prove useful, since images emerging from this work may reinforce the beginner's trust in arcane contour computations.

From chapter 5 on, this book deals with nonlinear dynamics in a self-contained way. Chapter 5 introduces some basic tools of Hamiltonian chaos needed to understand the chaotic regime of wave–particle interactions. Chapter 6 extends this introduction to the case of chaotic diffusion in a prescribed set of waves, with both its physical explanation and its analytical computation. In fact this dynamics is a good approximation to that occurring in many Hamiltonian systems. The concepts and techniques described in this chapter apply to issues as different as the chaos of magnetic field lines, the heating of particles by cyclotronic waves, and chaos of rays in geometrical optics. Chapter 7 shows how these ideas can be used to solve the warm beam–plasma instability and to hopefully close the previously mentioned quasilinear controversy; the theory of this chapter and the related part of the preceding one were found while this book was being drafted. The saturation of the warm beam–plasma instability is described through kinetic equations derived by taking into account the true chaotic character of the particle dynamics. This is made possible by the fact that particles interact by means of waves, and vice versa, a property which is absent in gas dynamics, which makes the derivation of the Boltzmann equation very problematic.

The final chapters focus on the long-time evolution of the wave–particle system, with further discussion of finite-N effects. Chapter 8 considers the more coherent dynamics of a single Langmuir wave with a cold or warm beam from a numerical point of view. Chapter 9 deals analytically with this problem through a

Gibbsian approach and is definitely beyond the undergraduate level. In the case of a negative slope for the distribution function, this approach enables us to provide evidence of a second-order phase transition in the dynamics, separating Landau damping from damping with trapping.

Some specialized or advanced questions and tools are treated in appendices. The Fokker–Planck equation is discussed in section E.3 and the Vlasov equation in appendix G. The similarity between these two equations should not conceal their different physical status: the Fokker–Planck equation relates to the stochastic motion of a particle, whereas the Vlasov equation relates to the deterministic evolution of $N \to \infty$ particles. Their different status is akin to the differences between a central limit theorem (Gaussian laws for fluctuations) and a law of large numbers (holding with probability 1).

Prerequisites and remarks

The contents of this book stand mainly under the plasma physics heading. However, chapters 5 and 6 provide general tools for Hamiltonian chaos, and the subsequent chapters may be viewed as illustrating the application of chaotic dynamics to high-dimensional systems. Chapter 3 is a unique example of the explicit solution to a high-dimensional Floquet problem. This chapter and appendix G may be viewed as a short introduction to the Vlasov equation, i.e. to kinetic theory as the regularized limit $N \to \infty$ thanks to the mean-field nature of the plasma dynamics—without introducing the Liouville equation and BBGKY hierarchy.

Three-quarters of this book can be read by students trained in Newtonian mechanics and elementary calculus. The technique most commonly used through this text is perturbation theory, which is deeply rooted in the scientific method. More advanced parts are indicated by an asterisk; the power of the Lagrangian or Hamiltonian formalism, the Laplace transform or Gibbs ensembles is only used there. In order to alleviate the reading, more technical aspects are set in appendices. While a detailed account of the extremely vast body of literature is far beyond the scope of our introductory text, we have tried to provide the reader with appropriate references for further study.

As our approach is quite direct, we indicate more traditional approaches in specific sections, where we stress their main points and refer to the relevant literature. Also, as we aim to argue convincingly with basic theoretical concepts, we leave aside the discussion of experimental works: plasma sensitivity to many excitations and the three-dimensional nature of our space plague experiments with many perturbations to the desired signals. However, nobody doubts that the microscopic description of a plasma is (quite reasonably) the mechanics of Coulomb-interacting particles, and one-dimensional models are common practice.

This textbook is intended for a one-semester course in plasma physics or in modern classical dynamics, in connection with statistical physics. Exercises provide the reader with technical and conceptual training: their level ranges from elementary questions to research projects. We remind the reader that a question may be deemed trivial only after a full, clear answer has been plainly cast in writing.

This book is the outcome of a collaboration between two quite different authors: one more interested in mathematics and details; and the other one more interested in physics and outlines. These opposite points of view generated a stimulating dialogue during the writing. We hope that the outcome of our work provides both global qualitative features and rigorous results. 'To show' and 'to prove' have somewhat elastic definitions in the physics community. Inspired guesses are of dramatic importance in the development of physics, while rigorous results are beacons in the ocean of scientific knowledge.

We thank Professor Hans Wilhelmsson for inviting us to write this book and for his work as a referee. We are indebted to Dr André Samain for his critical reading of the manuscript and for comments leading to important improvements. We thank Drs Didier Bénisti and Marie-Christine Firpo for commenting on large parts of the book; Didier Bénisti also provided us with many unpublished figures from his thesis. Professors Thomas M O'Neil, Patrick H Diamond and Marshall N Rosenbluth made useful comments to one of us (DE). The core of this work results from our collaborations and discussions with many colleagues, especially of Equipe Turbulence Plasma, agreeing or debating, whom we thank for these numerous fruitful exchanges.

Part of this book evolved from lectures to students at the universités d'Aix-Marseille (in plasma physics and in mathematical physics and modelling) and università degli studi di Roma la Sapienza (physics), whose reactions were welcome.

Last but not least, notre travail n'aurait pu être conduit sans le soutien et la patience de nos épouses Solange et Montserrat. Ce livre leur doit aussi d'exister.

Yves Elskens, Dominique Escande
Marseilles, June 2002

Chapter 1

Basic physical setting

This chapter defines the physical problem which is the central topic of this book, i.e. the interaction of particles with Langmuir waves, described by applying classical mechanics to the one-dimensional N-body description of the plasma. In section 1.1 we formulate the plasma model, which will be analysed in chapter 2 in terms of the wave–particle interaction. In section 1.2 we review the physical phenomenon central to the understanding of the particle motion, namely the existence of a resonance in this interaction. This will help in understanding why resonant particles must have a special status.

1.1 The original N-body system

1.1.1 Physical context

A plasma is an essentially neutral mixture of electrons and ions, for instance an ionized gas. The particles interact through long-range Coulomb forces. They have a tendency to shield each other which makes the electric potential produced in the plasma by a test particle fade away, like a Yukawa potential $(e^2/4\pi\epsilon_0 r)\exp(-r/\lambda_D)$ in a thermal plasma where

$$\lambda_D = \sqrt{\frac{\epsilon_0 k_B T}{\mathcal{N} e^2}} \tag{1.1}$$

is the Debye length, with T the plasma temperature, $\mathcal{N}$ its density, k_B the Boltzmann constant, e the electron charge (assuming singly charged ions) and ϵ_0 the vacuum permittivity. If the thermal plasma is contained in a vessel, the higher velocity of electrons makes them escape faster than ions if no confining potential is present. This leads to the electrostatic self-organization of the plasma which consists in the creation of electrostatic sheaths at the plasma boundary with a width of a few Debye lengths which confine the electrons. Both the shielding or screening and the sheath properties lead to requiring the size of the neutral mixture of electrons and ions to be much larger than λ_D for it to be considered as a true plasma.

Plasmas are highly complex media, especially when they are magnetized, and they exhibit both granular and collective aspects. A model which captures all this complexity is amenable neither to analytic treatment nor to physical intuition. Therefore plasmas must be described through a series of models which capture some aspects of this complexity and provide tools for dealing with more complex settings. Here we focus on one-component plasmas with a uniform ionic density (ions with infinite mass) and we concentrate on the motion of electrons due to a modulation of their density about the uniform neutralizing one. We only consider the electrostatic aspect of their motion, which means that their velocity is assumed to be much smaller than that of light, and to be parallel to a possible uniform magnetic field.

If a small electron density modulation is initially imposed with a scale larger than λ_D, we will see that the plasma responds collectively through a series of travelling sinusoidal potentials which are called Langmuir waves. At a fixed position each wave is seen to have a pulsation close to

$$\omega_p = \sqrt{\frac{\mathcal{N}e^2}{\epsilon_0 m_e}}, \tag{1.2}$$

which is traditionally called plasma frequency, where m_e is the electron mass. The product $\omega_p \lambda_D = \sqrt{\frac{k_B T}{m_e}}$ turns out to be the electron thermal velocity v_T. The scales λ_D and ω_p will show up naturally in our approach of Langmuir waves. These waves mainly involve electrons because of the higher ion inertia.

1.1.2 The plasma model

A further simplification of the plasma dynamics is its restriction to a single space dimension. This makes sense for the propagation of Langmuir waves along a magnetized plasma column with a large cross section. In an infinite or periodic one-dimensional plasma the electrons are self-confining, and their dynamics may be studied by accounting only for their mutual interactions. If the ions are massive enough, the plasma dynamics may be described without taking into account the ions either for their action on electrons or for their own dynamics. The last two simplifications rule out interesting plasma physics like the coupling of Langmuir waves with ion acoustic waves or their self-modulation at high intensity. A further defect of the one-dimensional description is the fact that Coulomb collisions are weaker in one space dimension than in three: two particles can cross and the force between them just changes sign at the moment of the crossing.

The plasma is described as a periodic one-dimensional mechanical system of N particles with same mass surface density m and same charge surface density q_* in Coulombian interaction, in a domain of length $L \gg \lambda_D$ with periodic boundary conditions. In three dimensions this corresponds to N parallel charged planes per spatial period L, each with a ratio of mass-to-charge densities m/q_*, moving in their common perpendicular direction. The energy surface density of the system is

$$H = K + V \tag{1.3}$$

with kinetic part

$$K = \sum_{r=1}^{N} \frac{p_r^2}{2m} \tag{1.4}$$

and potential part

$$V = \frac{1}{2N} \sum_{n=1}^{\infty} \sum_{l,r=1}^{N} q_*^2 V_n \cos k_n (x_r - x_l) \tag{1.5}$$

where x_r is the position of particle r and $p_r = m\dot{x}_r$. Boundary conditions are periodic on the interval of length L. The prefactor $1/N$ in the potential term avoids divergences that would occur in the plasma (mean-field) limit where N is large. This factor keeps the plasma frequency constant while $N \rightarrow \infty$.

V_n is the nth Fourier coefficient of the interparticle potential associated with wavevector $k_n = 2\pi n/L$. In the case of plasmas considered here, $V_n = n^{-2}\pi^{-2}L/\epsilon_0 = 4/(L\epsilon_0 k_n^2)$ for n running over odd positive integers and $V_n = 0$ for n even, defining the Coulomb plasma without a neutralizing background of Lenard (1961), Prager (1962) and others (see, e.g., the review by Choquard *et al* 1981).

Exercise 1.1.

(i) What are the dimensions of H, K, V, m, q_*, p_r, x_r, k_n and V_n, in the dimensional basis $[L, T, M, Q]$?
(ii) Plot the functions $V(x) = \sum_{n=1}^{\infty} V_n \cos k_n x$, $E(x) = -dV(x)/dx$ and $\epsilon_0 \rho(x) = dE(x)/dx$ corresponding to the contribution of a single particle to the potential, field and density.
(iii) Plot the analogous functions for the sums extending to all $n > 0$, even and odd.

Expressions (1.3)–(1.5) yield a compact way of defining the dynamics. Indeed the second-order equations of motion of x_r, $\ddot{x}_r = -\partial V/\partial x_r$, are generated by

$$\dot{x}_r = \partial H/\partial p_r = p_r/m \tag{1.6}$$

$$\dot{p}_r = -\partial H/\partial x_r = \frac{q_*}{N} \sum_{n=1}^{\infty} \sum_{l=1}^{N} q_* k_n V_n \sin k_n (x_r - x_l). \tag{1.7}$$

Then H is called the Hamiltonian of the system and p_r is called the conjugate momentum to x_r. Equation (1.7) may be equivalently rewritten

$$\dot{p}_r = \frac{q_*^2}{2N\epsilon_0} \sum_{l=1}^{N} \sigma \left(2\pi \frac{x_r - x_l}{L} \right) \qquad (1.8)$$

where $\sigma(y) = \text{sign}(\sin y)$ is the 2π-periodic sign function with values $-1, 0, 1$. This equation is more conveniently expressed using the quantities

$$E_n = \frac{q_* k_n V_n}{2iN} \sum_{r=1}^{N} e^{-ik_n x_r}. \qquad (1.9)$$

E_n characterizes the dynamics occurring on the spatial scale L/n. In terms of these quantities, the equations of motion become

$$\ddot{x}_r = \frac{q_*}{m} \sum_{n=-\infty}^{+\infty} E_n \exp(ik_n x_r) \qquad (1.10)$$

where $k_{-n} = -k_n$, $V_{-n} = V_n$, $V_0 = 0$ and E_n depends on time through (1.9). The N-body motion thus reduces to a collection of single-particle problems subjected to the self-consistent fields E_n with $n = \ldots, -2, -1, 1, 2, \ldots$ which may be interpreted as the Fourier components of the electric field due to all the particles. The time evolution of these Fourier components, which follows from (1.9) and (1.10), is not autonomous, as it results from the motion of all particles: they are not extra degrees of freedom. This electric field is a convenient object to characterize the action of $N - 1$ particles on one of them. However the field–particle dynamics is intrinsically self-consistent, since the field is created by the particles.

The periodic boundary conditions and the need for global neutrality of the plasma lead to a technical peculiarity of the model:

Exercise 1.2. Show that the contribution of a particle at x_r with charge q_* to the total field is the same as the contribution of a particle with charge $q'_* = -q_*$ at $x'_r \equiv x_r + L/2 \bmod L$, which we may call its 'ghost'. Show that $E(x + L/2) = -E(x)$ for any x. Show that the position of the ghost obeys the equation of motion $m\ddot{x}'_r = q'_* E(x'_r)$.

For all practical purposes, the plasma model behaves as if there were two species in the system, with a total number $N_{\text{tot}} = N_{q_*} + N_{-q_*} = 2N$ over the interval of length L. Or, in other words, there are N particles (some being electrons and the others being ghosts) in half the interval, with length $L/2$. With respect to the field, the ghosts are just as important as the electrons, because the field jumps at their location just as it does (in absolute value) at each electron location. As the acceleration of a particle changes suddenly wherever the field jumps, the positions of the ghosts are clearly important for understanding the dynamics of the plasma.

To formulate a model without ghosts (and restricted to a single species of particles), one must allow for a uniform neutralizing background. For this 'jellium' or one-component plasma, $V_n = 2/(L\epsilon_0 k_n^2)$ for $n \neq 0$. Then in (1.8) one replaces $\sigma(y)$ by $\sigma(y) - y/\pi$.

To conclude this section we note that one-dimensional models are very fruitful in physics, both for elementary and advanced studies. Many from various branches of physics are reviewed by Mattis (1993).

1.2 Wave–particle resonance: a paradigm of classical mechanics

This book focuses on the dynamics of electrons due to a modulation of their density about the uniform ion density. We have already introduced a central family of actors in this dynamics: the Langmuir waves which are collective motions of the particles. For N large, an individual particle has little action on the wave, and one may consider it as a test particle subjected to it. We now consider this test-particle dynamics, and we consider the motion of one particle in the presence of a static electrostatic potential which oscillates sinusoidally in the direction of a coordinate[1] q (the analysis is performed in the rest frame of a Langmuir wave). With an appropriate choice of units and origin for q, the equation of motion of an electron in this potential reads as

$$\ddot{q} = -A \sin q \tag{1.11}$$

with $A > 0$. This equation is the same as the one describing the motion of a pendulum with length l in a gravity field with acceleration g, when $A = g/l$, q is the angle from the vertical and $q = 0$ is the position of the stable equilibrium. The sum of the kinetic energy $\frac{1}{2}p^2$ and of the potential energy $-A\cos q$ corresponding to the equation of motion is the total energy

$$H(p, q) = \frac{1}{2}p^2 - A\cos q \tag{1.12}$$

where p is the linear momentum of the particle or the angular momentum of the pendulum (the mass of the electron m or the moment of inertia of the pendulum ml^2 being set equal to 1).

Exercise 1.3. Consider a particle with mass m and charge $Q > 0$ in the field of a travelling wave $E(x, t) = E_0 \sin(\theta_0 + \omega t - kx)$. Denoting the trajectory of the particle by $x_r(t)$, show that a Galileo transformation $y_r = x_r - (\omega/k)t$ and rescalings bring the particle equation of motion to the form (1.11). Express the

---

[1] We follow classical notations in mechanics, denoting coordinates by q and momenta by p. No confusion should arise with the charge q_* (which will not appear beyond chapter 2).

conclusions of the current section in terms of original variables. What changes if $Q < 0$?

The motion of the particle is a line in (p, q) space, termed the one-particle phase space. If the boundary conditions for q are periodic (as for the pendulum), this phase space is (equivalent to) the cylinder based on the circle of length $\Delta q = 2\pi$. If the boundary conditions are not periodic, i.e. if the potential is defined in infinite space, the phase space is just the plane. In either case, the phase space is two dimensional and the dynamics has one Hamiltonian degree of freedom.

Again H is the Hamiltonian of the system. Indeed the equations of motion are generated by

$$\dot{q} = \frac{\partial H}{\partial p} = p \qquad (1.13)$$

$$\dot{p} = -\frac{\partial H}{\partial q} = -A \sin q. \qquad (1.14)$$

As H does not depend on time, it is easy to check that the orbit corresponds to $H = $ constant in phase space (q, p). Here, the orbit may be computed by quadrature from the equations of motion by expressing p as a function $p_1(q)$ for a given energy and initial condition (q_0, p_0). The dynamics is said to be integrable. For the pendulum, which has a single degree of freedom, integrability stems from the existence of one constant of the motion: the energy. For more general Hamiltonians, integrability occurs when the number of commuting independent constants of the motion is equal to the number of Hamiltonian degrees of freedom (see Whittaker (1964) or Lichtenberg and Lieberman (1983) for a more precise definition of integrability). Equation (1.11) can be recovered from equations (1.13)–(1.14) by elimination of p.

As shown in figure 1.1, the nature of the motion depends on the value of the energy.

- For $H > A$, $p_1(q)$ never vanishes but oscillates along q with a definite sign; a given value of H corresponds to two opposite values of $p_1(q)$. This corresponds to an unbounded motion: the rotation of the pendulum and the passing motion for the electron.
- For $-A < H < A$, p vanishes for two values of q; $H = $ constant defines an ellipse-like curve in phase space ($p_1(q)$ is two valued) and the motion is periodic in q. This corresponds to a bounded motion: the libration for the pendulum and the trapped or bouncing motion for the electron in one of the (periodically repeated) potential wells.
- For $H = -A$, $p = 0$ and $q = 0 \bmod 2\pi$ for all times: this is the stable equilibrium point O of the pendulum and the family of stable equilibrium points O_i of the electron. Close to this point, $\cos q$ may be approximated by its expansion to second order in q. This yields the harmonic oscillator

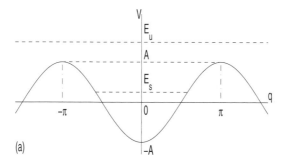

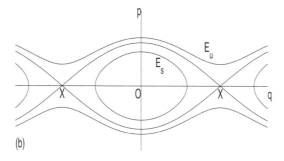

Figure 1.1. (*a*) Potential and (*b*) phase portrait of the pendulum Hamiltonian (1.12).

Hamiltonian with a bouncing or trapping pulsation

$$\omega_b = \sqrt{A}. \tag{1.15}$$

Therefore close to O the motion is almost harmonic and the orbit is like an ellipse. This motivates calling O an O-point.

- For $H = A$ and $p = 0$, $q = \pi \bmod 2\pi$ for all times: this is the unstable equilibrium point X of the pendulum, and the family of unstable equilibrium points X_i of the electron. Close to this point the orbits are branches of hyperbolae and the motion typically diverges from X. The rate at which they diverge is called the Lyapunov exponent of X

$$\lambda_X = \sqrt{A}. \tag{1.16}$$

- In phase space the set of points defined by $H(p, q) = A$ corresponds to two sinusoids ($p_1(q) = \pm 2A^{1/2} \cos(q/2)$) going through the X_i. This set is called the separatrix and corresponds to points of orbits asymptotic to the X_i for $t \to \pm\infty$. Close to the X_i the separatrix has the shape of an X, which motivates the name X-points for these points. It separates the domains of bounded and unbounded motions. It may be viewed as the set of bounded orbits with infinite period or as the set of unbounded orbits with vanishing mean velocity.

The set of bounded motion vanishes with A: one then recovers the free particle motion. Giving a finite value to A is like tearing the phase space of free motion considered as a rubber film (this yields the name tearing mode for the modes generating magnetic islands in configurations for magnetic confinement of plasmas). The free orbits are distorted and become the passing orbits. Inside the cut a new set of orbits is present: the trapped ones. The corresponding domain in phase space takes on the shape of a cat's eye and has a half-width

$$\Delta p_R = 2\sqrt{A} \tag{1.17}$$

termed the resonance or wave-trapping width. All the electrons with a bounded orbit have the same average velocity as the wave. The finite size of the separatrix defines a set of such orbits with positive measure (area). We will say that the corresponding electrons are in resonance with the wave. As shown later, the wave–particle resonance can be identified even if there is more than one wave and if the self-consistency of wave–particle interaction is taken into account. A similar resonant structure in phase space (in particular, width in p proportional to $A^{1/2}$) may be identified for many dynamics defined by a Hamiltonian similar to (1.12) with a part $H_0(p)$ which is nonlinear in p plus a term sinusoidal in q of amplitude $V(p)$. Therefore the wave–particle resonance is the paradigm of nonlinear resonances in classical mechanics (Chirikov 1979, Escande 1985). In this book the word *resonance* is also used for brevity to qualify this resonant term or the set of trapped orbits.

Far from resonance, the orbit is close to the free motion ($A = 0$). This can be stated more precisely by calculating the orbit to first order in A. We consider an orbit $(q(t), p(t))$ at (q_0, p_0) at $t = 0$. At zeroth order in A we get the free (or ballistic) motion $q^{(0)}(t) = q_0 + p_0 t$. At first order in A, with initial condition $(q^{(1)}(0), \dot{q}^{(1)}(0)) = (0, 0)$, we get the contribution

$$q^{(1)}(t) = \frac{A}{p_0^2}(\sin q^{(0)}(t) - \sin q_0 - p_0 t \cos q_0) \tag{1.18}$$

to $q(t)$. The orbit will be close to the free orbit if $|q^{(1)}(t)|$ remains small with respect to the wavelength 2π of the potential and if $|\dot{q}^{(1)}(t)|$ remains small with respect to the unperturbed velocity p_0.

The second condition, $|\dot{q}^{(1)}(t)| \ll |p_0|$, implies $|p_0| \gg \sqrt{A}$ which means that the orbit is far from the separatrix. As a consequence we find that there is a *locality* of the action of a resonance in phase space: it is only important for orbits whose velocity is close to that of the resonance (less than or of the order of the resonance half-width Δp_R).

The first condition, $|q^{(1)}(t)| \ll 2\pi$, cannot be satisfied over long times, because the last term in (1.18) generally grows linearly in time. This secular behaviour follows from the fact that the period of the orbit starting from (q_0, p_0) for $A = 0$ is generally not the period of the orbit starting from the same point for $A > 0$. However, one can avoid secularity by choosing a reference orbit $Q^{(0)}(t)$ with the same period as $(q(t), p(t))$.

Exercise 1.4. Express the period $T(E) = \int_{-\pi}^{\pi} (2E + 2A \cos q)^{-1/2} dq$ as a power series of A/E for $E > A$. Plot it as a function of E for fixed A and as a function of A for fixed E.

Exercise 1.5. For a given initial condition (q_0, p_0) and $0 < A \ll p_0^2/2$, propose a 'guiding-centre' reference orbit $Q^{(0)}(t)$ and a first-order correction $Q^{(1)}(t)$, such that $|Q^{(0)}(t) + Q^{(1)}(t) - q(t)|$ and $|\dot{Q}^{(0)}(t) + \dot{Q}^{(1)}(t) - p_0|$ remain $O(A^2)$ for all times. Hint: use exercise 1.4.

Exercise 1.6. Plot the potential and force generated by truncations $V(x) = \sum_{n=1}^{M} q_*^2 V_n \cos k_n x$ for various M and for $M \to \infty$, with the Coulomb values of V_n. Draw trajectories of Hamiltonian $H(x, p) = p^2/2 + V(x)$ in phase space (x, p) and in configuration space (x, t).

Chapter 2

From N-body dynamics to wave–particle interaction

This chapter shows that N electrostatically coupled particles are endowed with a collective vibrational motion which can be analysed in terms of Langmuir waves. Particles far from being resonant with these waves have a trivial oscillatory motion, and their set plays the role of a mere dielectric medium supporting their propagation. This is not the case for particles close to being resonant with the waves, and this chapter derives a Hamiltonian describing the self-consistent dynamics of these particles with the waves which appear as harmonic oscillators. Basically our approach consists in splitting the particle velocity distribution function into a non-resonant bulk and resonant tails (figure 2.1). Section 2.1 gives an intuitive derivation of this Hamiltonian to enable the reader to study the next chapters without entering the details of its full derivation from the N-body dynamics.

Section 2.2 introduces Langmuir waves as elementary vibrations of the plasma. We search for waves propagating quite freely, accompanied by a non-zero electric field. In section 2.2 we construct the corresponding particle motions, in the case where no particle is resonant with the waves, and we discuss the wave propagation characteristics. We show that the N-body plasma dynamics incorporates a $2M$-wave dynamics ($M \ll N$) and that these waves are independent (i.e. their contributions to the electric field just superpose linearly). Our calculation rests merely on the mechanical equations of motion of the particles and we identify small parameters which motivate our approximations.

In section 2.3 we show how these Langmuir waves interact with a small population of possibly resonant particles in the plasma and we obtain the self-consistent wave–particle equations upon which the next chapters on Langmuir waves rest. This self-consistent system of equations happens to derive from a Hamiltonian and we show in section 2.4 that this Hamiltonian can be obtained consistently within the Lagrangian formulation of mechanics, which yields a more controllable way of approximating the dynamics.

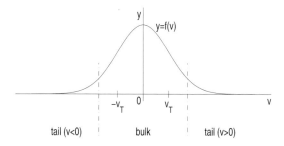

Figure 2.1. Splitting the particle velocity distribution function into a non-resonant bulk and resonant tails.

In the more technical second part of this chapter, we substantiate our approximations. Section 2.5 describes a reference state of the plasma, as simple as possible, similar to rest for a solid or a neutral fluid. As it is not possible to make the electric field vanish exactly, we discuss the thermal equilibrium states. Then in section 2.6 we explicitly check our approximations. The central assumptions are reviewed at the end of section 2.6.

Throughout the chapter, we argue that $N \gg 1$ but we do not rely on the formal limit $N \to \infty$, which is kinetic theory (discussed in appendix G). Indeed we want to establish the dynamics of some actual particles of the plasma and the kinetic limit loses formal track of individual particles. Furthermore, this allows us to keep permanently the intuitive mechanical image of the plasma, which is physically natural.

2.1   Intuitive derivation of the self-consistent Hamiltonian

For the following intuitive derivation of the self-consistent Hamiltonian, we take for granted the existence of Langmuir waves in the non-resonant plasma bulk, as due to the collective vibration of electrons, and we assume that their Bohm–Gross dispersion relation

$$\omega^2 = \omega_p^2 + 3k^2 v_T^2 + \cdots \tag{2.1}$$

is known. A mere extension of Hamiltonian (1.12) to R independent tail particles in the field of M propagating longitudinal waves is

$$H_{\text{nsc}} = \sum_{r=1}^{R} \frac{p_r^2}{2m} - \sum_{r=1}^{R} \sum_{j=1}^{M} W_j \cos(k_j x_r - \omega_{j0} t - \theta_{j0}) \tag{2.2}$$

where k_j and ω_{j0} are related through (2.1), and where W_j is the amplitude (incorporating the particle charge) of the potential of wave j with phase θ_{j0}. If we want to deal with Langmuir waves as mechanical objects, it is natural to consider

them as harmonic oscillators corresponding to the vibrating bulk electrons. The
Hamiltonian for these oscillators is

$$H_{\text{harm}} = \sum_{j=1}^{M} \omega_{j0} I_j \tag{2.3}$$

where the wave action I_j is proportional to its energy. Then the angle conjugate
to I_j evolves like $\theta_j = \omega_{j0} t + \theta_{j0}$. Since the electrostatic energy of a wave is
proportional to the square of its amplitude, there is some constant c_j such that
$W_j = c_j \sqrt{I_j}$. Therefore it is natural to take the full wave–particle self-consistent
Hamiltonian as

$$H_{\text{wp}} = \sum_{r=1}^{R} \frac{p_r^2}{2m} + \sum_{j=1}^{M} \omega_{j0} I_j - \sum_{r=1}^{R} \sum_{j=1}^{M} c_j \sqrt{I_j} \cos(k_j x_r - \theta_j). \tag{2.4}$$

Combining the equations of motion of p_r and I_j shows that the linear momentum

$$P = \sum_{r=1}^{R} p_r + \sum_{j=1}^{M} k_j I_j \tag{2.5}$$

is a constant. Since the first term in the right-hand side is the particles'
contribution to total momentum, the second term is interpreted as the waves'
contribution to momentum (a wave momentum is proportional to W_j^2, as is the
Poynting vector in electrodynamics).

The following exercise provides further insight into the calculations in this
chapter and their final outcome.

Exercise 2.1.

(i) Considering our 'particles in one-dimensional space' as 'charged sheets in
three-dimensional space', let $[q_*] = [QL^{-2}]$ and $[m] = [ML^{-2}]$. Then H
here is an energy per unit surface, $[H] = [MT^{-2}]$. What are the dimensions
of p, P, I_j and c_j?

(ii) Let $E(x) = \sum_{j=-\infty}^{\infty} F_j \exp(\mathrm{i} k_j x)$. In view of (1.5), and recalling that
the electrostatic field energy density (in volume) is $|E(x)|^2 \epsilon_0/2$, note that
the contribution of mode j to the volume density of electrostatic energy will
read $N|F_j|^2 \epsilon_0$. Take proper care of the fact that $E(x)$ is defined locally in
space, while F_j is a mode coefficient (Parseval identities for Fourier series
may be useful).

(iii) Gathering the definition of parameters ε and β_j, and of variable I_j
from (2.107), (2.44) and (2.46), check that in the Hamiltonian (2.110) the
coefficient of the coupling term is just the amplitude of the electrostatic
potential times q_* and the wave contribution $\omega_{j0} I_j$ is just the electrostatic

energy density times L. These two remarks provide a way to compute directly c_j in (2.4); check your result with (2.53).

At this point the reader may either jump to section 2.7 to learn the notation used in the remainder of this book or follow the explicit derivation of the self-consistent Hamiltonian.

2.2 Langmuir waves without resonant particles

We consider a plasma close to a spatially uniform state but whose velocity distribution function may be evolving due to collisions; and we focus on its small amplitude oscillations, with a wavenumber bounded by condition $|n| \leq M$ for some M to be determined ('ultraviolet cut-off'). To this end we split the motion of each particle into two components,

$$x_r = \xi_r + \eta_r \tag{2.6}$$

where ξ_r is a guiding-centre motion to be defined later, and η_r is the small oscillation associated with collective fields. Accordingly we split the electric field E_n defined by (1.9) into a bulk field

$$E_{bn} = \frac{q_* k_n V_n}{2iN} \sum_{r=1}^{N} e^{-ik_n \xi_r} \tag{2.7}$$

corresponding to E_n where x_r is substituted with ξ_r, and a collective part

$$E_{cn} = E_n - E_{bn}. \tag{2.8}$$

2.2.1 Decomposition of the field and particle motion—relevant small parameters

We define η_r from x_r by imposing a zero time-average and by requiring

$$\ddot{\eta}_r = \frac{q_*}{m} \sum_{n=-M}^{M} E_{cn} e^{ik_n x_r}. \tag{2.9}$$

The guiding-centre motion is then governed by

$$\ddot{\xi}_r = \ddot{x}_r - \ddot{\eta}_r = \frac{q_*}{m} \sum_{|n|>M} E_n e^{ik_n x_r} + \frac{q_*}{m} \sum_{|n|\leq M} E_{bn} e^{ik_n x_r} \tag{2.10}$$

where all terms in the right-hand side are small. Indeed, in the first term the field E_n has a large wavenumber and we show after (2.20) that it is small for $N \to \infty$.

The factor E_{bn} in the last term of (2.10) is controlled by the spatial distribution of ξ_r in (2.7). We require a quasi-uniform spatial distribution[1] such that for $|n| \leq M$

$$N^{-1} \sum_{r=1}^{N} e^{ik_n \xi_r} = O(N^{-1/2}) \qquad (2.11)$$

ensuring that $\ddot{\xi}_r \sim N^{-1/2}$, which enables us to define a small parameter ε_{gc} below. The guiding-centre motion is thus quasi-ballistic. Condition (2.11) is reminiscent of the thermal scaling (2.77) discussed later but the system departs from equilibrium in two (related) respects: the particles have a fast motion with a small amplitude around their guiding centres; and the long-wavelength components of the fields have above-thermal amplitude oscillations.

It is convenient to express E_{cn} as

$$E_{cn} = F_n e^{-i\psi_n} + F_{-n} e^{i\psi_{-n}} \qquad (2.12)$$

where F_n is a slow (real) amplitude[2] and ψ_n is a fast (real) phase, while $\dot{\psi}_n$ is slow. We rewrite (2.9) as

$$\ddot{\eta}_r = \frac{q_*}{m} \sum_{n=-M}^{M} F_n (e^{ik_n x_r - i\psi_n} + e^{-ik_n x_r + i\psi_n}). \qquad (2.13)$$

The requirement of small amplitude oscillations (compared to mode wavelengths) is expressed by

$$\varepsilon_{ampl} = k_M \max(|\eta_r|) \ll 1. \qquad (2.14)$$

This motivates our bounding $|n|$ by M. Therefore our first approximation is to replace x_r by ξ_r in (2.13), which corresponds to the usual neglect of mode coupling[3] for long wavelengths. This reduces (2.13) to

$$\ddot{\eta}_r = \frac{q_*}{m} \sum_{n=-M}^{M} F_n (e^{ik_n \xi_r - i\psi_n} + e^{-ik_n \xi_r + i\psi_n}). \qquad (2.15)$$

---

[1] A limiting regime of our assumptions is obtained analytically, for N large enough and given M, with a multibeam state (see exercise 2.7). In the limit $N \to \infty$, thermal states are close to such reference states. As such a multibeam reference state has vanishing fields for $|k_n| \leq k_M$, only the departures η_r of particles from this repartition (and quasi-resonant particles in section 2.3) generate the fields.

[2] It will turn out to generate the wave envelope. More generally, one could write $E_{cn} = \sum_\chi F_{n\chi} e^{-i\psi_{n\chi}}$ with some appropriate index χ. As we argue in section 2.2.3 that the contribution we retain is the most robust one, we simplify the following calculations by only keeping this one. These terms occur in complex pairs in (2.12) because $k_{-n} = -k_n$ and $\ddot{\eta}_r$ is real valued.

[3] Expanding the η_r contribution to the right-hand side of (2.13), using (2.13) again, yields a nonlinear expression in terms of modes $F_n e^{-i\psi_n}$. Mode coupling is intimately related to nonlinear wave evolution equations, such as those which occur, e.g., in strong turbulence regimes. Here we only consider weak turbulence.

As the right-hand side of (2.15) does not involve η_r, one can integrate it twice with respect to time t to obtain η_r. However, the amplitude F_n and phase $k_n \xi_r - \psi_n$ depend on time. We define the Doppler-shifted frequency of mode n in the frame of particle r as

$$\Omega_{rn}(t) = k_n \dot{\xi}_r - \dot{\psi}_n. \tag{2.16}$$

2.2.2 Collective dynamics

In (2.16) $\dot{\xi}_r$ is nearly constant, evolving with a time-scale τ_{gc} determined by $\ddot{\xi}_r$. We assume

$$\varepsilon_{gc} = \sup_{r,n} |\tau_{gc} \Omega_{rn}|^{-1} \ll 1. \tag{2.17}$$

As the guiding-centre acceleration scales like $N^{-1/2}$ in the limit $N \to \infty$, such a condition is easy to satisfy. Special (multibeam, exercise 2.7) reference states even ensure $\varepsilon_{gc} = 0$. Equation (2.17) precludes the existence of particles close to resonance with the modes (section 2.3 will deal with these quasi-resonant particles). The time evolution of Ω_{rn} is also ruled by the time evolution of $\dot{\psi}_n$ but the next subsection shows that this evolution is slower than that of $\dot{\xi}_r$ in the absence of resonant particles. Therefore condition (2.17) is the condition for a slow evolution of the Ω_{rn}.

   Equation (2.15) describes the particle's oscillations in a family of slowly modulated waves. It is integrated asymptotically with the small ordering parameter ε_{gc}, i.e. we write a formal series of powers of ε_{gc} and identify its terms recursively. Recalling that η_r was defined so that its time average vanishes (secular drifts being incorporated into the guiding-centre motion) yields

$$\eta_r = \frac{q_*}{m} \sum_{n=-M}^{M} \mathcal{F}_{rn} (e^{ik_n \xi_r - i\psi_n} + e^{-ik_n \xi_r + i\psi_n}) \tag{2.18}$$

with the (two-time-scale) expansion in ε_{mode} given by $\mathcal{F}_{rn} = \mathcal{F}_{rn}^{(0)} + \mathcal{F}_{rn}^{(1)} + \cdots$. The dominant O(1) term and the O(ε_{mode}) term read:

$$\mathcal{F}_{rn}^{(0)} = -\Omega_{rn}^{-2} F_n \tag{2.19}$$

$$\mathcal{F}_{rn}^{(1)} = -iF_n^{-1} \frac{d}{dt}(\Omega_{rn}^{-3} F_n^2) \tag{2.20}$$

where the singularities as $F_n \to 0$ are fictitious, being cancelled by a factor F_n or F_n^2. We note that the occurrence of small denominators in (2.19) is ruled out by condition (2.17). Differentiating (2.18) twice with respect to time yields (2.15) up to O(ε_{mode}^2) corrections, which would be eliminated by subsequent terms in the expansion of $\mathcal{F}$.

   Equation (2.18) expresses η_r as a superposition of small-wavenumber terms $e^{ik_n \xi_r}$. Then for large wavenumber ($|s| > M$) the field component $E_s = q_* k_s V_s/(2iN) \sum_{r=1}^{N} e^{-ik_s \xi_r} e^{-ik_s \eta_r}$ must be small as (2.90) will show.

For $|n| \leq M$, inserting into the definition (1.9) of E_n the decomposition of the particle motion (2.6) and expanding to first order in η_r leads to the approximate equation

$$E_n = \frac{k_n V_n q_*}{2iN} \sum_{r=1}^{N} e^{-ik_n \xi_r} (1 - ik_n \eta_r) \qquad (2.21)$$

which, on using (2.8) and (2.18) for η_r, becomes

$$E_{cn} = -\frac{k_n^2 V_n q_*^2}{2Nm} \sum_{r=1}^{N} \sum_{m=-M}^{M} (\mathcal{F}_{rm} e^{-i\psi_m} + \mathcal{F}_{r,-m} e^{i\psi_{-m}}) e^{i(k_m - k_n)\xi_r}. \qquad (2.22)$$

Now we introduce the so-called dielectric function

$$\epsilon(k_n, \omega, t) = 1 - \frac{k_n^2 V_n q_*^2}{2Nm} \sum_{r=1}^{N} (k_n \dot{\xi}_r - \omega)^{-2} \qquad (2.23)$$

where the time dependence (through $\dot{\xi}_r$) is slow and will be left implicit in the following. Note that $\epsilon(-k_n, -\omega) = \epsilon(k_n, \omega)$. In (2.22) we express E_{cn} and $\mathcal{F}_{rm}$ in terms of F_m using (2.8), (2.19) and (2.20):

$$\Re\left[\left(\epsilon(k_n, \dot{\psi}_n) F_n + \frac{i}{2F_n} \frac{d}{dt}\left(\frac{\partial \epsilon}{\partial \omega} F_n^2\right)\right) e^{-i\psi_n}\right] = \frac{k_n^2 V_n}{2} \Re(\alpha_n) \qquad (2.24)$$

where we define

$$\alpha_n \equiv \frac{q_*^2}{mN} \sum_{r=1}^{N} \sum_{m=-M, m \neq n}^{M} e^{i(k_m - k_n)\xi_r - i\psi_n} \left[\frac{F_m}{\Omega_{rm}^2} + \frac{i}{F_m} \frac{d}{dt}\left(\frac{F_m^2}{\Omega_{rm}^3}\right)\right]. \qquad (2.25)$$

The left-hand side of (2.24) collects terms of size $O(E)$, where E is the typical amplitude of the E_n, while the right-hand side collects terms of size $O(E/\sqrt{N})$ because of the spatial quasi-uniformity of guiding centres. As we consider non-thermal fields here, we neglect the right-hand side in (2.24). Then, since ψ_n is fast while $\dot{\psi}_n$ and F_n are slow, the left-hand side of (2.24) vanishes only if

$$\epsilon(k_n, \dot{\psi}_n) F_n + \frac{i}{2F_n} \frac{d}{dt}\left(\frac{\partial \epsilon}{\partial \omega} F_n^2\right) = 0. \qquad (2.26)$$

This final form motivates calling ϵ the dielectric function. Indeed if the plasma supports an applied field E_{an} in addition to its self-field E_n, then E_{an} will occur in the right-hand side as does α_n in (2.24). The actual field in the plasma is then estimated by $F_n \sim \epsilon^{-1} E_{an}$.

Exercise 2.2. Plot $\epsilon(k, \omega)$ versus ω for various choices of n and some simple choices of the distribution of velocities $\dot{\xi}_r$ (also known as waterbags): the distribution is piecewise constant. Pay special attention to the existence and positions of real zeros.

2.2.3 Bohm–Gross modes

For a given k_n, if $\dot{\psi}_n$ is a root ω of the dispersion relation

$$\epsilon(k_n, \omega) = 0, \qquad (2.27)$$

(2.26) admits a solution with arbitrary time-independent F_n. This means that an excitation at wavenumber k_n, i.e. an eigenmode of the medium, can propagate with a significant amplitude.

Thus we discuss the zeros of the dielectric function ϵ, which is real valued for real (k_n, ω). The dispersion relation (2.27) admits $2N$ complex roots for each wavenumber k_n. It is clear from (2.23) that $\epsilon(k_n, \omega) = 1$ at $|\omega| = \infty$; that $\epsilon(k_n, \omega) < 1$ for real ω; that $\epsilon(k_n, \omega)$ is a continuous function of ω except for its poles $\epsilon(k_n, k_n\dot{\xi}_r) = -\infty$ $\forall r$; and hence that the dispersion relation always has at least two real roots[4], the largest one $\omega_{n,\mathrm{BG}+} > \sup_r(k_n\dot{\xi}_r)$ and the smallest one $\omega_{n,\mathrm{BG}-} < \inf_r(k_n\dot{\xi}_r)$. The symmetry $k_{-n} = -k_n$ implies that $\omega_{-n,\mathrm{BG}+} = -\omega_{n,\mathrm{BG}-}$ and we write $\omega_n \equiv \omega_{n,\mathrm{BG}+}$ for any $n \neq 0$.

Depending on the values of the velocities $\dot{\xi}_r$, the other roots may be real or complex[5]. But a real root (or the real part of a complex root[6]) is generally close to a resonant value $k_n\dot{\xi}_r$ so that linearization hypotheses may be broken. These roots are very sensitive to the instantaneous values of $\dot{\xi}_r(t)$, so that their average effect will be negligible. We shall discuss in more detail a similar situation in relation with van Kampen modes in section 3.8.3.

In the following, we focus on the ω_n. Both modes, denoted BG+ and BG−, with the same wavelength L/n contribute to the Fourier components $E_n = F_{n,\mathrm{BG}+}e^{-i\psi_{n,\mathrm{BG}+}} + F_{n,\mathrm{BG}-}e^{-i\psi_{n,\mathrm{BG}-}}$ and $E_{-n} = E_n^*$ of the field. Taking advantage of the fact that there are only two modes BG$\pm$, we alleviate notation and write $F_{\pm n} = F_{n,\mathrm{BG}\pm}$ and $\psi_{\pm n} = \psi_{n,\mathrm{BG}\pm}$. Note that $F_{n,\mathrm{BG}-} = F_{-n,\mathrm{BG}+}$ and $\psi_{n,\mathrm{BG}-} = -\psi_{-n,\mathrm{BG}+}$ because $E_{-n} = E_n^*$, and $\omega_{n,\mathrm{BG}-} = -\omega_{-n,\mathrm{BG}+}$. In summary, the Langmuir wave contribution to the electric field is

$$E_{\mathrm{waves}}(x) = \sum_{n=1}^{M}(E_n e^{ik_n x} + E_n^* e^{-ik_n x}) = \sum_{n=-M, n\neq 0}^{M} 2F_n \cos(k_n x - \psi_n) \quad (2.28)$$

where the mode index n runs from $-M$ to M (excluding 0). The mode of excitation of the plasma corresponding to ω_n is called the Bohm–Gross mode with this pulsation. When seen as a travelling electrostatic wave, it is the Langmuir wave with pulsation ω_n of the plasma without resonant particles.

---

[4] For a symmetric (even) distribution of guiding-centre velocities, these roots are opposite.
[5] For the multibeam case of exercise 2.7, one finds explicitly all the beam modes (in the same way as for the beam–wave interaction in chapter 3): they all lie inside the circle with centre $(\omega_{n,\mathrm{BG}-} + \omega_{n,\mathrm{BG}+})/2$ and radius $|\omega_{n,\mathrm{BG}+} - \omega_{n,\mathrm{BG}-}|/2$; many of them have zero real part.
[6] Complex roots ω are worse. Indeed, since the dielectric function (2.23) is a rational function with real coefficient, complex roots occur in conjugate pairs, generating instabilities which ultimately break the small oscillation hypothesis.

Expanding ϵ in (2.26) about ω_n to first order yields

$$(\dot{\psi}_n - \omega_n)\frac{\partial \epsilon}{\partial \omega}F_n + \frac{i}{2}F_n^{-1}\frac{d}{dt}\left(\frac{\partial \epsilon}{\partial \omega}F_n^2\right) = 0 \qquad (2.29)$$

or, separating the real and imaginary parts,

$$\frac{d}{dt}\left(F_n^2\frac{\partial \epsilon}{\partial \omega}(k_n, \omega_n)\right) = 0 \qquad (2.30)$$

$$\dot{\psi}_n = \omega_n. \qquad (2.31)$$

From (2.30) we note that

$$F_n^2\frac{\partial \epsilon}{\partial \omega}(k_n, \omega_n) = C \qquad (2.32)$$

where C is a constant. The left-hand side corresponds to the classical wave energy (proportional to its action) in a dielectric plasma. In the kinetic limit $N \to \infty$, (2.32) yields for F_n

$$\left(\frac{F_n(t)}{F_n(0)}\right)^2 = \frac{\int_\Lambda (k_n v - \omega_n(t))^{-3} f(x, v, 0)\, dx\, dv}{\int_\Lambda (k_n v - \omega_n(t))^{-3} f(x, v, t)\, dx\, dv}. \qquad (2.33)$$

Equations (2.31)–(2.32) show that the mode pulsation and amplitude are slaved to f, the coarse-grained velocity distribution function defined by

$$N^{-1}\sum_{l=1}^{N}\bullet = \int_{-\infty}^{\infty}\bullet f(v, t)\, dv = \int_\Lambda \bullet f(x, v, t)\, dx\, dv. \qquad (2.34)$$

Since it averages over many particles, f has a slower evolution than the typical evolution of the velocity of individual particles. At thermal equilibrium f has no evolution even though the individual velocities do evolve. As a result equations (2.31)–(2.32) show that the eigenmodes evolve slowly (they are said to be adiabatic), and even more slowly than the velocity of particles, as anticipated in the previous subsection. For N large, $\varepsilon_{\mathrm{gc}}$ is close to 0; then F_n may be considered as a constant and the derivation of (2.24) is simplified by restraining $\mathcal{F}$ to $\mathcal{F}^{(0)}$.

Our identification of Bohm–Gross modes ensures that ω_n is real and as 'large' as possible. Then condition (2.14) and equations (2.18)–(2.19) determine how small the field amplitudes F_n must be for a given minimum value of $|\Omega_{rn}|$. Physically, this restriction is related to the linearization of the motion which precludes particles from being trapped in the wave as we discuss in section 2.6. This is the counterpart for particles of the locality in velocity of the action of a wave on particles (section 1.2).

Finally, in the kinetic limit $N \to \infty$ (appendix G), the sum over all particles in (2.23) may be replaced by integral (2.34). Then (2.27) yields the dispersion relation of Langmuir waves, first written by Bohm and Gross (1949a),

$$1 - \omega_p^2\int_{\mathbb{R}}\frac{f_0(v)dv}{(k_n v - \omega)^2} = 0 \qquad (2.35)$$

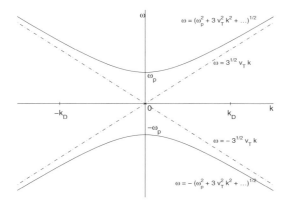

Figure 2.2. Bohm–Gross dispersion relation.

in a plasma with a space-independent velocity distribution function $f_0(v)$ (normalized to unity) and with plasma frequency[7]

$$\omega_p = \sqrt{\frac{2q_*^2}{Lm\epsilon_0}}. \tag{2.36}$$

By expanding the integrand in (2.35) in powers of kv/ω, these equations yield, to second order in k, the approximation usually called the Bohm–Gross dispersion relation (2.1) where the mean velocity of the particles $\langle v_r \rangle_{f_0}$ is assumed to be zero, and where $v_T^2 = \langle v_r^2 \rangle_{f_0}$. Then (2.1) yields a lower bound to the phase velocities of Langmuir waves:

$$v_{BG} = \sqrt{3}v_T. \tag{2.37}$$

Exercise 2.3. Derive (2.1). Comment on our requirement that there are no resonant particles ($f_0(\omega/k)$ and its derivative $f_0'(\omega/k)$ must vanish). Estimate the first neglected term in (2.1). How does the graph change if you change v_T?

Figure 2.2 displays the Bohm–Gross dispersion relation; a straight line going through the origin corresponds to a fixed velocity, which enables us to visualize the velocities susceptible to resonate with the waves easily. The closeness of the slower-wave phase velocity to the thermal particle velocity implies that some waves may easily be resonant with some particles (even non-relativistic ones). A simple way to make all particles within the distribution f_0 non-resonant with any of the M Langmuir modes is to impose the restriction that the largest particle velocity be smaller than the smallest phase velocity, i.e. the (Langmuir) mode number M (the phase velocity is a decreasing function of the wavenumber in the

---

[7] To compare (2.36) with (1.2), note that here we scale the coupling constant by N and introduce 'ghost' particles (exercise 1.2), so that $\mathcal{N} = 2/L$.

sector $k > 0$, $\omega > 0$: exercise 2.4). This can be done easily for many velocity distribution functions. For instance, if the distribution f_0 is uniform between, say $-v_{mode}$ and v_{mode}, then $v_{BG} = v_{mode}$, which leaves almost all the phase velocity range free from particles. A similar situation is obtained by cutting the tails of a Maxwellian distribution.

Exercise 2.4. For waves following the Bohm–Gross dispersion relation, determine the phase and group velocities

$$v_{\phi n} = \omega_n / k_n \tag{2.38}$$

$$v_{gn} = \frac{d\omega}{dk} = -\frac{\partial\epsilon/\partial k}{\partial\epsilon/\partial\omega}(k_n, \omega_n) \tag{2.39}$$

and plot them versus k_n and versus ω_n. Allowing k and ω to be negative, how do the various branches of your graphs correspond? Discuss the limit $k_n/k_D \to 0$. Show that $v_{\phi,n}$ satisfies $N^{-1} \sum_r (v_{\phi,n} - \dot{\xi}_r)^{-2} = 2mq_*^{-2}V_n^{-1}$, so that $dv_{\phi,n}/dV_n > 0$ for $n > 0$ and the phase velocity is a decreasing function of k_n.

In view of the Poisson equation and of the longitudinal motion of the particles, Bohm–Gross modes appear as charge-density modes. They are plasma analogues of the acoustic modes of neutral fluids but, in contrast with these, they are dispersive. In particular, in the long-wavelength limit $k_n/k_D \to 0$, the phase velocity diverges while the group velocity tends to zero.

2.3 Coupled motion of quasi-resonant particles with Bohm–Gross modes

At this point we have shown the Langmuir waves are the collective vibration modes of the particles but we have only considered the case where resonant particles are absent. We now turn to the more general case where $R \ll N$ particles are close to resonance with the Langmuir modes. In order to avoid small denominators in (2.19) and (2.20), these particles cannot be included in our previous definition of collective vibrations. Therefore we keep their granular character and we now derive the equations describing their coupled motion with the Bohm–Gross modes. We number the $N - R$ non-resonant particles (the bulk plasma) with indices 1 to N' and the quasi-resonant (or tail) particles with indices from $N'' = N'+1$ to N. We derive a mode evolution similar to that of section 2.2, but respecting the granular character of the quasi-resonant particles.

Accordingly, we split (1.9) into two parts and deal with the non-resonant one as in the previous section. To first order in η_r we find

$$E_n = -\frac{iq_*k_n V_n}{2N} \sum_{r=N''}^{N} e^{-ik_n x_r} - \frac{iq_*k_n V_n}{2N} \sum_{r=1}^{N'} e^{-ik_n \xi_r}(1 - ik_n \eta_r). \tag{2.40}$$

A treatment similar to that in section 2.2 yields an equation similar to (2.24):

$$
\left[(\dot{\psi}_n - \omega_n)\beta_n^{-2} F_n + \frac{i}{2 F_n} \frac{d}{dt}(\beta_n^{-2} F_n^2) \right] e^{-i\psi_n}
$$
$$
+ \left[(\omega_{-n} - \dot{\psi}_{-n})\beta_{-n}^{-2} F_{-n} + \frac{i}{2 F_{-n}} \frac{d}{dt}(\beta_{-n}^{-2} F_{-n}^2) \right] e^{i\psi_{-n}}
$$
$$
= -\frac{i q_* k_n V_n}{2N} \sum_{r=N''}^{N} e^{-ik_n x_r} \tag{2.41}
$$

where, in the right-hand side, we take into account the resonant particles (other contributions are shown to be negligible in section 2.6). Frequencies $\omega_{\pm n}$ are the Bohm–Gross frequencies of the bulk plasma, which we define as zeros of the dielectric function

$$
\epsilon(k_n, \omega) = 1 - \frac{q_*^2 k_n^2 V_n}{2Nm} \sum_{r=1}^{N'}(k_n \dot{\xi}_r - \omega)^{-2} = 1 - \frac{1}{N} \sum_{r=1}^{N'} \frac{\omega_p^2}{(k_n \dot{\xi}_r - \omega)^2} \tag{2.42}
$$

with the summation restricted to the N' bulk particles. We keep the denominator N in (2.42) though the sum runs only over N' particles because N scales the potential coefficients. Choosing a denominator N' would not affect our results significantly: it would only change the normalizations of coupling constants in the final self-consistent equations.

In (2.41) the presence of tail particles, in general, makes $\dot{\psi}_n$ different from ω_n, and F_n non-constant. In order to derive equation (2.41), we needed again Ω_{rn} to be a slow quantity. However, now the slowness of $\dot{\psi}_n$ is no longer granted by that of f as defined on the bulk particles, because tail particles force the mode to evolve. As a result, we must introduce a new condition, i.e. for $|n| \le M$, F_n and $\dot{\psi}_n$ are slow variables. They evolve on a time-scale τ_{mode} and we assume that

$$
\varepsilon_{\text{mode}} = \sup_{r,n} |\tau_{\text{mode}} \Omega_{rn}|^{-1} \ll 1. \tag{2.43}
$$

Coupling coefficients $\beta_{\pm n}$ are defined for $-M \le n \le M, n \ne 0$, by

$$
\beta_n = \left(\frac{\partial \epsilon}{\partial \omega}(k_n, \omega_n) \right)^{-1/2}. \tag{2.44}
$$

The fact that $\omega_n \equiv \omega_{n,\text{BG+}} > \sup_r(k_n \dot{\xi}_r)$ ensures that $\beta_n^2 > 0$ and we retain the positive determination of the square root.

Exercise 2.5. Show that $\omega_n = \omega_p$ and $\beta_n = \sqrt{\omega_p/2}$ for $n > 0$ for a cold plasma, i.e. in the limit $\epsilon = 1 - \omega_p^2/\omega^2$. Determine $v_{\phi n}$ and v_{gn}.

It will prove convenient to rescale β_n to

$$
\kappa_n \equiv \beta_n q_* V_n^{1/2} = k_n^{-1} \left(-\frac{1}{Nm} \sum_{r=1}^{N'}(k_n \dot{\xi}_r - \omega_n)^{-3} \right)^{-1/2} \tag{2.45}
$$

and to introduce the rescaled mode complex representation

$$Z_n = -2i\sqrt{\frac{N}{k_n^2 V_n}}\,\beta_n^{-1}F_n e^{-i\psi_n} = X_n + iY_n = \sqrt{2I_n}\,e^{-i\theta_n} \qquad (2.46)$$

where (X_n, Y_n) are the 'Cartesian' components of the mode n and (I_n, θ_n) are its intensity-phase (or action-angle) components, in the 'laboratory frame'. With these variables, (2.41) reduces to

$$\kappa_n^{-1}(\dot{Z}_n + i\omega_n Z_n) + \kappa_{-n}^{-1}(\dot{Z}_{-n}^* - i\omega_{-n}Z_{-n}^*) = iN^{-1/2}\sum_{r=1}^{N'} e^{-ik_n x_r}. \qquad (2.47)$$

In the left-hand side, for $0 < n \le M$, both ω_n and ω_{-n} are Bohm–Gross frequencies (equal to each other for a symmetric bulk velocity distribution). The mode Z_n oscillates at ω_n and has a phase velocity ω_n/k_n, larger than all bulk particles; Z_{-n}^* oscillates at $-\omega_{-n}$ and has a phase velocity $\omega_{-n}/(-k_n)$, smaller than all guiding-centre velocities $\dot{\xi}_r$. Therefore, the second term in the left-hand side of (2.47) cannot resonate with the first term and can be averaged out. This leaves for the modes indexed by $-M \le n \le M, n \ne 0$:

$$\dot{X}_n = \omega_n Y_n + N^{-1/2}\sum_{r=N''}^{N} \kappa_n \sin k_n x_r \qquad (2.48)$$

$$\dot{Y}_n = -\omega_n X_n + N^{-1/2}\sum_{r=N''}^{N} \kappa_n \cos k_n x_r. \qquad (2.49)$$

Substituting (2.46) in the equations of motion (1.10) for the quasi-resonant particles, we obtain the system

$$\dot{x}_r = \frac{p_r}{m} \qquad (2.50)$$

$$\dot{p}_r = -N^{-1/2}\sum_{n=-M,n\ne0}^{M} \kappa_n k_n (Y_n \cos k_n x_r + X_n \sin k_n x_r) \qquad (2.51)$$

where the effect of Fourier modes with $|n| > M$ is neglected, as it is for the bulk plasma. Equations (2.48)–(2.51) provide a closed set of equations with $R + 2M$ pairs of conjugate variables. They can be recovered through equations $\dot{x}_r = \partial_{p_r} H_{\mathrm{sc}}^{R,2M},\ \dot{p}_r = -\partial_{x_r} H_{\mathrm{sc}}^{R,2M},\ \dot{X}_n = \partial_{Y_n} H_{\mathrm{sc}}^{R,2M}$ and $\dot{Y}_n = -\partial_{X_n} H_{\mathrm{sc}}^{R,2M}$, where

$$H_{\mathrm{sc}}^{R,2M} = \sum_{r=N''}^{N} \frac{p_r^2}{2m} + \sum_{n=-M,n\ne0}^{M} \omega_n \frac{X_n^2 + Y_n^2}{2}$$

$$+ \frac{1}{\sqrt{N}} \sum_{r=N''}^{N} \sum_{n=-M,n\neq0}^{M} \kappa_n (Y_n \sin k_n x_r - X_n \cos k_n x_r) \quad (2.52)$$

$$= \sum_{r=N''}^{N} \frac{p_r^2}{2m} + \sum_{n=-M,n\neq0}^{M} \omega_n I_n$$

$$- \frac{1}{\sqrt{N}} \sum_{r=N''}^{N} \sum_{n=-M,n\neq0}^{M} \kappa_n \sqrt{2I_n} \cos(k_n x_r - \theta_n). \quad (2.53)$$

Thus $H_{\mathrm{sc}}^{R,2M}$ is the Hamiltonian of the self-consistent wave–particle dynamics; and (p_r, x_r) and (X_n, Y_n) are pairs of conjugate variables. It is easily checked that $\dot{I}_n = -\partial_{\theta_n} H_{\mathrm{sc}}^{R,2M}$ and $\dot{\theta}_n = \partial_{I_n} H_{\mathrm{sc}}^{R,2M}$ when $H_{\mathrm{sc}}^{R,2M}$ is written in terms of the pair of (conjugate) variables (I_n, θ_n). Therefore $H_{\mathrm{sc}}^{R,2M}$ is again the Hamiltonian in this case. This Hamiltonian is appropriate for studying, in a *self-consistent* way, the evolution of the long-wavelength modes of the system (1.3) and of the particles of this system which may interact significantly with them. It distinguishes between three contributions: the free motion (kinetic energy) of the R relevant particles, the (harmonic) oscillation of the $2M$ modes and the coupling between relevant particles and large-scale modes. As (1.3), our Hamiltonian (2.53) is invariant under translations in time and in space (with $x = \theta_n/k_n$ defining the position of the O points for mode n): total energy E_{sc} and momentum P_{sc}

$$E_{\mathrm{sc}} = H_{\mathrm{sc}}, \quad (2.54)$$

$$P_{\mathrm{sc}} = \sum_{r=N''}^{N} p_r + \sum_{n=-M,n\neq0}^{M} k_n I_n, \quad (2.55)$$

are conserved.

In system (2.48)–(2.51) and in (2.52) or (2.53) we can still eliminate some modes and include their quasi-resonant particles in the bulk. In particular, our derivation takes into account M modes with positive phase velocities and M modes with negative phase velocities but particles cannot be quasi-resonant with both sets of modes simultaneously. Therefore, as far as the evolution of modes with $\omega_n/k_n > 0$ is concerned, we can neglect the other set of modes. Including fast particles with $\dot{x}_r < 0$ into the bulk for the modes with $\omega_n/k_n > 0$ will not change ϵ and β_n very much ($|\Omega_{rn}|$ being large). The resulting self-consistent Hamiltonian (2.109)–(2.110) reads like (2.52)–(2.53), with a running mode index $1 \leq n \leq M$.

The overall coupling factor $N^{-1/2}$, which goes to zero as $N \to \infty$, calls for two comments:

(i)   If the mode variables (X_n, Y_n) are O(1), i.e. if the field amplitudes F_n scale as $N^{-1/2}$ as in thermal equilibrium, then a test particle motion becomes ballistic in the kinetic limit.

(ii) The actual mode–particle interaction involves amplitudes $F_n \gg F_{Tn}$ where F_{Tn} is the thermal energy (2.77), and the scaling $(X_n, Y_n) = O(N^{1/2})$ balances the prefactor $N^{-1/2}$ in (2.50)–(2.51). Then some of the R particles can be trapped into wave resonances. In turn, these trapped particles significantly affect the mode evolution provided their number is large enough to balance the prefactor in (2.48)–(2.49): the natural scaling is to keep the fraction of quasi-resonant particles constant, i.e. $R/N = \mathcal{N}_b/\mathcal{N}$ where $\mathcal{N}_b$ is the density of, say, an injected beam, and $\mathcal{N}$ is the density of the whole beam–plasma system. This is the right scaling to investigate, e.g., the quasilinear regime of weak electrostatic turbulence as we do in chapter 7.

There is neither direct mode–mode coupling nor direct particle–particle coupling in H_{sc}: the only coupling occurs between particles and modes. The specific coupling between mode n and particle r is controlled by two factors: the particle charge q_* and the mode 'sensitivity' β_n. The latter depends indirectly on the potential coefficient V_n through the explicit value of the phase velocity ω_n/k_n which enters Ω_{rn}; a factor V_n is also incorporated in the definition of (X_n, Y_n).

Variables (I_n, θ_n) are the adiabatic variables for the mode 'oscillators'; if the bulk plasma evolves, their relation to the fields E_n evolves accordingly (on time-scale τ_{gc} as shown in section 2.6).

The elimination of large-wavenumber components from the field has a major consequence on the particle dynamics. Whereas the original Coulomb force is singular at its point particle sources (like r^{1-d} in d space dimensions), the forces associated with our $2M$ modes are smooth. This implies more regularity in the particle motion. So we focus on large-scale motion by eliminating a small-scale, less regular motion.

As the limit $N \to \infty$ of our mechanical approach is the Vlasov–Poisson dynamics (see appendix G), the Hamiltonian (2.52)–(2.53) coincides with the one obtained by Mynick and Kaufman (1978), with the coefficients ω_n, κ_n obtained in the same way from the dielectric function $\epsilon(k_n, \omega)$.

Finally, note that particles of the bulk plasma can also be included among the R distinguished particles considered in the self-consistent Hamiltonian: we only need that the eliminated $N - R$ particles be non-resonant and impose no restriction on the remaining R particles. Hence our approach includes naturally the usual test particles introduced *a posteriori* in traditional treatments.

2.4   Lagrangian formulation*

The previous derivation of the self-consistent wave–particle dynamics is straightforward, inasmuch it rests on identifying special types of approximate solution to the equations of motion. But in this kind of approach one may lose specific properties of the mechanical system under consideration. There are more systematic ways to investigate effective dynamics with a reduced number of degrees of freedom, preserving some structures inherent to mechanical systems. One such structure is the detailed force balance (reaction opposing the action in two-body interactions). Another such structure, which implies the detailed balance, is that the equations of motion follow from a Lagrangian least-action principle. A third such structure is the Hamiltonian structure with pairs of conjugate variables. These three structures are quite intimately related mathematically.

Thus we now show how the original N-body mechanical problem can be reduced to an effective problem with fewer degrees of freedom in a way which preserves the Lagrangian structure of mechanics. The formulation is a little more abstract than the direct one of section 2.3 but it is safer and illustrates a general procedure of theoretical physics.

Hamiltonian (1.3) is conjugate to the Lagrangian

$$\mathcal{L} = K - V = \sum_{r=1}^{N} \frac{m\dot{x}_r^2}{2} - \frac{1}{2N} \sum_{n=1}^{\infty} \sum_{l,r=1}^{N} q_*^2 V_n \cos k_n (x_r - x_l) \qquad (2.56)$$

which is a function of the coordinates x_r and the velocities $\dot{x}_r$. The Lagrangian generates the equations of motion in the form

$$\frac{d}{dt}\frac{\partial \mathcal{L}}{\partial \dot{x}_r} - \frac{\partial \mathcal{L}}{\partial x_r} = 0. \qquad (2.57)$$

The conjugate momentum

$$p_r \equiv \frac{\partial \mathcal{L}}{\partial \dot{x}_r} \qquad (2.58)$$

to coordinate x_r is generally a function of all variables $(x_l, \dot{x}_l)$. However, for the special form (2.56), it reduces to the well known $p_r = m\dot{x}_r$. One obtains the Hamiltonian H from $\mathcal{L}$ by a Legendre transformation

$$H(\mathbf{x}, \mathbf{p}) = \sum_{r=1}^{N} p_r \dot{x}_r - \mathcal{L}(\mathbf{x}, \dot{\mathbf{x}}) \qquad (2.59)$$

and the equations of motion follow from (2.57) and (2.58).

To reduce the full Lagrangian (2.56) to an effective wave–particle Lagrangian, we express now the bulk particle motion $x_r = \xi_r + \eta_r$ in terms of the mode variables X_n, Y_n (and their time derivatives). As in the previous sections, the guiding-centre motion is quasi-ballistic and we take it as given.

Consider first the kinetic term in (2.56). In

$$K = \sum_{r=1}^{N'} \frac{m}{2}\dot{\xi}_r^2 + \sum_{r=1}^{N'} m\dot{\xi}_r\dot{\eta}_r + \sum_{r=1}^{N'} \frac{m}{2}\dot{\eta}_r^2 + \sum_{r=N'+1}^{N} \frac{m}{2}\dot{x}_r^2 \qquad (2.60)$$

the first term does not depend on the η_r, and the second term is linear in $\dot{\eta}_r$, with $m\dot{\xi}_r$ also constant (up to $O(\varepsilon_{gc})$). Hence these two terms do not contribute to the

Lagrange evolution equation for η_r and may be neglected. To express the third term using the mode variables, we rewrite (2.15) using (2.45)–(2.46) as

$$\ddot{\eta}_r = \frac{1}{m} \sum_{n=-M}^{M} \kappa_n k_n N^{-1/2} \Re(\mathrm{i} Z_n \mathrm{e}^{\mathrm{i} k_n \xi_r}). \tag{2.61}$$

The scale separation ($|(\mathrm{d}/\mathrm{d}t)(Z_n \mathrm{e}^{\mathrm{i}\omega_n t})| \ll |\omega_n Z_n|$) yields, to first order in $\varepsilon_{\mathrm{mode}}$, the time integral

$$\dot{\eta}_r = \frac{1}{m} \sum_{n=-M}^{M} \kappa_n k_n N^{-1/2} \Re \left[((k_n \dot{\xi}_r - \omega_n)^{-1} Z_n \right.$$
$$\left. + \mathrm{i}(k_n \dot{\xi}_r - \omega_n)^{-2}(\dot{Z}_n + \mathrm{i}\omega_n Z_n)) \mathrm{e}^{\mathrm{i} k_n \xi_r} \right]. \tag{2.62}$$

Therefore $\dot{\eta}_r^2$ is approximated to the same order by

$$\dot{\eta}_r^2 = \frac{1}{2Nm^2} \sum_{n=-M}^{M} \kappa_n^2 \left((\dot{\xi}_r + \omega_n/k_n)^{-2}(X_n^2 + Y_n^2) \right.$$
$$\left. - 2k_n^2(k_n \dot{\xi}_r - \omega_n)^{-3}(X_n \dot{Y}_n + \omega_n X_n^2 - \dot{X}_n Y_n + \omega_n Y_n^2) \right) \tag{2.63}$$

and the kinetic energy of the bulk particles oscillations reads

$$\sum_{r=1}^{M} \frac{m}{2} \dot{\eta}_r^2 = \frac{1}{2} \sum_{n=-M}^{M} (\beta_n^2 + \omega_n)(X_n^2 + Y_n^2) + \frac{1}{2} \sum_{n=-M}^{M} (X_n \dot{Y}_n - \dot{X}_n Y_n) \tag{2.64}$$

where we used (2.23), (2.44) and $\epsilon(k_n, \omega_n) = 0$.

Now consider the potential term

$$V = \sum_{n=-M}^{M} \frac{q_*^2 V_n}{4} \rho_n \rho_n^* \tag{2.65}$$

where the contribution of the large wavenumbers was eliminated as in section 2.3. Here we define

$$\rho_n = N^{-1/2} \sum_{r=1}^{N} \mathrm{e}^{-\mathrm{i} k_n x_r} \tag{2.66}$$

which reduces to

$$\rho_n = N^{-1/2} \sum_{r=N'+1}^{N} \mathrm{e}^{-\mathrm{i} k_n x_r} + N^{-1/2} \sum_{r=1}^{N'} \mathrm{e}^{-\mathrm{i} k_n \xi_r} - \mathrm{i} k_n N^{-1/2} \sum_{r=1}^{N'} \eta_r \mathrm{e}^{-\mathrm{i} k_n \xi_r} \tag{2.67}$$

to first order in η_r. In (2.67) the second term is negligible due to quasi-uniformity of guiding centres. We estimate the third term by integrating (2.62):

$$
\eta_r = \frac{1}{m} \sum_{n=-M}^{M} \kappa_n k_n N^{-1/2} \Re[e^{ik_n \xi_r}(-i(k_n\dot{\xi}_r - \omega_n)^{-2}Z_n
$$
$$
+ 2(k_n\dot{\xi}_r - \omega_n)^{-3}(\dot{Z}_n + i\omega_n Z_n))] \tag{2.68}
$$

to order $\varepsilon_{\text{mode}}$. Then

$$
\sum_{r=1}^{N'} \eta_r e^{-ik_n \xi_r} = -i\frac{N^{1/2}}{k_n V_n q^2}(\kappa_n Z_n + \kappa_{-n}Z^*_{-n}) - N^{1/2}k_n^{-1}\kappa_n^{-1}(\dot{Z}_n + i\omega_n Z_n)
$$
$$
- N^{1/2}k_{-n}^{-1}\kappa_{-n}^{-1}(\dot{Z}^*_{-n} - i\omega_{-n}Z^*_{-n}) \tag{2.69}
$$

neglecting the terms corresponding to $k_s \neq k_n$ as we did in section 2.3. Thus the leading terms in (2.67) are

$$
\rho_n = N^{-1/2} \sum_{r=N'+1}^{N} e^{-ik_n x_r} - \frac{\kappa_n}{q^2 V_n}Z_n - \frac{\kappa_{-n}}{q^2 V_n}Z^*_{-n}
$$
$$
+ i\kappa_n^{-1}(\dot{Z}_n + i\omega_n Z_n) - i\kappa_{-n}^{-1}(\dot{Z}^*_{-n} - i\omega_{-n}Z^*_{-n}) \tag{2.70}
$$

so that the potential (2.65) reads

$$
V = \tfrac{1}{2} \sum_{n=-M}^{M} (\beta_n^2 + 2\omega_n)(X_n^2 + Y_n^2) + \sum_{n=-M}^{M} (X_n\dot{Y}_n - \dot{X}_nY_n)
$$
$$
+ \sum_{n=-M}^{M} \sum_{r=N'+1}^{N} \kappa_n N^{-1/2}(Y_n \sin k_n x_r - X_n \cos k_n x_r) \tag{2.71}
$$

where we have kept the dominant term in the collective variables Z_n and the first corrections due to quasi-resonant particles and to the slow evolution of the mode envelopes. We neglected (or averaged off) the wave-pairing terms $Z_n Z_{-n}$, which oscillate at $\omega_n + \omega_{-n}$, as we did when separating the two parts of (2.47) in the derivation of (2.48) and (2.49).

Finally, let $C = (1/2)\sum_{n=-M}^{M} X_n Y_n$ and recall that the evolution equations do not change if one adds the total derivative $\dot{C}$ to the Lagrangian. Gathering (2.71), (2.64) and the fourth term in (2.60) yields the Lagrangian

$$
\mathcal{L}_{\text{sc}}^{R,2M} = \sum_{n=-M}^{M} \left(\dot{X}_n Y_n - \omega_n \frac{X_n^2 + Y_n^2}{2} \right) + \sum_{r=N'+1}^{N} \frac{m}{2}\dot{x}_r^2
$$
$$
+ \sum_{r=N'+1}^{N} \sum_{n=-M}^{M} N^{-1/2}\kappa_n(X_n \cos k_n x_r - Y_n \sin k_n x_r). \tag{2.72}
$$

As the conjugate momentum to X_n turns out to be

$$\frac{\partial \mathcal{L}_{\text{sc}}^{R,2M}}{\partial \dot{X}_n} = Y_n \tag{2.73}$$

the Legendre transform $H = \sum_{r=N'+1}^{N} \dot{x}_r p_r + \sum_{n=-M}^{M} \dot{X}_n Y_n - \mathcal{L}_{\text{sc}}^{R,2M}$ yields the self-consistent Hamiltonian (2.52), (2.53).

Exercise 2.6. Show that (2.72) is invariant under the transformations $x'_r = x_r + a$, $X'_n + iY'_n = (X_n + iY_n)e^{-ik_n a}$ for any $a \in \mathbb{R}$. This implies the conservation of total momentum (2.55) by Noether's theorem (see, e.g., Goldstein 1980). Similarly, the invariance of (2.72) under time translations implies the conservation of energy (2.54).

2.5   Reference states of the plasma*

The first step in analysing the plasma dynamics is to investigate some of its simplest solutions. Because of the strength of Coulomb interactions, the plasma tends to minimize the electric fields in itself, so that it appears as nearly neutral. Otherwise, strong accelerations due to the field cause the plasma to evolve rapidly. Therefore, we start with a reference state of the plasma where the fields are (nearly) zero.

2.5.1   Non-existence of zero-field states

Unfortunately, with a finite number of particles N, one cannot realize a plasma with an electric field vanishing everywhere. Indeed, if we require that $E_n = 0$ for all n, then (1.9) reduces with $w_r = e^{-2\pi i x_r/L}$ to

$$\sum_{r=1}^{N} w_r^n = 0 \tag{2.74}$$

for all $n = 1, 2, \ldots$. This is an infinite system of algebraic equations for the finite family of unknowns w_r ($1 \leq r \leq N$): it admits a single solution, $w_r = 0 \; \forall r$, which violates condition $|w_r| = 1$. From a physicist's viewpoint, it is clear that, with a finite number N of particles, each particle is a singularity of the field (a point at which the field has a jump, in one space dimension), so that it is impossible to ensure a vanishing field over the whole spatial domain. This dead-end actually tells us how to amend our requirement on the state to define more natural weak-field near-equilibrium states.

Exercise 2.7. Consider, for some large N, a configuration of particles moving in monokinetic beams, say N_s particles (with $\sum_s N_s = N$) equally spaced (so that $x_r = y_s + rL/N_s \bmod L$) on a beam with $v_r = v_s$. Such a *multibeam* configuration is characterized by the beam populations N_s, beam velocities v_a and reference positions y_s. Show that, for all $1 \leq n \leq \min(N_s)$, the field component E_n vanishes identically. Physically, the particle contributions to E_n interfere destructively.

2.5.2 Thermal distribution of the electric field

Actual plasmas are not generated by placing particles carefully at given positions with given velocities. They are produced macroscopically and one only controls some of their average properties. In the absence of external perturbations, one first describes them in a thermal equilibrium state following an approach which we further discuss in sections 9.1 and 9.8.

To describe this statistical equilibrium, one is first interested in the macroscopic constants of the motion. These are total energy, $E = H$, given by (1.3), total momentum $P = \sum_{r=1}^{N} p_r$, thanks to periodic boundary conditions and a centre-of-mass reference position $\sum_r (mx_r - p_r t)/(Nm)$, as the dynamics (1.10) is invariant under Galileo transformations. If our plasma model involves no interaction with its surroundings, the mathematical description of the statistical equilibrium assumes that the microscopic state (x, p) in the $2N$-dimensional phase space is given by the so-called microcanonical probability distribution.

However, one generally considers a system interacting with its surroundings. Its ability to exchange energy brings it to a situation where its energy fluctuates around an equilibrium value. The quantity controlling its total energy is the temperature T and the system temperature is equal to the temperature of the environment, conventionally called a bath. Similarly, keeping the plasma in a vessel enables it to exchange momentum via the wall pressure, and fixes its centre-of-mass position. This more common physical situation is described mathematically by the canonical probability distribution in phase space

$$d\mathbb{P}_{\text{can}}(x, p) = \frac{1}{C_N \mathscr{Z}(T, N, L)} e^{-H(x,p)/(k_B T)} d^N x \, d^N p \qquad (2.75)$$

where the partition function $\mathscr{Z}$ normalizes $\mathbb{P}_{\text{can}}$ to unity. The constant C_N appears here for dimensional and fundamental reasons but its value is irrelevant for the following. The temperature T is the absolute temperature (positive for normal, stable states of matter) and we set the Boltzmann constant k_B to unity by measuring T as an energy.

A major characteristic of the Gibbs canonical distribution is that, at a finite temperature T, it gives small weight to high-energy microstates. It also gives little weight to the low-energy states if H admits an absolute minimum, which is the case for (1.3) since $K \geq 0$ and (1.5) reads:

$$V = 2N \sum_{n=1}^{\infty} \frac{|E_n|^2}{k_n^2 V_n} \geq 0 \qquad (2.76)$$

thanks to (1.9). By tuning the temperature, one preferentially selects states with energy near a given value.

To compute the canonical distribution of the fields E_n, one computes (very symmetrical) $2N$-dimensional integrals (see, e.g., Lenard (1961, 1963), Prager (1962) or Elskens and Antoni (1997)). The fields E_n have a joint Gaussian distribution, with nearly independent components[8], with their phases Arg E_n distributed uniformly on $[0, 2\pi]$ and

$$
\begin{aligned}
\langle |E_n|^2 \rangle^{1/2} = F_{\mathrm{T}n} &= \left(\frac{k_n^2 V_n}{4N} \right)^{1/2} \left(\frac{1}{2T} + \frac{1}{q_*^2 V_n} \right)^{-1/2} \\
&= N^{-1/2} q_*^{-1} T k_{\mathrm{D}} (1 + \lambda_{\mathrm{D}}^2 k_n^2)^{-1/2} \\
&= \frac{m \lambda_{\mathrm{D}} \omega_{\mathrm{p}}^2}{N^{1/2} q_*} (1 + \lambda_{\mathrm{D}}^2 k_n^2)^{-1/2}.
\end{aligned}
\tag{2.77}
$$

Short-range components (with $|k_n \lambda_{\mathrm{D}}| \gtrsim 1$) are significantly smaller than long-range components, i.e. the Debye length is a natural cut-off for the Fourier components of the electric field. The real-space equivalent to the scaling (2.77) is the exponential decay of the two-body correlation function, i.e. the conditional probability density f to find a particle at position x_2 given that there is a particle at x_1,

$$
\frac{N-1}{N} f(x_2 | x_1) \simeq L^{-1} - \frac{k_{\mathrm{D}}}{2N} \mathrm{e}^{-k_{\mathrm{D}} |x_1 - x_2|}
\tag{2.78}
$$

for $|x_1 - x_2| < L/2$.

Exercise 2.8. Compute the expectation of the total field energy (2.76). As this expectation is finite, conclude that $\lim_{k_n \to \infty} \langle |E_n|^2 \rangle = 0$. Moreover, assuming $\lambda_{\mathrm{D}} \ll L$, show that, for any k_*,

$$
\left\langle \sum_{k_n = k_*}^{\infty} |E_n|^2 \right\rangle \simeq \frac{m^2 \lambda_{\mathrm{D}} L \omega_{\mathrm{p}}^4}{2\pi q_*^2 N} \left(\frac{\pi}{2} - \arctan(k_* \lambda_{\mathrm{D}}) \right).
\tag{2.79}
$$

We will be interested in states for which the fields significantly exceed these equilibrium values. Our reference states will be such that the long-range part of the fields are small, while we impose no specific restriction on the short-range part as they are typically small. Thus we focus on field components $|n| \leq M$, for some large M to be determined later.

In a sense, the limit $N \to \infty$ is an artifact to set all thermal fields equal to zero. However, actual plasmas support finite thermal fields, generating significant noise levels in probes and other measurement devices.

---

[8] For finite N, the fields E_n are not independent but their correlation coefficients vanish like N^{-1}.

2.6 Physical scalings and error estimates*

The derivation of self-consistent equations (2.48)–(2.51) in the previous sections rests on neglecting various terms on physical grounds. We now estimate them explicitly to validate our derivation mathematically. As the derivative of small quantities may be large (but irrelevant in averaging methods[9]), we focus here on evolution equations rather than on the Lagrangian approach.

First recall the 'natural' scalings of physical quantities. The plasma frequency ω_p (2.36) and the thermal velocity v_T, at temperature $T = m v_T^2 = m \langle \dot{\xi}^2 \rangle$, determine $\lambda_D = k_D^{-1} = v_T / \omega_p$. The grain parameter, which is the reciprocal of the average number of particles in a Debye length,

$$N_D = N \frac{\lambda_D}{L} \qquad (2.80)$$

controls the relaxation of bulk particle distribution and the time evolution of $\dot{\xi}_r$. Typically, one finds (see, e.g., Rouet and Feix 1991) the time scaling $N_D^2 \omega_p^{-1}$ for distributions close to Maxwellians, while a subset of 'test' particles in the bulk relaxes to the 'overall' bulk distribution with a rate scaling as

$$\varepsilon_{gc} \simeq N_D^{-1} \qquad (2.81)$$

or $\tau_{gc} \sim N_D \omega_p^{-1}$. Condition (2.17) is then satisfied automatically in the kinetic limit. The variations in the bulk-particle velocity distribution influence the mode propagation only through the dielectric function (2.23). The zeros ω_n and the derivatives β_n^{-2} of $\epsilon(k_n, \omega)$ then also depend on time generically[10] with scale τ_{gc}.

The thermal distribution of non-resonant particles in space also determines a scale for fields E_n (section 2.5.2). This distribution found analytically in the Gibbs canonical ensemble was checked to describe correctly the fields in numerical simulations (see, e.g., Rouet and Feix (1991, 1996) and references therein). Equation (2.77) provides an important estimate for the scaling of F_n since the probability for a large deviation ($|E_n^2| > \mathcal{E}_n^2$) is exponentially small in the limit $N \to \infty$ for any finite amplitude $\mathcal{E}_n$: as the fields in which we are interested do not need to vanish in the limit $N \to \infty$, they are above the thermal estimates.

Our key assumption is that bulk particles and modes are weakly coupled: the motion of these particles with respect to the modes' phase velocities is fast, so that this coupling acts perturbatively. This will not only provide control on solutions

[9] Consider, e.g., $\epsilon \sin(x/\epsilon^2)$ for small ϵ.

[10] Generically means that the set of ϵ depending on time with scale τ_{gc} is the countable intersection of open dense sets. An example of a generic set is the complementary set to a Cantor set. Since a Cantor set may have a finite measure, a non-generic set is not necessarily of zero measure. Therefore saying that a property is generic is not as strong as saying it has probability 1.

of the linear equation (2.15) for η_r; it also ensures that nonlinear effects in (2.13) are small. Thus we assume

$$|q_* F_n k_n^{-1}| \leq m \left(\dot{\xi}_r - \frac{\dot{\psi}_n}{k_n} \right)^2 \varepsilon_{coupling} \tag{2.82}$$

with $\varepsilon_{coupling} \ll 1$, for any mode $-M \leq n \leq M$ and non-resonant particle $1 \leq r \leq N'$.

If we assume the Bohm–Gross relation (2.1), small wavenumbers have larger phase velocities, scaling as ω_p/k_n since $|\omega_n| \geq \omega_p$ by (2.1), and (2.37) is a lower bound on phase velocities. With (2.1), a requirement $\omega_M/k_M = \mu v_T$ yields $k_M = k_D(\mu^2 - 3)^{-1/2}$; for $\mu > 2$, this reads:

$$k_M < k_D \tag{2.83}$$

and $M < L/(2\pi \lambda_D) = N/(2\pi N_D)$. For a Maxwell distribution, the condition $\mu > 2$ places 92% of the particles in the bulk (which entails that the fraction of tail particles is not very large). Denoting the maximum of $|\dot{\xi}_r|$ in the bulk by λv_T, (2.82) reads:

$$\left| \frac{q_* F_n}{k_n m v_T^2} \right| \leq (\mu - \lambda)^2 \varepsilon_{coupling}. \tag{2.84}$$

Condition (2.82) allows for *large* field amplitudes in the long-wavelength (large-phase-velocity) range. A more stringent condition, such as $|\dot{\xi}_r - \dot{\psi}_n/k_n| \geq v_0 = O(v_T)$ would yield $|q_* F_n| \ll k_n m v_0^2$, which is a usual weak-turbulence requirement; even the latter is easily satisfied simultaneously with the scaling $|F_n| \gg F_{Tn}$.

Note also that the nonresonance condition (2.82) ensures that *direct* mode–mode coupling, i.e. coupling via nonlinear effects in the *bulk* particles' motion, is weak. Indeed, these nonlinear effects are related to the $O(k_n^2 \eta_r^2)$ terms neglected in (2.21). Assumption (2.82) yields an estimate for the amplitude of oscillations of bulk particles, which determines our small parameter ε_{ampl} in (2.14):

$$\varepsilon_{ampl} = \sup_r |k_m \eta_r| \leq 2\varepsilon_{coupling} \sum_{n=1}^{M} \frac{k_M}{|k_n|} \sim 2\varepsilon_{coupling} M \ln M. \tag{2.85}$$

Direct mode coupling is thus neglected if

$$\varepsilon_{coupling} M \ln M \ll 1. \tag{2.86}$$

However, this condition places no constraint on couplings occurring through quasi-resonant particles (as in the turbulent regime of chapter 7).

Besides, to ensure that non-resonant particles do not contribute strong poles in the dielectric function and its derivative β_n^{-2}, we also require

$$|k_n \dot{\xi}_r - \dot{\psi}_n| \geq \omega_0 \gg \tau_{mode}^{-1} \tag{2.87}$$

where $\omega_0 \simeq \omega_p(1 - \mu^2/6)$ is the minimum value of $|k_n \dot{\xi}_r - \omega_n|$.

These estimates imply that the large-wavenumber components of the field $|s| > M$ are small. Indeed,

$$
\begin{aligned}
E_s &= \frac{\omega_p^2 m}{k_s q_* N \mathrm{i}} \sum_{r=1}^{N} \mathrm{e}^{-\mathrm{i}k_s x_r} = \frac{\omega_p^2 m}{k_s q_* N \mathrm{i}} \sum_{r=1}^{N} \mathrm{e}^{-\mathrm{i}k_s \xi_r} \mathrm{e}^{-\mathrm{i}k_s \eta_r} \\
&= \frac{\omega_p^2 m}{k_s q_* N \mathrm{i}} \sum_{r=1}^{N} \mathrm{e}^{-\mathrm{i}k_s \xi_r} + \frac{\omega_p^2 m}{k_s q_* N \mathrm{i}} \sum_{r=1}^{N} \mathrm{e}^{-\mathrm{i}k_s \xi_r} (\mathrm{e}^{-\mathrm{i}k_s \eta_r} - 1) \quad (2.88)
\end{aligned}
$$

where the first term is $\mathrm{O}(N^{-1/2})$ as guiding centres have a uniform distribution, up to thermal fluctuations. The second term is estimated as

$$
\begin{aligned}
\left| N^{-1} \sum_{r=1}^{N} \mathrm{e}^{-\mathrm{i}k_s \xi_r} (\mathrm{e}^{-\mathrm{i}k_s \eta_r} - 1) \right| \\
\leq N^{-1} \sum_{r=1}^{N} |\mathrm{e}^{-\mathrm{i}k_s \eta_r} - 1| \\
\leq N^{-1} \sum_{r=1}^{N} |k_s \eta_r| = N^{-1} \sum_{r=1}^{N} \left| \sum_{n=-M}^{M} \frac{2q_*}{m} \mathcal{F}_{rn} k_n \cos(k_n \xi_r - \psi_n) \right| \\
= \sum_{n=-M}^{M} \mathrm{O}(\varepsilon_{\text{coupling}}) \quad (2.89)
\end{aligned}
$$

using (2.18) and (2.82). Hence

$$
E_s = \mathrm{O}(N^{-1/2}) + \frac{\omega_p^2 m}{2 k_s q_*} \mathrm{O}(M \varepsilon_{\text{coupling}}). \quad (2.90)
$$

Now we estimate the scales of the coupling terms in the evolution equations (2.48)–(2.51), rewritten as

$$
m \ddot{x}_r = 2q_* \sum_{n=-M}^{M} F_n \cos(k_n x_r - \psi_n) + m A_r' \quad (2.91)
$$

$$
\frac{\mathrm{i} \mathrm{e}^{-\mathrm{i}\omega_n t}}{\beta_n} \frac{\mathrm{d}U_n}{\mathrm{d}t} = -\frac{\mathrm{i}q_* k_n V_n}{2N} \sum_{r=N''}^{N} \mathrm{e}^{-\mathrm{i}k_n x_r} + \alpha_n' - \frac{\mathrm{i} \mathrm{e}^{\mathrm{i}\omega_{-n} t}}{\beta_{-n}} \frac{\mathrm{d}U_{-n}^*}{\mathrm{d}t} \quad (2.92)
$$

where the remainders A_r' and α_n' collect the contributions neglected in previous approximations.

As in the previous sections we first discuss the mode equations, which we simplify by introducing envelopes

$$
U_n = \frac{F_n}{\beta_n} \mathrm{e}^{-\mathrm{i}(\psi_n - \omega_n t)}. \quad (2.93)
$$

Following (2.43), the right-hand side in (2.92) determines (a lower estimate for) the time-scale τ_{mode} and (an upper estimate for) the small parameter $\varepsilon_{\text{mode}}$:

$$\tau_{\text{mode}}^{-1} = E^{-1} O(\varrho N^{-1/2} m\omega_{\text{p}}^3 (q_* k_n)^{-1}, \alpha'), \qquad \varepsilon_{\text{mode}} = (\omega_{\text{p}} \tau_{\text{mode}})^{-1} \quad (2.94)$$

where E is an order of magnitude for E_n (i.e. for F_n) and where we have introduced the dimensionless quantity

$$\varrho = N^{-1/2} \sum_{r=N''}^{N} e^{-ik_n x_r} \tag{2.95}$$

associated with the source term in (2.48) and (2.49). This ϱ is of the order of $(R/N)^{1/2}$ if the quasi-resonant particles are almost uniformly distributed in space; it becomes $O(N^{-1/2} R)$ if most quasi-resonant particles become bunched in their interaction with the modes. Condition (2.83) generally ensures that quasi-resonant particles are not numerous, i.e. that ϱ is small indeed (a few per cent for a Maxwell distribution function).

The right-hand side of (2.92) yields the estimate $\beta_n \dot{U}_n = O(E\omega_{\text{p}} \varepsilon_{\text{mode}})$ or, separating the real part from the imaginary one

$$\dot{F}_n = E\omega_{\text{p}} O(\varepsilon_{\text{mode}}), \qquad \dot{\psi}_n - \omega_n = \omega_{\text{p}} O(\varepsilon_{\text{mode}}). \tag{2.96}$$

The last term in (2.92) is the forcing of mode n by mode $-n$. Its contribution to $\dot{U}_n$ is non-resonant, with relative frequency $\omega_n + \omega_{-n}$. Hence it contributes only secularly to driving U_n and this secular contribution is controlled by the time evolution of U_{-n}, which occurs at the same scale τ_{mode} as for U_n.

We now estimate the remainders in (2.92) to check that the evolution of the field resulting from the motion of the bulk particles is consistent with our two-scale assumptions (2.17) and (2.43).

The coupling coefficient β_n defined by (2.44) is easily estimated. For the long-wavelength modes ($|k_n| < k_D$) and bulk particles ($\dot{\xi}_r = O(v_T)$), $|\Omega_{rn}| = |\omega_n - k_n \dot{\xi}_r|$ cannot be smaller than $O(\omega_{\text{p}})$, as shown by (2.87). Thus we estimate $\partial \varepsilon / \partial \omega(\omega_n) \sim \omega_n^{-1}$ and $\beta_n = O(\omega_n^{1/2}) = O(\omega_{\text{p}}^{1/2})$.

Estimating explicitly the error terms through our computations yields the following remainder for (2.92):

$$\alpha_n' = \alpha_{n,1}' + \alpha_{n,2}' + \alpha_{n,3}' \tag{2.97}$$

with

$$\alpha_{n,1}' = c_1 \frac{i}{2} E_n (\omega_n - \dot{\psi}_n)^2 \frac{\partial^2 \varepsilon}{\partial \omega^2} \tag{2.98}$$

$$\alpha_{n,2}' = -\frac{q_* k_n V_n}{2N} \sum_{r=1}^{N'} e^{-ik_n \xi_r} c_2 k_n^2 \eta_r^2 \tag{2.99}$$

$$\alpha_{n,3}' = -\frac{i}{2} k_n^2 V_n \alpha_n (1 + O(\varepsilon_{\text{mode}}^2, \varepsilon_{\text{gc}}^2)) \tag{2.100}$$

where c_1 and c_2 are O(1) and α_n is defined by (2.25). The $\alpha'_{n,1}$ term accounts for the expansion of ϵ near ω_n in the left-hand side of (2.29) and (2.41); it is easily estimated using (2.94) as $E_n O(\varepsilon^2_{\text{mode}})$.

The $\alpha'_{n,2}$ term accounts for the expansion with respect to η_r in (2.21) and (2.40). Physically, its main effect is related to the fact that the period of nonlinear oscillations of bulk particle motion in the modes depends on the oscillation amplitude. The small-amplitude assumption (2.14) for $|n| \leq M$ and spatial quasi-uniformity of guiding centres yield

$$\alpha'_{n,2} \sim \frac{m\omega^3_{\text{p}}}{q_*} \frac{k_n}{k^2_M} N^{-1/2}\varepsilon^2_{\text{ampl}}.$$ (2.101)

The $\alpha'_{n,3}$ term is defined by incorporating into the right-hand side of (2.25) the error estimates for the expansion of $\mathcal{F}$ in (2.18). In the square brackets of (2.25), the d/dt term is $O(\varepsilon_{\text{mode}}, \varepsilon_{\text{gc}})$ smaller than the first term. Since spatial quasi-uniformity ensures that, for $k_m \neq k_n$,

$$\alpha_n = \frac{q^2_*}{m\omega^2_0} \sum_{m=-M,m\neq n}^{M} O(E_m N^{-1/2}),$$ (2.102)

the estimate (2.100) scales like

$$\alpha'_{n,3} = EO(M^{1/2}N^{-1/2}).$$ (2.103)

Thus $\alpha'_{n,1} \sim \varepsilon_{\text{mode}}\alpha'_{n,3}$ and $\alpha'_{n,2} \sim \varepsilon_{\text{ampl}}M^{-1/2}\alpha'_{n,3}$ are negligible with respect to $\alpha'_{n,3}$ and finally

$$\varepsilon_{\text{mode}} = (M/N)^{1/2} = (k_M\lambda_D)^{1/2}(2\pi N_D)^{-1/2}$$ (2.104)

which is small in the plasma limit $N_D \rightarrow \infty$ provided that $k_M\lambda_D$ is finite. Note that N/M is just the number of particles per wavelength of the shortest mode retained. Hence, $\varepsilon_{\text{mode}}$ is also small if one restricts the modes to the long-wavelength ones ($k_M \ll k_D$).

Now turn to the error estimate in the particle equation of motion (2.91) and recall that the physical process causing relaxation of the modes is their coupling with (many) particles. For modes with wavenumbers over k_D, the particles with velocities near the mode phase velocity are numerous and F_n is readily estimated by (2.77). Then (2.79) implies that in (2.91)

$$A'_r = O((\lambda_D L/N)^{1/2}\omega^2_{\text{p}})$$ (2.105)

in the quadratic mean, so that the remainder $m A'_r$ is negligible with respect to the leading term in (2.91) if the long-wavelength mode amplitudes are not thermal[11].

---

[11] Here thermal is meant as due to the natural fluctuations of the bulk particles. In chapter 4 we will define a different thermal level due to tail particles.

Thus our scale separation involves four small parameters:

(i) $\varepsilon_{\text{mode}} = (M/N)^{1/2}$: Long-wavelength modes with thermal amplitudes may be neglected as modes are uncoupled in the reduced Hamiltonian H_{sc}. Furthermore the contribution of these modes to the force in (2.91) is of the size of (2.105), i.e. they are negligible. Then, *even if N_D is not large*, the parameter $\varepsilon_{\text{mode}}$ is small provided that $k_M \ll k_D$, i.e. only modes with wavelength much over the Debye length are excited. The smallness of $\varepsilon_{\text{mode}}$ follows from the weakness of coupling between fast modes and slow bulk particles as seen in (2.82). This improves over the usual Vlasovian derivation of the self-consistent dynamics, since kinetic theory implies the limit $N_D \rightarrow \infty$.

(ii) $\varepsilon_{\text{coupling}}$: This scales the mode amplitudes so that direct mode coupling is negligible and bulk-particle oscillations remain in the linear regime (in particular, it also controls $\varepsilon_{\text{ampl}}$). Its small value does not prevent large-wavelength modes becoming highly suprathermal. For $k_n \simeq k_D$ however, the phase velocity is close to v_T and the amplitude of the mode must be small in order to avoid too large a number of resonant particles. It must satisfy condition (2.86).

(iii) A third parameter of interest

$$\eta \equiv R/N \ll 1 \qquad (2.106)$$

indicates the extent to which the resonant source is perturbative for the waves. The regime $\eta = O(1)$ would lead to strong turbulence regimes. However, this parameter is not crucial to our derivation of the wave–particle dynamics: it occurs only in (2.94), as $N^{-1/2}\varrho$ would scale like η in regimes where a large fraction of quasi-resonant particles would become bunched in the wave. This would apply to the saturation regime of a cold beam–plasma instability, but if the wave amplitude E in (2.94) is large $\tau_{\text{mode}}\omega_p$ would still be large.

(iv) A fourth parameter, $\varepsilon_{\text{gc}} \leq \varepsilon_{\text{mode}}$, which is automatically small for a reference state in thermal equilibrium in the limit $N_D \rightarrow \infty$, characterizes the (slow) evolution of the distribution of guiding-centre velocities $\dot{\xi}_r$, without resorting to an intermediate kinetic equation for the particle motion and without introducing 'test' particles.

Finally, note that our error estimates remain valid if the long-wavelength mode amplitudes are thermal. This situation is relevant in discussing the initial conditions for instabilities such as the weak-warm-beam (also called bump-on-tail) instability (which motivated the introduction of the model with $M = 1$). However, in the low-amplitude regime, the source term in (2.48) and (2.49) is comparatively more important and, as the particles are close to a uniform distribution and quite independent, the source is then relatively 'noisy'.

2.7   Final form of the Hamiltonian

In the following chapters, we associate with the total number $N \gg 1$ of particles in the length L of the plasma the small parameter[12]

$$\varepsilon = N^{-1/2} q_* k_n V_n^{1/2} = \frac{2q_*}{\sqrt{L\epsilon_0 N}} = \omega_p \sqrt{\frac{2m}{N}}. \qquad (2.107)$$

The M wavenumbers of interest may occasionally be restricted to a smaller range than the full $0 < k_j \ll k_D$. In particular, the single-wave model $M = 1$ occurs naturally in the study of instabilities: if several unstable modes coexist in the plasma, the fastest-growing one typically becomes dominant over the other ones.

From now on, we denote by N the number of particles interacting with the waves (R until now). Therefore we write ε as

$$\varepsilon = \omega_p \sqrt{\frac{2m\eta}{N}} \qquad (2.108)$$

with η defined by (2.106).

After setting $m = 1$, i.e. identifying particle momentum with velocity, and rewriting $\omega_{n0} \equiv \omega_n$, the self-consistent Hamiltonian reads:

$$H_{sc}^{N,M} = \sum_{r=1}^{N} \frac{p_r^2}{2} + \sum_{j=1}^{M} \omega_{j0} \frac{X_j^2 + Y_j^2}{2}$$

$$+ \varepsilon \sum_{r=1}^{N} \sum_{j=1}^{M} k_j^{-1} \beta_j (Y_j \sin k_j x_r - X_j \cos k_j x_r) \qquad (2.109)$$

$$= \sum_{r=1}^{N} \frac{p_r^2}{2} + \sum_{j=1}^{M} \omega_{j0} I_j$$

$$- \varepsilon \sum_{r=1}^{N} \sum_{j=1}^{M} k_j^{-1} \beta_j \sqrt{2I_j} \cos(k_j x_r - \theta_j) \qquad (2.110)$$

where β_j is defined by (2.44). It generates the evolution equations

$$\dot{x}_r = p_r \qquad (2.111)$$

$$\dot{p}_r = \varepsilon \Re \left(\sum_{j=1}^{M} i\beta_j Z_j e^{ik_j x_r} \right) \qquad (2.112)$$

$$= -\varepsilon \sum_{j=1}^{M} i\beta_j \sqrt{2I_j} \sin(k_j x_r - \theta_j) \qquad (2.113)$$

---

[12] In the following ω_p may be considered as a quantity of order 1.

$$\dot{Z}_j = -\,\mathrm{i}\omega_{j0}Z_j + \mathrm{i}\varepsilon\beta_j k_j^{-1}\sum_{r=1}^{N}\mathrm{e}^{-\mathrm{i}k_j x_r} \tag{2.114}$$

where $Z_j = X_j + \mathrm{i}Y_j$, and

$$\dot{\theta}_j = \omega_j - \varepsilon\sum_{j=1}^{M} k_j^{-1}\beta_j(2I_j)^{-1/2}\cos(k_j x_r - \theta_j) \tag{2.115}$$

$$\dot{I}_j = \varepsilon\sum_{j=1}^{M} k_j^{-1}\beta_j(2I_j)^{1/2}\sin(k_j x_r - \theta_j). \tag{2.116}$$

As said previously, the total energy $E_{\mathrm{sc}} = H_{\mathrm{sc}}^{N,M}$ and the total momentum $P_{\mathrm{sc}} = \sum_{r=1}^{N} p_r + \sum_{j=1}^{M} k_j I_j$ are conserved.

2.8   Historical background and notes

The collective behaviour of electrons in a plasma was first described as eigenmodes by Bohm, Gross and Pines (Bohm and Gross 1949a, b, Bohm and Pines 1951, Pines and Bohm 1952). The self-consistent equations for the wave–particle system were introduced by Onishchenko *et al* (1970) and by O'Neil *et al* (1971) to analyse the nonlinear regime of the interaction of a small beam with an electrostatic mode in a plasma. The fact that they derive from a Hamiltonian was first noted by Mynick and Kaufman (1978) and a consistently Hamiltonian derivation of these equations was established by Tennyson *et al* (1994). All these derivations start from the kinetic theory of the plasma, i.e. from the Vlasov–Poisson system of equations. A direct derivation of the self-consistent model, within the framework of classical mechanics, was finally introduced by Escande (1991). Our presentation is based on the subsequent formulation by Antoni *et al* (1998).

The self-consistent Hamiltonian is a paradigm for the interaction of particles with collective degrees of freedom. The single-wave Hamiltonian ($M = 1$) is a general model for electrostatic instabilities (Crawford and Jayaraman 1999) and also captures essential features of the interaction of vortices with finite-velocity flow in hydrodynamics (del-Castillo-Negrete 2000, del-Castillo-Negrete and Firpo 2002). It is also worth noting that the single-wave model has close connections, to be further clarified, with systems of coupled nonlinear (dissipative) oscillators, such as those first studied by Kuramoto (1984).

The core of this derivation is not the long-range nature of the Coulomb interaction but the collective dynamics. Indeed, the key to the construction of the self-consistent wave–particle Hamiltonian is the mean-field limit, which enables a particle to interact with a large number of partners—in contradistinction with systems on a lattice, where the number of interacting neighbours is constant in the thermodynamic limit; and with gas models like the hard spheres, where the number of interacting neighbours is also limited by geometry (at most six in two space dimensions). The fact that the Coulomb potential does not decay rapidly for $r \to \infty$ was important only to ensure the finiteness of the plasma frequency $\lim_{k_n \to 0} k_n^2 V_n q_*^2 / m$. As an extreme case one directly obtains the single-wave model in analysing the Langmuir modes of the dynamical XY model, which has the two-body interaction $V(x) = \cos x - 1$ (Antoni and Ruffo 1995).

This universality of the wave–particle self-consistent Hamiltonian model is not so surprising. In fact, the derivation here relied much on the fact that the potential Fourier coefficients V_n are positive. Their scaling with k_n as k_n^{-2} enabled us to express our results in terms of the plasma frequency ω_p but in the weak-intensity regime the plasma waves are not coupled to each other, so that our analysis is easily extended to other repulsive ($V_n \geq 0$) potentials with a mean-field (i.e. $1/N$ pairwise) interaction.

Chapter 3

Dynamics of the small-amplitude wave–particle system

This chapter deals with the self-consistent dynamics in the case of waves with a small amplitude. It provides the analogue of the Landau and van Kampen description of eigenmodes in Vlasovian plasmas, with the advantage of emphasizing mechanical intuition at each step of the calculations. Because of the small wave amplitude, particle trajectories may be considered as nearly free. Such an approximation is correct as long as nonlinear effects like trapping or chaos may be neglected. This translates into upper bounds in time for the description of the dynamics equal to the trapping times of the various particles in the global wave potential (see section 1.2) or equal to the time of spreading of the position due to chaotic transport (see section 6.8.2). Indeed for tail particles, linearizing the motion is a lot more demanding than for bulk ones: a new limiting time-scale is involved.

The treatment proceeds with the following steps. In section 3.1 we exhibit a reference state similar to that of section 2.2 where the particles stay on multibeam arrays and the waves keep a zero amplitude. Then we give the explicit and complete solution of the small perturbations close to this equilibrium. This solution is made up of two parts: one corresponding to the dynamics with small waves (section 3.4); and the (simpler) so-called ballistic motion related to non-oscillatory perturbations to the beams (section 3.3). A new feature with respect to the Bohm–Gross modes is that a given wavenumber is no longer related to a unique frequency.

The waves may be damped or unstable due to their interaction with the particles. Three limiting cases are considered as far as the particle velocity distribution is concerned. In the cases of a monokinetic beam and of two monokinetic beams (section 3.7), the origin of stable and unstable behaviour is easily identified. In the case of many beams with a smooth coarse-grained velocity distribution function of the particles (section 3.8), one recovers the growth rate of unstable Langmuir waves introduced by Landau in 1946 if the slope of the distribution function of tail particles is positive. If it is negative, a given wavenumber corresponds to a family of beam modes which are reminiscent of the van Kampen modes of a Vlasovian plasma. For a typical initial condition a Langmuir wave is found to damp with the Landau damping rate in the limit $N \to \infty$. This coherence with the Vlasovian approach agrees with the limit theorem of appendix G. Finally (section 3.9), a second-order perturbation calculation shows that an unstable or damped wave tends to synchronize particles with it.

3.1 Reference state with a vanishing electric field

The self-consistent Hamiltonian (2.109) describes waves and particles as a mutually interacting system. As a consequence, the particles yield a source term in (2.114) and, in general, the electric field cannot vanish for all times. This is the signature of a spontaneous emission of waves by the particles. For the N-body dynamics, exercise 2.7 showed that a Fourier component of the field could vanish for all times if there was a destructive interference of the electric fields produced by individual particles. This interference was obtained by setting the particles on beam arrays. Such a property still holds after the elimination of the individual degrees of freedom of the bulk. As exercise 2.7 showed that one cannot ensure destructive interference for all field components, the self-consistent Hamiltonian (2.110) is even more appropriate since it involves only a finite number of field components. In this chapter, we consider only waves ($1 \leq m \leq M$) with wavenumbers in the form

$$k_m = 2\pi \frac{\mu + m}{L} \tag{3.1}$$

where the integer $\mu \geq 0$ is chosen such that the slowest mode (with index M) has a velocity larger than the largest velocity in the background plasma.

Setting the bulk particles on beam arrays meant in exercise 2.7 setting their collective field to zero. Here we set the tail particles on beam arrays (Escande 1991, Escande *et al* 1996): the particles of the self-consistent dynamics are the elements of a set of b beams where each beam is an array (see figure 3.1).

Beam s has a velocity v_s and N_s particles ($\sum_{s=1}^{b} N_s = N$) which are uniformly distributed on the spatial period L of the plasma. Let x_{ns} be the position of its nth particle (in the following (x_l, p_l) still refer to the lth particle in Hamiltonian (2.109), with $l = n$ for $s = 1$, $l = n + \sum_{r=1}^{s-1} N_r$ for $s > 1$). Assuming vanishing waves, we get

$$x_{ns}(t) = v_s t + x_{ns}(0) = v_s t + n \frac{L}{N_s} + \phi_s. \tag{3.2}$$

The source term in (2.114), related to the emission of waves by particles, is proportional to

$$\sum_{l=1}^{N} e^{ik_j x_l(t)} = \sum_{s=1}^{b} \sum_{n=1}^{N_s} e^{ik_j x_{ns}(t)}. \tag{3.3}$$

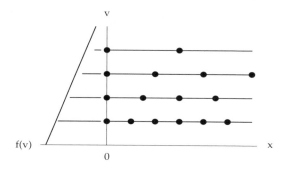

Figure 3.1. Reference multibeam state.

Relation (3.2) makes this vanish provided the ratio $(\mu + j)/N_s$ is not an integer whatever j is. For this to be satisfied, a sufficient condition is

$$N_s > (\mu + M) \qquad \forall s \in [1, b]. \tag{3.4}$$

Thus if the waves have zero amplitude initially, their amplitudes stay zero, which implies free motion of the particles, preserving the regular beam arrays. Such a family of states, parametrized by one phase ϕ_s and one velocity v_s per beam, forms a $2b$-dimensional manifold $\mathcal{S}$ in the $2(N + M)$-dimensional phase space, on which orbits are quasi-periodic with periods $T_s = L/v_s$. In the simplest case of such states (which is sufficient for the following analysis), all velocities v_s are integer multiples of a given Δp, so that the motion is periodic with period $T = L/\Delta p$. This state with zero electric field and a set of beam arrays is, hereafter, called the reference state.

3.2   Small perturbation to the reference state

Let us consider an orbit on manifold $\mathcal{S}$ defined by

$$x_l(t) = x_{ns}(t), \qquad p_l(t) = v_s \tag{3.5}$$
$$X_j(t) = Y_j(t) = 0. \tag{3.6}$$

Let $(\ldots, \delta x_l, \delta p_l, \ldots, \delta X_j, \delta Y_j, \ldots)$ be a perturbation of this orbit. Its evolution is determined by the linearized equations deduced from (2.114)

$$\delta \dot{x}_l = \delta p_l \tag{3.7}$$

$$\delta \dot{p}_l = \varepsilon \Re \left(\sum_{j=1}^{M} i \beta_j Z_j e^{i k_j x_l} \right) \tag{3.8}$$

$$\dot{Z}_j = -i \omega_{j0} Z_j + \varepsilon \beta_j \sum_{l=1}^{N} \delta x_l e^{-i k_j x_l} \tag{3.9}$$

where $Z_j = \delta X_j + i\delta Y_j$ is the complex amplitude of the electrostatic wave.

Such an approximation of the particle dynamics is meaningful for times such that the wave–particle interaction does not disturb the free-particle dynamics too much. In particular for particles resonant with a wave, the time must be smaller than the bounce time in the wave's potential well:

$$t \ll \omega_b^{-1} = (\varepsilon k\beta)^{-1/2}|Z|^{-1/4} \qquad (3.10)$$

following (1.15) with the wave's complex amplitude (2.46) in the Hamiltonian (2.110).

Denoting by $U(t)$ the vector $(\ldots, \delta x_l, \ldots, \delta p_l, \ldots, Z_j, \ldots)$, the set of differential equations (3.7)–(3.9) may be formally written:

$$\dot{U}(t) = [M_0 + \varepsilon M_1(t)]U(t) \qquad (3.11)$$

where M_0 is a constant matrix and M_1 a matrix with quasi-periodic coefficients of periods T_s. In the case of a single period T, equation (3.11) belongs to the Floquet class of differential equations. Then its solutions are of the type

$$U(t) = V(t)e^{\alpha t} \qquad (3.12)$$

where $V(t)$ is a vector of period T and α a complex number. Such equations are met, for instance, in the study of the perturbations of planetary motion and are generally not explicitly solvable with elementary functions even for the simplest one, the Mathieu equation. In contrast, the high-dimensional quasi-periodic Floquet problem defined by equations (3.7)–(3.9) is explicitly solvable. Its full solution will turn out to be a superposition of wavelike and ballistic solutions.

Indeed, the linearized dynamics is drastically simplified by the facts that the reference state has zero wave amplitudes; that a particle motion can deviate from ballistic only due to the waves; and that a wave can change amplitude and phase only due to interaction with particles. Mathematically, the occurrence of x_l in (3.8) and (3.9) only through the trigonometric functions $e^{ik_j x_l(t)}$, which makes (3.11) a Floquet problem, invites one to introduce the Fourier representation of the perturbation $(\delta x_l, \delta p_l)$ (see appendix A). Indeed, we have a complete representation of the initial perturbation for each beam s in the form

$$\delta p_{ns}(t) = \sum_{m \in \mu_s} A_{ms}(t)e^{ik_m x_{ns}(t)} \qquad (3.13)$$

$$\delta x_{ns}(t) = -i \sum_{m \in \mu_s} C_{ms}(t)e^{ik_m x_{ns}(t)} \qquad (3.14)$$

where μ_s is defined as the set of integers

$$\mu_s = \begin{cases} \{j \in \mathbb{Z} : |\mu + j| \le (N_s - 1)/2\} & \text{if } N_s \text{ is odd} \\ \{j \in \mathbb{Z} : 1 - N_s/2 \le \mu + j \le N_s/2\} & \text{if } N_s \text{ is even.} \end{cases} \qquad (3.15)$$

Thus μ_s has just N_s elements[1]. The reciprocal change of variables reads (see appendix A)

$$A_{ms}(t) = N_s^{-1} \sum_{n=1}^{N_s} \delta p_{ns}(t) e^{-ik_m x_{ns}(t)} \tag{3.16}$$

$$C_{ms}(t) = iN_s^{-1} \sum_{n=1}^{N_s} \delta x_{ns}(t) e^{-ik_m x_{ns}(t)}. \tag{3.17}$$

This yields $(N_s - 1)/2$ complex coefficients and one real $(A_{-\mu,s})$ if N_s is odd, so that the number of independent real numbers necessary to define the A_{ms} is equal to the number of particles in beam s. A similar counting holds for the C_{ms}.

Exercise 3.1. Extend the counting of independent real variables to the case of even N_s.

Exercise 3.2. For any $m \in \mu_s$, show that $A^*_{m-\mu,s} = A_{-m-\mu,s}$ and $C^*_{m-\mu,s} = -C_{-m-\mu,s}$ as δx_l and δp_l are real. For $c \in \mathbb{Z}$, let $m' = m + cN_s$ and $k_{m'} = 2\pi(m' + \mu)/L$. With $\kappa_s = e^{-\pi i c N_s(\phi_s + v_s t)/L}$, show that $A_{m',s} = \kappa_s^2 A_{m,s}$ and $C_{m',s} = \kappa_s^2 C_{m,s}$ if one extends (3.16)–(3.17) to arbitrary integer m'.

The evolution equations for the Fourier variables A_{ms}, C_{ms} are readily found as

$$\dot{C}_{ms} = -ik_m v_s C_{ms} + iA_{ms} \tag{3.18}$$

$$\dot{A}_{ms} = -ik_m v_s A_{ms} + i \sum_{j=1}^{M} \frac{\varepsilon \beta_j}{2N_s} \sum_{n=1}^{N_s} (Z_j e^{-i(k_m - k_j) x_{ns}(t)} - Z^*_{-j} e^{-i(k_m + k_j) x_{ns}(t)}) \tag{3.19}$$

$$\dot{Z}_j = -i\omega_{j0} Z_j + i\varepsilon \beta_j \sum_{s=1}^{b} \sum_{n=1}^{N_s} \sum_{m \in \mu_s} C_{ms} e^{i(k_j - k_m) x_{ns}(t)}. \tag{3.20}$$

The right-hand side of (3.19) and (3.20) involves trigonometric sums, which may depend periodically on time t. However, only $m = j$ remains in (3.20) thanks to the orthonormality identity (A.3). Moreover, in (3.19), the Z_j contributions can be simplified by a reasoning similar to that in section 3.1, on replacing condition (3.4) by the more stringent

$$N_s > 2(\mu + M) \qquad \text{for } 1 \le s \le b. \tag{3.21}$$

[1] This tedious definition of μ_s is due to the definition (3.1) of k_m. If N_s is odd, in (3.13)–(3.14) k_m sweeps the range $[-\pi(N_s - 1)/L, \pi(N_s - 1)/L]$ by steps of $2\pi/L$. If N_s is even, k_m sweeps the range $[-2\pi(N_s/2 - 1)/L, \pi N_s/L]$. See also exercise 3.2.

Then (3.19) reduces to

$$\dot{A}_{ms} = -ik_m v_s A_{ms} + i\frac{\varepsilon}{2} \sum_{j=1}^{M} \beta_j (Z_j \delta_{j,m} - Z_j^* \delta_{j,N_s-2\mu-m}) \qquad (3.22)$$

where $\delta_{m,j}$ is the Kronecker symbol (see section A.1).

Exercise 3.3. Show that, for any s and m, there is at most one contribution in (3.22) from the modes, in the form of either Z_j or Z_j^*.

As the set μ_s includes the set $\mu_{sM} = \{j \in \mathbb{Z} : 1 \le j \le M$ or $1 \le N_s - 2\mu - j \le M\}$ defining the Langmuir modes of interest, we denote by $\mu_s' = \mu_s \setminus \mu_{sM}$ its complement. The discrete Fourier expansion of initial conditions may be split into two sums:

$$\delta p_{ns}(t) = \sum_{j=1}^{M} (A_{js} e^{ik_j x_{ns}(t)} + A_{js}^* e^{-ik_j x_{ns}(t)}) + \sum_{m \in \mu_s'} A_{ms} e^{ik_m x_{ns}(t)} \qquad (3.23)$$

$$\delta x_{ns}(t) = i \sum_{j=1}^{M} (C_{js} e^{ik_j x_{ns}(t)} + C_{js}^* e^{-ik_j x_{ns}(t)}) + i \sum_{m \in \mu_s'} C_{ms} e^{ik_m x_{ns}(t)}. \qquad (3.24)$$

This splitting of particle perturbations into two families of components enables evidencing two behaviours in the particle–wave dynamics. The expansion related to the k_j corresponds to the resonant interaction of tail particles with Langmuir waves described in the previous chapter, while the set of the k_m with $m \in \mu_s'$ will be found to be off resonance with Langmuir waves.

Exercise 3.4. As the wave–particle dynamics is Hamiltonian, the final equations (3.18), (3.20) and (3.22) should also derive from a Hamiltonian. Assume N_s to be odd and define $\mu_s^+ = \{m \in \mathbb{Z} : 1 \le m + \mu \le (N_s - 1)/2\}$. Let $A_{ms}' = \Re A_{ms}$, $A_{ms}'' = \Im A_{ms}$, $D_{ms}' = 2N_s C_{ms}'' = 2N_s \Im C_{ms}$ and $D_{ms}'' = -2N_s C_{ms}' = -2N_s \Re C_{ms}$ for $m \in \mu_s^+$. Let $D_{-\mu,s} = -iN_s C_{-\mu,s}$. Define for $1 \le s \le b$

$$H_{-\mu,s} = \frac{N_s}{2} A_{-\mu,s}^2, \qquad (3.25)$$

for $1 \le s \le b$ and $m \in \mu_s' \cap \mu_s^+$

$$H_{ms} = N_s A_{ms} A_{ms}^* + k_m v_s (A_{ms}' D_{ms}'' - A_{ms}'' D_{ms}'), \qquad (3.26)$$

and for $1 \le j \le M$

$$H_j = \sum_{s=1}^{b} [N_s A_{js} A_{js}^* + k_j v_s (A_{js}' D_{js}'' - A_{js}'' D_{js}')]$$
$$+ \frac{\omega_{j0}}{2}(X_j^2 + Y_j^2) + \frac{\varepsilon}{2}\beta_j \sum_{s=1}^{b}(D_{js}' Y_j - D_{js}'' X_j). \qquad (3.27)$$

Considering each D_α as the generalized coordinate conjugate to the generalized momentum A_α, show that the Hamiltonian

$$H_{\text{modes}} = \sum_{s=1}^{b} \left[H_{-\mu,s} + \sum_{m \in \mu_s^+} H_{ms} \right] + \sum_{j=1}^{M} H_j \qquad (3.28)$$

generates the evolution equations (3.18), (3.20) and (3.22). Note that the expression of H_{modes} as a sum of bilinear terms ensures that the evolution equations are linear. Moreover, as no term couples distinct values of m, the resulting evolution equations decouple modes with different m.

Exercise 3.5. Does each of the H_{ms} and H_j admit invariants, say P_{ms} and P_j, similar to P? Do other invariants exist?

3.3   Ballistic solutions

We first discuss the case $m \in \mu_s'$. Then

$$\dot{C}_{ms} = -ik_m v_s C_{ms} + iA_{ms} \qquad (3.29)$$
$$\dot{A}_{ms} = -ik_m v_s A_{ms} \qquad (3.30)$$

so that

$$A_{ms}(t) = A_{ms}(0)e^{-ik_m v_s t} \qquad (3.31)$$
$$C_{ms}(t) = (C_{ms}(0) + it A_{ms}(0))e^{-ik_m v_s t}. \qquad (3.32)$$

These Fourier components being insensitive to the modes Z_j preserve the ballistic free motion of the particles. This ballistic free motion allows for two kinds of perturbation:

- sliding the positions, with $A_{ms}(0) = 0$—this implies a genuinely oscillating time dependence at the beam frequency $k_m v_s$; and
- shifting the reference velocity, with $C_{ms}(0) = 0$—this implies oscillations at frequency $k_m v_s$ for the velocity along with the drifting-oscillating time dependence for positions (see exercise 3.8).

Exercise 3.6. Show that the resulting motion of particles reads:

$$\delta p_{ns}(t) = \delta p_{ns}(0) = \sum_{m \in \mu_s'} A_{ms}(0)e^{ik_m x_{ns}(0)} \qquad (3.33)$$
$$\delta x_{ns}(t) = \delta x_{ns}(0) + \delta p_{ns}(0)t$$
$$= \sum_{m \in \mu_s'} (iC_{ms}(0) + A_{ms}(0)t)e^{ik_m x_{ns}(0)}. \qquad (3.34)$$

For each beam s, the set μ'_s contains $N_s - 2M$ indices, so that the dimension of the subspace of ballistic solutions is

$$d_{\text{bal}} = \sum_{s=1}^{b} (2N_s - 4M) = 2N - 4Mb. \qquad (3.35)$$

3.4   Wavelike solutions

Now we turn to the case $1 \le m = j \le M$, for which the evolution equations read:

$$\dot{C}_{js} = -ik_j v_s C_{js} + iA_{js} \qquad (3.36)$$

$$\dot{A}_{js} = -ik_j v_s A_{js} + i\frac{\varepsilon}{2}\beta_j Z_j \qquad (3.37)$$

$$\dot{Z}_j = -i\omega_{j0} Z_j - i\varepsilon\beta_j \sum_{s=1}^{b} N_s C_{js}. \qquad (3.38)$$

Exercise 3.7. Show that the case $m = N_s - 2\mu - j$ is redundant. Hint: see exercise 3.2.

Equations (3.36)–(3.38) form a system of $2b + 1$ homogeneous linear first-order differential equations with constant coefficients, for the $2b + 1$ complex variables (C_{js}, A_{js}, Z_j). They involve only the modulation of the beams with the Fourier wavenumber k_j. The major advantage of this system is that waves with distinct wavenumbers $k_{j'} \ne k_j$ are uncoupled from the wave of interest.

In the case of zero coupling, Z_j is purely oscillatory with pulsation ω_{j0} and defines a propagating Langmuir wave. We still characterize Langmuir waves by their wavenumber but in contrast to the Bohm–Gross case they are not endowed with a unique frequency and their amplitude may vary. We look for a similar solution for non-zero coupling. Let

$$C_{js} = c_s e^{\alpha t}, \qquad A_{js} = a_s e^{\alpha t}, \qquad Z_j = z e^{\alpha t} \qquad (3.39)$$

where c_s, a_s, z and α are complex constants. Then (3.36)–(3.38) reduce to the linear system

$$\alpha \mathbf{C} = -i\mathbf{M} \cdot \mathbf{C} \qquad (3.40)$$

for the vector $\mathbf{C} \in \mathbb{C}^{2b+1}$

$$\mathbf{C} = \begin{pmatrix} c_1 \\ \cdots \\ c_b \\ a_1 \\ \cdots \\ a_b \\ z \end{pmatrix} \qquad (3.41)$$

where M is a $(2b + 1) \times (2b + 1)$ square matrix with real elements $m_{r',r}$, whose only non-zero elements are, for $1 \le s \le b$: $m_{s,s} = m_{b+s,b+s} = k_j v_s$, $m_{s,b+s} = 1$, $m_{b+s,2b+1} = -\varepsilon\beta_j/2$, $m_{2b+1,s} = -\varepsilon\beta_j N_s$, $m_{2b+1,2b+1} = \omega_{j0}$. Non-zero solutions of (3.40) are obtained if and only if $\alpha = -i\sigma$ where σ is an eigenvalue of M and C is an associated eigenvector. Given $\sigma = \sigma_{jr} = i\alpha_{jr}$, one readily finds the eigenvector

$$a_{jrs} = \frac{\varepsilon}{2}\beta_j(k_j v_s - \sigma)^{-1} \tag{3.42}$$

$$c_{jrs} = \frac{\varepsilon}{2}\beta_j(k_j v_s - \sigma)^{-2} \tag{3.43}$$

$$z = 1 \tag{3.44}$$

where we use the normalization freedom on C to impose (3.44). With the index $1 \le r \le 2b + 1$, the eigenvectors form a square matrix [C] such that $-i\mathsf{M} \cdot [\mathsf{C}] = \mathrm{diag}(\alpha_{jr}) \cdot [\mathsf{C}]$. These coefficients show that the beams which are most excited by the perturbation are those with velocity v_s closest to ω_{jr}/k_j.

Substituting these expressions into the equations yields the condition for a non-zero solution, namely the dispersion relation

$$\alpha = -i\omega_{j0} + \frac{i}{2}\varepsilon^2\beta_j^2 \sum_{s=1}^{b} N_s(\alpha + ik_j v_s)^{-2} \tag{3.45}$$

i.e. $\sigma \equiv i\alpha = \chi(\sigma)$ with

$$\chi(\sigma) = \omega_{j0} - \frac{\varepsilon^2\beta_j^2 N}{2} \sum_{s=1}^{b} \frac{N_s}{N}(\sigma - k_j v_s)^{-2}. \tag{3.46}$$

Multiplying both sides of (3.45) by the product of all denominators yields a polynomial of degree $2b + 1$ in α, which is the characteristic polynomial of $-i\mathsf{M}$. Let α_{jr} (for $1 \le r \le 2b+1$) be its (generically distinct) $2b+1$ roots. Note that for all (j, r, s), $\alpha_{jr} + ik_j v_s$ is non-zero since its vanishing would imply a pole on the right-hand side of (3.45). As the coefficients of $i\alpha$ in this odd-degree polynomial are real, at least one of the α_{jr} is purely imaginary $-i\omega_{jC}$ (with $\omega_{jC} > \omega_{j0}$ and $\gamma_{jC} = 0$); the others appear as b pairs of solutions $\pm\gamma_{jr} - i\omega_{jr}$ (with γ_{jr} real or imaginary). We shall return to this in section 3.7.

Exercise 3.8. In the limit $\beta_j \to 0$ what do the roots σ_{jr} and the corresponding eigenvectors become? Show that one root remains simple and that in the limit there are b double roots with only b eigenvectors of M: each of the double roots corresponds to an upper triangular Jordan block $\begin{pmatrix} \alpha & 1 \\ 0 & \alpha \end{pmatrix}$. Show that each double root and the corresponding vector space correspond to two ballistic modes with only one beam excited, while the simple root corresponds to a pure wave non-interacting with the beams.

Exercise 3.9. The Hamiltonian (3.27) generating these evolution equations can be written in the form

$$H_j = \sum_{r=1}^{4b+2} \lambda_r u_r^2 \tag{3.47}$$

where the coefficients λ_r take only values $+1$, 0 and -1 and the variables $u_r \in \mathbb{R}$ are linear combinations of $\Re A_{js}$, $\Im A_{js}$, $\Re C_{js}$, $\Im C_{js}$, X_j and Y_j. Show that the signature of H_j is $(2(b+1)+, 2b-)$, i.e. that one can choose the u_r so that $\lambda_r = +1$ for $1 \le r \le 2(b+1)$ and $\lambda_r = -1$ for $2(b+1)+1 \le r \le 2(2b+1)$. Noting that the u_r 'mix' generalized coordinates and momenta, this does not warrant that there are many unstable eigenmodes, but it allows for as many as b unstable eigenmodes[2].

This construction provides us, for each of the M waves with wavenumbers and frequencies (k_j, ω_{j0}), with $2b+1$ wavelike solutions of the linear system (3.7)–(3.9). Generically the M distinct dispersion relations (3.45) define M times $2b+1$ different complex growth rates $\alpha_{jr} = \gamma_{jr} - i\omega_{jr}$. Thus our construction yields $(2b+1)M$ linearly independent solutions[3]. The linear superposition of these solutions yields the following motion for the lth particle:

$$\delta p_{ns}(t) = \varepsilon \Re \left(i \sum_{j=1}^{M} \beta_j \sum_{r=1}^{2b+1} \frac{\zeta_{jr}}{\alpha_{jr} + ik_j v_s} e^{\alpha_{jr} t + ik_j x_{ns}} \right) \tag{3.48}$$

$$\delta x_{ns}(t) = \varepsilon \Re \left(i \sum_{j=1}^{M} \beta_j \sum_{r=1}^{2b+1} \frac{\zeta_{jr}}{(\alpha_{jr} + ik_j v_s)^2} e^{\alpha_{jr} t + ik_j x_{ns}} \right) \tag{3.49}$$

along with the evolution of the jth wave

$$Z_j(t) = \sum_{r=1}^{2b+1} \zeta_{jr} e^{\alpha_{jr} t}. \tag{3.50}$$

Each Langmuir wave turns out to be the superposition of $2b+1$ eigenmodes whose amplitude is defined by the ζ_{jr}.

3.5   Initial value problem

Now we check whether system (3.7)–(3.9) admits other solutions than a linear combination of (3.33)–(3.34) and (3.48)–(3.50). Index r takes on $2b+1$ values,

---

[2] If the signature of H_j were fully positive or fully negative, this would imply that all eigenvalues in the dispersion relation would be imaginary. A deeper discussion would bring in the concept of 'negative energy modes', useful to the understanding of some instabilities.

[3] Even if two wavelike solutions with $j \ne j'$ satisfy $\omega_{jr} = \omega_{j'r'}$, the corresponding wavelike modes are independent, because they are obtained by solving independent systems (3.40).

ζ_{jr} corresponds to two real dimensions, and there are M values for j. Therefore the space of solutions determined by (3.48)–(3.50) has generically dimension

$$d_{\text{wav}} = 2M(2b + 1). \tag{3.51}$$

The space of all solutions found so far to the linearized dynamics has a dimension which is the sum of those of the subspaces of wavelike and ballistic solutions. According to (3.35) and (3.51), this is

$$d_{\text{wav}} + d_{\text{bal}} = 2N + 2M \tag{3.52}$$

that is the dimension of the differential system (3.7)–(3.9). Hence this system is completely solved.

In order to make explicit the solution of (3.7)–(3.9) completely, the $(2b + 1)M$ constants ζ_{jr} must be expressed in terms of initial conditions. For a given wavenumber k_j, in view of (3.42)–(3.43), this amounts to solving the linear system of $2b + 1$ equations with $2b + 1$ unknowns ζ_{jr}

$$A_{js}(0) = \frac{i}{2}\varepsilon\beta_j \sum_{r=1}^{2b+1} \frac{\zeta_{jr}}{\alpha_{jr} + ik_j v_s} \tag{3.53}$$

$$C_{js}(0) = -\frac{\varepsilon}{2}\beta_j \sum_{r=1}^{2b+1} \frac{\zeta_{jr}}{(\alpha_{jr} + ik_j v_s)^2} \tag{3.54}$$

$$Z_j(0) = \sum_{r=1}^{2b+1} \zeta_{jr}. \tag{3.55}$$

Initial data $(\delta x_{ns}(0), \delta p_{ns}(0))$ yield the $(A_{js}(0), C_{js}(0))$ through (3.16) and (3.17). Then the ζ_{jr} are computed from $(\ldots Z_j(0), \ldots, A_{js}(0), C_{js}(0), \ldots)$ through (3.53)–(3.55). Note that the wave variable Z_j is not merely the amplitude ζ_{jr} of a wavelike eigenmode: the physical wave results from a linear superposition of $2b + 1$ wavelike modes.

As the α_{jr} are eigenvalues of the matrix $-iM$, and the terms in (3.53)–(3.55) are their eigenvectors, solving this system reduces to finding the left eigenvectors of this matrix. These are the (row) vectors $C' = (c'_1, \ldots, c'_b, a'_1, \ldots, a'_b, z') \in \mathbb{C}^{2b+1}$ such that $C'\alpha_{jr} = -iC' \cdot M$, actually

$$c'_{jrs} = \varepsilon\beta_j N_s (\sigma_{jr} - k_j v_s)^{-1} z'_{jr} \tag{3.56}$$

$$a'_{jrs} = -\varepsilon\beta_j N_s (\sigma_{jr} - k_j v_s)^{-2} z'_{jr}. \tag{3.57}$$

It is convenient to normalize the right and left eigenvectors so that $C' \cdot C = 1$ if C' and C correspond to the same eigenvalue. This yields the normalization

$$z'_{jr} = \left[1 + \varepsilon^2\beta_j^2 \sum_s N_s (\sigma_{jr} - k_j v_s)^{-3}\right]^{-1} \tag{3.58}$$

and the solution to (3.53)–(3.55)

$$\zeta_{jr} = z'_{jr} Z_j(0) + \sum_s (c'_{jrs} C_{js}(0) + a'_{jrs} A_{js}(0)).$$ (3.59)

The dominant terms in (3.59) are those for which $|\sigma_{jr} - k_j v_s|$ is small, i.e. those corresponding to beams moving at a velocity close to the mode phase velocity $\Re\sigma_{jr}/k_j$, within a range of the order of $|\Im\sigma_{jr}/k_j|$.

Exercise 3.10. What are the physical dimensions of each entry of $\mathbf{M}$, an eigenvector $\mathbf{C}$ and an eigenvector $\mathbf{C}'$? Show that the square matrix $[\mathbf{C}']$, whose rows are the left eigenvectors, verifies the equations:

- $[\mathbf{C}'] \cdot [\mathbf{C}] = \mathbf{I}$ and $[\mathbf{C}] \cdot [\mathbf{C}'] = \mathbf{I}$, i.e. the matrices are reciprocal to each other;
- $[\mathbf{C}'] \cdot \mathbf{M} = i[\mathbf{C}'] \cdot \text{diag}(\alpha_{jr})$.

Exercise 3.11. Write explicitly the coefficients ζ_{jr} in terms of initial data $(\delta x_{ns}(0), \delta p_{ns}(0), Z_j(0))$.

3.6 Dispersion relation for wavelike modes

The wavelike modes are characterized by the rate constants α_{jr}. We note that $\alpha = 0$ is not a solution of the dispersion relation (3.45), because this would imply $\omega_{j0} + (\varepsilon^2 \beta_j^2/2) \sum_s N_s (k_j v_s)^{-2} = 0$, which is impossible if the frequencies ω_{j0} are positive. Thus all wavelike solutions are genuinely time dependent.

We also note that index r takes on $2b + 1$ values. This fact has two interpretations. First, it follows from the algebra of dispersion relation (3.45), which has $2b + 1$ complex solutions. But it also corresponds to the dimension counting for the non-ballistic modes, i.e. the number of equations and unknowns in system (3.53)–(3.55).

Among the rate constants α_{jr}, one at least plays a special role. Indeed, we noted that at least one α_{jr} is purely imaginary, $\alpha_{jC} = -i\omega_{jC}$, solving

$$\omega_{jC} - \omega_{j0} = \frac{\varepsilon^2 \beta_j^2}{2} \sum_s N_s (k_j v_s - \omega_{jC})^{-2}.$$ (3.60)

Thus $\omega_{jC} \geq \omega_{j0}$, with equality in the limit $\varepsilon \to 0$ of free waves. Moreover, the right-hand side of (3.60) is a sum of positive functions diverging at $\omega = v_s/k_j$, so that $\omega_{jC}/k_j < v_s$ for all beam velocities larger than ω_{j0}/k_j. Whether (3.45) admits other pure imaginary solutions depends on the set of beam velocities v_s, numbers of particles N_s, wavenumber k_j and free wave frequency ω_{j0}.

Complex solutions read $\alpha_{jr} = \gamma_{jr} - i\omega_{jr}$, with

$$\gamma_{jr} = \varepsilon^2 \beta_j^2 \sum_{s=1}^b N_s \frac{\gamma_{jr}(k_j v_s - \omega_{jr})}{[\gamma_{jr}^2 + (k_j v_s - \omega_{jr})^2]^2}$$ (3.61)

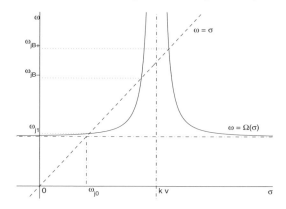

Figure 3.2. Graphical solution of the dispersion relation for one beam and one wave.

$$\omega_{jr} = \omega_{j0} + \frac{\varepsilon^2 \beta_j^2}{2} \sum_{s=1}^{b} N_s \frac{(k_j v_s - \omega_{jr})^2 - \gamma_{jr}^2}{[\gamma_{jr}^2 + (k_j v_s - \omega_{jr})^2]^2}. \tag{3.62}$$

3.7   Physical interpretation: cold beams

Now that we have an explicit general solution of the linearized dynamics, let us investigate the specific properties of its various components. As the general solution is a linear combination of the fundamental solutions, it is sufficient to discuss these solutions separately. Since we are mostly interested in the evolution of the wave, we focus on the wavelike modes, for the special cases of one beam and two beams and in the limit of many beams. We discuss the ballistic modes inasmuch they are related with the wavelike modes.

3.7.1   Case of a single beam

The simplest case corresponds to a single beam ($b = 1$) of particles interacting with the wave (Akhiezer and Fainberg 1951). Let v be the velocity of the beam. Then (3.45) reads, with $\sigma = i\alpha = \omega + i\gamma \in \mathbb{C}$,

$$\sigma = \omega_{j0} + q(\sigma - kv)^{-2} \tag{3.63}$$

where $q = N\varepsilon^2 \beta_j^2/2 \geq 0$. The nature of each solution to this algebraic equation for σ is easily understood from a graph. The right-hand side of (3.63) defines a function $\Omega(\sigma)$, displayed in figure 3.2, the fixed points of which are the solutions to (3.63). For all $\sigma \in \mathbb{R}$, one has $\Omega(\omega) > \omega_{j0}$ and Ω diverges at $k_j v$.

Let $q_M = 4(k_j v - \omega_{j0})^3/27$. Assume that $\omega_{j0} < k_j v$, as in figure 3.2. Then, if $0 < q < q_M$, $\Omega(\sigma)$ admits three real fixed points, while if $q > q_M$, $\Omega(\sigma)$ admits only one real fixed point and two complex conjugate ones, as displayed in figure 3.3(a). The weak-coupling regime $0 < q < q_M$ allows for three oscillating modes of the beam–wave system. Denote their frequencies by ω_{jr} ($r \in \{1, 2, 3\}$) with $\omega_{j0} \leq \omega_{j1} \leq \omega_{j2} \leq \omega_{j3}$:

- Consider first the mode with frequency ω_{j1}. In the limit $\varepsilon \to 0$, ω_{j1} tends to ω_{j0}. The motions (3.48), (3.49) of the particles generated by this mode are oscillations with an amplitude proportional to $\varepsilon \beta_j z$, which vanishes in the limit $\varepsilon \to 0$, while the amplitude z of the wave oscillations does not vanish. This mode thus describes the propagation of the wave, mildly perturbed by the beam with which it interacts.

- The other two modes have frequencies ω_{j2}, ω_{j3} with lowest approximations $\omega_j = k_j v \pm \sqrt{q/(k_j v - \omega_{j0})} + \cdots$ in the limit $q \to 0$. They generate particle motions (3.48), (3.49) with amplitude for δp proportional to $\varepsilon \beta_j z / \sqrt{q} = z\sqrt{2/N}$, which does not vanish in the limit $\varepsilon \to 0$. These modes describe resonant oscillations of the beam coupled to the wave.

For $q/q_M = O(1)$, characteristics of the modes 'mix'. In particular, as $q \to q_M$, the eigenvalues ω_{j1} and ω_{j2} tend to each other. For $q > q_M$, the corresponding roots σ of (3.63) have become complex conjugate. Then $\omega_{j3} > k_j v > \omega_{j0}$ corresponds to the purely oscillating mode. Physically, one cannot hope to observe this oscillating mode easily, as one of the complex conjugate roots has a positive growth rate γ: a typical initial perturbation of the reference state will have a non-vanishing component following this eigenmode, which will ultimately overwhelm the other components.

Thus for any value of q there exists at least one purely oscillating eigenmode of the beam–wave system but this mode for $q > q_M$ is not the continuation of the free-wave mode existing for $q = 0$.

Let us now turn to the case $\omega_{j0} > k_j v$. Then $q_M < 0 < q$, so that Ω admits one real and two complex conjugate fixed points (see figure 3.3(b)). The analysis follows closely the previous one but now the damped-unstable pair of conjugate modes corresponds to the beam modes (with $\omega \simeq k_j v$) while the oscillating mode is the continuation of the free-wave mode.

The physics involved in the instability is easily grasped from the conservation law (2.5) for momentum, $P = \sum_n m v_n + k_j |Z_j|^2 / 2$. In the reference state, $P = Nmv$. The wave can only contribute a positive momentum (for $k_j > 0$), whereas the particles can reduce their velocity to yield momentum or increase their velocity to gain momentum. A growing instability supplies momentum to the wave, thus slows down the particles.

Anticipating that the wave–particle interaction will bring the particle velocities closer to the mode velocity ω_j / k_j (see section 3.9), it is worth noting here that, for a wavelike eigenmode, $Z = z e^{\alpha t}$, so that the phase velocity of the wave is not the free velocity ω_{j0}/k_j. In our single-beam case, it turns out that the damped and the unstable eigenmodes have the same phase velocity $\Re \sigma_{u,s}/k_j$, easily computed since the sum of the three roots of (3.63) is a coefficient of the

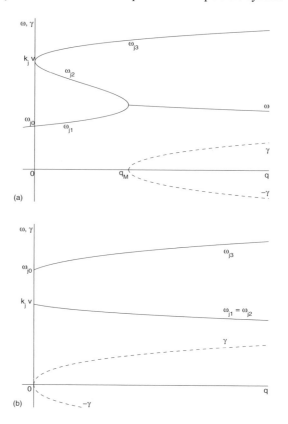

Figure 3.3. Real and imaginary parts of solutions $\sigma = \omega + i\gamma$ of the single-beam dispersion relation (3.63), vsersus the coupling constant q for fixed $k_j v - \omega_{j0}$: (*a*) case $k_j v > \omega_{j0}$; (*b*) case $k_j v < \omega_{j0}$.

cubic equation

$$\sigma_u + \sigma_s + \omega_{jC} = 2\Re \sigma_{u,s} + \omega_{jC} = 2k_j v + \omega_{j0} \tag{3.64}$$

where ω_{jC} is the only real root. As we noted, $\omega_{jC} > \omega_{j0}$ for $q > 0$, which implies that the damped and the unstable eigenmodes verify

$$\Re \sigma_{u,s} < k_j v. \tag{3.65}$$

Thus the eigenmode can synchronize the particles with the wave, as the latter propagates more slowly than the beam. This stresses the fact that the wavelike eigenmodes of the beam–wave system are not mere modulations of the free wave but result from an explicitly, consistently coupled set of degrees of freedom.

Finally, recall that in plasmas and wave–particle systems, there are many wave degrees of freedom, with many wavevectors k_j. As we saw that waves

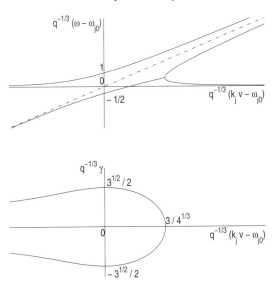

Figure 3.4. Real and imaginary parts of solutions $\sigma = \omega + i\gamma$ of the single-beam dispersion relation (3.63), versus the detuning $k_j v - \omega_{j0}$ for fixed coupling constant q.

with distinct wavevectors k_j are not coupled in the linear dynamics, it appears that in most many-wave systems there exist some waves (k_j, ω_{j0}) such that their parameters satisfy $q_j > q_{jM}$: these waves generate unstable eigenmodes. The experimental (or numerical) observation, starting from typical initial data, then enables one to observe easily the most rapidly growing ones. In our model, this is best discussed by considering a fixed coupling constant q and by varying the beam–wave frequency detuning $k_j v - \omega_{j0}$. The real and imaginary parts of the three roots of (3.63) are plotted in figure 3.4. The growth rate of the unstable eigenmode reaches a maximum for the resonant case, $k_j v = \omega_{j0}$. Then the oscillating mode has eigenfrequency $\omega_{jC} = \omega_{j0} + q^{1/3}$ while the damped and unstable modes have $\omega = \omega_{j0} - q^{1/3}/2$ with $\gamma = \pm\sqrt{3}q^{1/3}/2$.

3.7.2   Case of two beams

The case with two beams ($b = 2$) interacting with a single wave sheds some light on the many-beam case discussed later. The dispersion relation (3.45) now reads

$$\sigma = \omega_{j0} + q_1(\sigma - k_j v_1)^{-2} + q_2(\sigma - k_j v_2)^{-2} \qquad (3.66)$$

and its graphical solution is plotted in figure 3.5 for a typical choice of $q_1 = N_1 \varepsilon^2 \beta_j^2/2$, $q_2 = N_2 \varepsilon^2 \beta_j^2/2$. Let $v = k_j(v_2 - v_1)$. For $v = 0$, one recovers the single-beam case with $q = q_1 + q_2$. But for $v \neq 0$, the dispersion relation (3.66) admits five complex solutions. For v 'large enough', these solutions are of

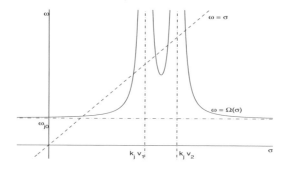

Figure 3.5. Graphical solution of the dispersion relation for two beams and one wave.

the same type as those discussed in the previous subsection and one may consider that the beams do not perturb strongly each other's interaction with the wave.

The limit $v_1 = v_2$ amounts to considering a single beam as a superposition of two beams with the same velocity (and $N = N_1 + N_2$). To understand this limit, writing the dispersion relation (3.66) as the quintic equation

$$
\begin{aligned}
0 &= q_1(\sigma - k_j v_1 - v)^2 + q_2(\sigma - k_j v_1)^2 \\
&\quad - (\sigma - \omega_{j0})(\sigma - k_j v_1)^2(\sigma - k_j v_1 - v)^2 \\
&= [q - (\sigma - \omega_{j0})(\sigma - k_j v_1)^2] \\
&\quad \times \left[(\sigma - k_j v_1 - v)^2 + \frac{q_2(2\sigma - 2k_j v_1 - v)}{q - (\sigma - \omega_{j0})(\sigma - k_j v_1)^2} \right]
\end{aligned}
\qquad (3.67)
$$

one sees that its roots are, in the limit $v \rightarrow 0$, those of (3.63) with $v = v_1 + O(v_2 - v_1)$ along with two more roots,

$$
\sigma = k_j \frac{q_2 v_1 + q_1 v_2}{q} \pm i \frac{\sqrt{q_1 q_2}}{q} k_j(v_2 - v_1) + \cdots.
\qquad (3.68)
$$

As these two roots are complex, one of them describes an unstable mode but its instability growth rate (as the damping rate of the stable mode) scales like v, hence vanishes in the limit $v \rightarrow 0$, while the frequency of these modes is sandwiched between $k_j v_1$ and $k_j v_2$, associated with beam intrinsic oscillations. Furthermore, for $v \rightarrow 0$, the ratios $z/|\delta p|$ and $|\delta p|/|\delta x|$ given by (3.42), (3.43) and (3.16), (3.17) scale like v, i.e. the intensity of the wave perturbation goes to zero, and particles reach far ahead and trail far behind their reference motions for a small velocity perturbation. These additional roots thus correspond to ballistic modes.

This shows that one can continuously change the 'nature' of eigenmodes from ballistic to wavelike and conversely. Extending this analysis to the case of many beams enables one to discuss the 'warm-beam' case and to show, as O'Neil and Malmberg (1968) did for the Vlasov–Poisson model, how the thermal spreading of the beam particles' velocities makes the wave–particle system

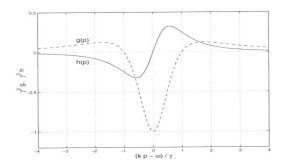

Figure 3.6. Contribution from a particle with velocity p to the growth rate γ (full line) and frequency ω (dotted line) of the Floquet exponent for the mode with complex frequency $\sigma = \omega - i\gamma$.

unstable: two monokinetic beams with nearby velocities behave like a 'warm' single beam.

3.8 Many cold beams or a warm beam

We now consider the case where many beams are present, defined by $b \to \infty$ (hence $N \to \infty$) and a vanishing velocity difference Δp between neighbouring beams. We first study (3.61), (3.62) by assuming the real part γ_{jr} of the Floquet exponent α_{jr} stays finite in this limit and discuss their physical interpretation in terms of individual wave–particle interaction. With the help of a more mathematical analysis, we identify one unstable mode (with $\gamma = \gamma_L > 0$), which is the Landau mode in the kinetic approach. In the next section, we discuss how this mode emerges from the wavelike eigenmodes of section 3.4 and show how the other wavelike eigenmodes generate one damped eigenmode (with damping rate $-\gamma_L$) and many singular van Kampen–Case eigenmodes. We revisit the initial value problem in this new setting and find, in particular, the classical Landau damping.

3.8.1 Landau unstable mode

The function

$$h(p) = \frac{\gamma(k_j p - \omega_{jr})}{[\gamma^2 + (k_j p - \omega_{jr})^2]^2} \tag{3.69}$$

plotted in figure 3.6 expresses the contribution to γ_{jr} of a particle with velocity p. It vanishes for p equal to the phase velocity ω_{jr}/k_j of the mode of interest, reaches extrema for velocities $\omega_{jr}/k_j \pm \gamma/(\sqrt{3}k_j)$ and goes to zero when $|p| \to \infty$. Hence the main contribution to γ_{jr} comes from particles with a velocity typically in the range $[(\omega_{jr} - \gamma_{jr})/k_j, (\omega_{jr} + \gamma_{jr})/k_j]$. Now, for γ_{jr} to be physically meaningful, the rise time of the corresponding unstable mode must verify (3.10), i.e. $\gamma_{jr} \gg \omega_{bj} = (\varepsilon k_j \beta_j)^{1/2} |Z_j|^{1/4}$. Therefore trapped particles, whose velocity is typically ω_{bj}/k_j away from γ_{jr}/k_j, contribute little to (3.61) and the instability is not due to a resonance between the wave and particles in the sense of section 1.2. The instability may be termed a weakly resonant effect since it involves particles whose velocity has a mismatch of order γ_{jr}/k_j with the phase velocity.

Now we turn to the mathematical formulation. Consider a sequence of self-consistent Hamiltonian systems with increasing N describing the same bulk and tail (with $N\varepsilon^2/\omega_p^2 = 2\eta = 2n_t/n_0 = 2\mathcal{N}_b/\mathcal{N} \ll 1$). The sequence of corresponding multi-beams is chosen such that $v_{s+1} - v_s = O(\Delta p) = O(N^{-1/2})$ and $N_s/N = O(N^{-1/2})$. Let $f_t(v)$ be a coarse-grained normalized velocity distribution function[4] for the tail particles ($\int f_t(v)\,dv = 1$), and write $f_t'(p) = \partial f_t/\partial p$.

We consider modes C_{jr} with $|\omega_{jr} - \omega_{j0}| \ll k_j \Delta p_f$, where Δp_f is the characteristic scale over which f_t changes ($|f_t'(p)| \simeq f_t(p)/\Delta p_f$). If $\Delta p \ll \gamma_j/k$, γ_{jr} may be evaluated by going to the continuous limit of (3.61) and integrating by parts[5]. Then (3.61) becomes the fixed-point equation $\gamma_j = \varphi(\gamma_j)$, with

$$\varphi(\gamma) = -\frac{\varepsilon^2 \beta_j^2}{2k_j}\gamma_j \int N\frac{f_t'(p)}{\gamma^2 + (k_j p - \omega_j)^2}\,dp. \tag{3.70}$$

Assume that f_t varies on a scale

$$\Delta p_f \gg \gamma_j/k \tag{3.71}$$

i.e. $f_t \gg \frac{\pi}{2}\varepsilon^2 \beta_j^2 k_j^{-3} N f_t'^2$ near ω_{j0}/k_j. Then (B.17) implies that $\lim_{\gamma \to 0} \varphi(\gamma) = \frac{\pi}{2}\varepsilon^2 \beta_j^2 k_j^{-2} N f_t'(\omega_j/k_j)$. As φ is a continuous function of γ, which has a finite positive limit for $\gamma \to 0+$ and goes to 0 as $\gamma \to +\infty$, it has a fixed point $\gamma_j > 0$. Therefore, in the limit $\varepsilon \to 0$ or $f_t' \to 0$, (3.70) reduces to

$$|\gamma_j| = \gamma_{jL} \equiv \frac{\pi}{2}\frac{\varepsilon^2 \beta_j^2}{k_j^2} N f_t'(\omega_j/k_j). \tag{3.72}$$

This solution of (3.70) is well defined if and only if f_t' is positive. Moreover, if we consider modes C_{jr} with $|\omega_{jr} - \omega_{j0}| \ll k_j \Delta p_f$, the slope in (3.72) may be evaluated at $p = \omega_{j0}/k_j$ just as well as at ω_{jr}/k_j. Condition (3.71) implies that $|\gamma_j| \ll \omega_p$. However, the finiteness of (3.72) ensures that, for a perturbation amplitude ζ_{jr} small enough, the assumption $\gamma_j \gg \omega_{bj}$ is satisfied.

---

[4] This coarse-grained distribution function for N particles is not the Vlasovian mean-field distribution function corresponding to N infinite for the full plasma. It only describes the beam particles.

[5] Here one uses the fact that $\lim_{p \to \pm\infty} f_t(p) = 0$ and that $\gamma_j > 0$ so that $[\gamma_j^2 + (k_j p - \omega_j)^2]^{-1} \leq \gamma_j^{-2}$ is bounded for $p \in \mathbb{R}$.

The corresponding frequency ω_j is given by (3.62). The contribution of a particle with velocity p in this sum is weighted by function

$$g(p) = \frac{(k_j p - \omega_j)^2 - \gamma^2}{[\gamma^2 + (k_j p - \omega_j)^2]^2} \qquad (3.73)$$

plotted in figure 3.6. It also displays a width of order γ_j / k_j, which enables us to use the continuous limit for $\Delta p \ll \gamma_j / k_j$:

$$\omega_j = \omega_{j0} + \frac{\varepsilon^2 \beta^2 N}{2k_j} \int \frac{k_j p - \omega_j}{\gamma_j^2 + (k_j p - \omega_j)^2} f'_t \, dp. \qquad (3.74)$$

With condition (3.71), at the continuous limit γ_j may be neglected in (3.74), which yields the fixed-point equation

$$\omega_{jL} = \psi(\omega_{jL}) \qquad (3.75)$$

with the function

$$\psi(\omega) = \omega_{j0} + \frac{\varepsilon^2 \beta^2 N}{2k_j} \mathrm{PP} \int \frac{f'_t}{k_j p - \omega} \, dp \qquad (3.76)$$

where PP denotes the principal part of the integral[6]. The integral gives the small shift of the mode's frequency (compared to the free wave) due to the contribution of the tail particles to the total (bulk + tail) mode frequency.

Expression (3.72), which is called the Landau growth rate of the wave, was first obtained by Landau (1946) by using a Vlasovian description of the plasma. Equation (3.72) implies that in a plasma whose velocity distribution has a tail with a range of positive slope, there are unstable eigenmodes in this phase velocity range for any realization with a non-zero initial amplitude of these modes. These unstable modes are characterized by unstable wave envelopes at the corresponding wavenumbers. The modes with the opposite value of γ_{jr} will be discussed later.

3.8.2 Eigenmodes and initial value problem

Now we reconsider the connection between Landau modes and wavelike eigenmodes. As assumed in its derivation, the Landau type of solution corresponds to a γ_{jr} which does not vanish when $\Delta p \to 0$. The denominator in (3.61) shows that $O(\gamma_{jL}/(k_j \Delta p))$ beams contribute to this growth rate. Now,

---

[6] It reads

$$\mathrm{PP} \int \frac{f'_t}{k_j p - \omega_j} \, dp = \lim_{a \to 0+} \int_a^\infty \frac{f'_t(\omega_j/k_j + w) - f'_t(\omega_j/k_j - w)}{k_j w} \, dw \qquad (3.77)$$

which is typically positive if $f'_t(\omega_{j0}/k_j) > 0$, assuming $\omega_j \simeq \omega_{j0}$.

recall that this denominator is also the denominator in (3.48), (3.49), so that for these beams a Landau mode causes significant perturbations of the particle motion. Given initial data $(\delta x_{ns}(0), \delta p_{ns}(0), Z_j(0))$, the Landau mode is excited according to (3.59) and (3.56)–(3.58) with a coefficient

$$z'_{jL} = \frac{1}{1 - \chi'(\sigma_{jL})} \tag{3.78}$$

where $\sigma_{jL} \equiv \omega_{jL} + i\gamma_{jL}$ and $\chi(\sigma) = \psi(\sigma) + i\varphi(\sigma)$ is defined by (3.46). Calculations along the previous discussion yield

$$\chi'(\sigma) \simeq \frac{\varepsilon^2 \beta_j^2 N}{2k_j^3} \int \frac{f_t''(v)}{v - \sigma/k_j} \, dv = O\left(\frac{\gamma_{jL}}{k_j \Delta p_f}\right). \tag{3.79}$$

Therefore, $z'_{jL} \simeq 1$ and the contribution of the Landau mode to the evolution of the wave envelope is

$$Z_j(t) = \zeta_{jL} e^{(\gamma_{jL} - i\omega_{jL})t} + 2b \text{ other terms} \tag{3.80}$$

where $\zeta_{jL} = Z_j(0) + \sum_s a'_{jLs} A_{js}(0) + \sum_s c'_{jLs} C_{js}(0)$. For the elementary case with $\delta x_{ns}(0) = 0$, $\delta p_{ns} = 0$ for all particles, i.e. a wave launched in a quiet plasma, this reduces to $\zeta_{jL} = Z_j(0)$.

Exercise 3.12. Express ζ_{jL} in terms of the perturbation of initial particle data. How does your result tend to the expression of section G.2 in the continuous limit?

However, to solve the initial value problem, we need to identify the other modes as well. Indeed, the eigenvalues of M are the roots of equation

$$\sigma - \chi(\sigma) = 0 \tag{3.81}$$

where the finite-N function χ is the ratio of a degree $2b - 1$ polynomial to a degree $2b$ polynomial. In the previous section we found one solution σ_{jL} in the limit $b \to \infty$ as the particle distribution has a smooth limit f_t. Now we must locate the other roots.

We first note that the Landau mode is isolated. More precisely, it is the only eigenmode with $\Re\alpha > 0$ and $\Re\alpha \not\to 0$ in the continuous limit $b \to \infty$. Indeed, the Landau eigenvalue is a solution of (3.81) and in its neighbourhood $|\chi'(\sigma)| \ll 1$. Therefore the implicit function theorem ensures that, for $b \to \infty$, this equation admits no other solution in the vicinity of σ_{jL}.

Then we note that σ_{jL}^* is also a root of (3.81), since this is a rational equation with real coefficients, and the same argument as that for the Landau root implies that σ_{jL}^* is also isolated. However, this second root describes a time-damping wave—which may be interpreted as a wave unstable towards the reversed time evolution. Its weight in the eigenmode decomposition is $z'_{jL}{}^* =$

$(1 - \chi'(\sigma^*_{jL}))^{-1} \to 1$ in the continuous limit. Thus the unstable and the damped Landau modes' joint contribution is

$$Z_j(t) = 2Z_j(0)e^{-i\omega_{jL}t} \cosh \gamma_{jL}t + \text{terms with the } (A_{js}(0), C_{js}(0))$$
$$+ (2b - 1) \text{ non-Landau terms.} \tag{3.82}$$

This expression agrees with the reversible nature of Hamiltonian dynamics but the explicit terms add up to twice the initial data: hence the non-Landau terms must not be neglected. These will be discussed in the next section along with particle aspects of the eigenmodes.

3.8.3 Van Kampen modes*

So far we have identified no eigenmode in the continuous limit if $\partial f_t/\partial p < 0$: then the Floquet exponents α_{jr} have real parts γ_{jr} which must vanish when Δp goes to zero; otherwise, (3.72) would yield a negative absolute value. Thus for a given value of j there exist wavelike modes with phase velocities ω_{jr}/k_j in the range of the beam velocities. Unless there would exist modes with $|\gamma|$ large enough to violate (3.71), their number must be $2b + 1$ if $\partial_p f_t(\omega_{j0}/k_j) < 0$, and $2b - 1$ if $\partial_p f_t(\omega_{j0}/k_j) > 0$.

In the continuous limit, these solutions accumulate on the imaginary axis in the complex α-plane, as do the two 'nearly ballistic' solutions (3.68) in the two-beam case (section 3.7.2). These many $(O(b) = O(\Delta p^{-1}))$ modes, whose frequencies form a continuum in the limit $\Delta p \to 0$, correspond to the van Kampen modes found in Vlasovian theory.

Figure 3.7 displays, in the complex σ-plane, the eigenvalues and weights z' computed numerically (to 28 decimal places) for three cases with $b = 40$ beams. Here, $\omega_{j0} = 0$ and the beams are equally spaced by Δp to produce a coarse-grained tail distribution function with constant slope f'_t. The beam velocities are poles of the function $\sigma - \chi(\sigma)$ and the eigenvalues are zeros of $\sigma - \chi(\sigma)$. The Landau point is denoted by a star. The full and dotted lines will be estimated analytically by (3.96) later.

In the unstable case (figure 3.7(a)), one eigenvalue is found close to the Landau point, with coefficient $z'_{jL} = 0.85$. Most other eigenvalues (besides its conjugate) are found to accumulate close to the beam velocities, with coefficients z'_{jr} depending smoothly on ω_{jr}. The size of these coefficients scales like $\Delta p/(2\pi)$, with a nearly constant real part close to $-\gamma_{jL}$ and an imaginary part linear in $\omega_{jr} - \omega_{jL}$ as we discuss later; the Landau mode coefficient z'_{jL} is much larger than the other coefficients and creates a hook in the line of $\Re[\Delta p/(2\pi z'_{jr})]$.

In the neutral case (figure 3.7(b)) and the Landau-damped case (figure 3.7(c)), the Landau value gives rise to no special eigenvalue. All eigenvalues exhibit the same smooth dependence on ω_{jr} and their coefficients scale similarly, with again $\Re[\Delta p/(2\pi z'_{jr})] \simeq -\gamma_{jL}$ up to corrections vanishing for $\Delta p \to 0$.

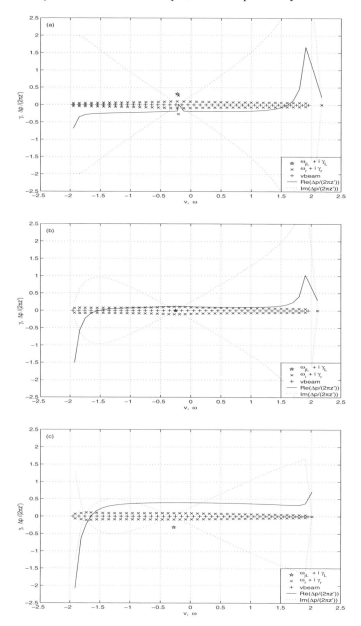

Figure 3.7. Eigenvalues and scaled coefficients of eigenvectors in the complex plane for $\omega_{j0} = 0$, $k_j = 1$, $\varepsilon^2\beta_j^2 N = 2$, $f_{\mathrm{t}}(v) = f_{\mathrm{t}}(0) + vf_{\mathrm{t}}'(0)$ on $[-2, 2]$, for $b = 40$ beams equally spaced by $\Delta p = 0.1$. Beam velocities $(+)$, eigenvalues $(\times)$, Landau value (star), weight coefficients $1/z'$ scaled by $\Delta p/(2\pi)$ (full and dotted lines are guides to the eye). (a) Landau unstable case; (b) Landau neutral case; (c) Landau damped case.

In all three graphs, the ends of the distribution break the very regular pattern and it seems that the approximations used later cannot apply when ω_{jr}/k_j is not in the bulk of the support of the distribution function f_t. Mathematically, this is because near these points one cannot find a 'wide' interval over which one may approximate f_t by a linear function. Physically this means that, with respect to a mode with phase velocity near the end of the distribution support, the multibeam system does not appear very different from a cold beam with a velocity near the average of f_t, as we also observed for the two-beam case for $v \to 0$.

Exercise 3.13. Solve numerically the dispersion relation (3.45) and compute the coefficients z'_{jr}. Compare four techniques: direct search for the fixed points of χ, search for the zeros of the characteristic polynomial (numerator of the fraction to which $\sigma - \chi(\sigma)$ reduces), search for the eigenvalues of M and full diagonalization of the matrix M. Compute the eigenvectors C_{jr}. Discuss their dependence on b, $\varepsilon\beta$, N, on the beam velocity interval ends and on the specific shape of the function f_t. In particular, how do the Landau eigenvalue and eigenmode emerge from the family of other eigenvalues and eigenmodes as df_t/dp passes through 0? Warning: beware of the numerical intricacies generated by the closeness between zeros and poles of $\sigma - \chi(\sigma)$. Use, e.g., complex conjugation symmetry and $[C] \cdot [C'] = I$ to test the accuracy of your results.

Now we discuss analytically the data displayed in this figure. The techniques we use rest on classical calculus. The reader may wish to first read section 4.2, where a lighter form of the argument is applied (taking advantage of averaging) and then return to the current section. However, a proper understanding of section 4.2 requires the further conceptual effort inherent to a statistical approach.

The right-hand side of (3.45) becomes singular when the phase velocity ω_{jr}/k_j is close to the velocity v_s of a beam and when γ_{jr} goes to zero. The more beams one considers, the more values these phase velocities take, as each ω_{jr} lies between two successive beam frequencies $k_j v_s$ and $k_j v_{s+1} = k_j v_s + k_j \Delta p$. It is not very convenient to try to follow individual eigenmodes in the limit $\Delta p \to 0$, because the number of modes depends on Δp. It is more natural to consider a reference phase velocity, hence a frequency ω_{jr}. Considering then an eigenvalue in the accumulating continuum, with an imaginary part near ω_{jr}, one uses the dispersion relation (3.45) to associate a real part $\pm\gamma_{jr}$ with this imaginary part ω_{jr}. Then one may estimate the right-hand side of (3.45) by isolating the most resonant terms, as we shall also discuss in section 4.2.

It is convenient to space the beam velocities equally, setting

$$v_s = v_0 + s\Delta p \tag{3.83}$$

and to decompose the eigenvalue $\sigma_{jr} = (s_{jr} + u_{jr})k_j\Delta p$ with $s_{jr} \in \mathbb{Z}$, and $u_{jr} = w_{jr} + i\gamma_{jr}/(k_j\Delta p)$. Requiring $|w_{jr}| \leq 1/2$ ensures that s_{jr} identifies the beam with the velocity closest to the phase velocity of eigenmode r. Provided that $f_t'(\omega_{jr}/k_j) \neq 0$, a genuine small parameter is

$$\delta = \frac{f_t'(\omega_{jr}/k_j)\Delta p}{f_t(\omega_{jr}/k_j)}. \tag{3.84}$$

In the continuous limit

$$\frac{N_s}{N\Delta p} = f_t(\omega_{jr}/k_j) + f_t'(\omega_{jr}/k_j)(v_s - \omega_{jr}/k_j)$$

$$+ O((v_s - \omega_{jr}/k_j)^2 f_t''(\omega_{jr}/k_j)) \tag{3.85}$$

and the sum in (3.45) may be decomposed into a singular part, which is summed exactly using section A.2, and a regular part converging to an integral. This procedure is akin to the definition of the Cauchy principal part. Explicitly, one finds

$$\sigma_{jr} - \omega_{j0} - W(\sigma_{jr}/k_j) = \frac{\varepsilon^2 \beta_j^2 N}{2k_j^2 \Delta p} f_t(\omega_{jr}/k_j) \frac{\pi^2 - \pi\delta \sin \pi u_{jr} \cos \pi u_{jr}}{\sin^2 \pi u_{jr}} \tag{3.86}$$

where

$$W(\sigma_{jr}/k_j) = \frac{\varepsilon^2 \beta_j^2 N}{2k_j^2} \int_{\Delta p/2}^{\infty} [f_t(v_{s_{jr}} + v) + f_t(v_{s_{jr}} - v) - 2f_t(v_{s_{jr}})$$

$$- 2(\sigma_{jr}/k_j - v_{s_{jr}})f_t'(v_{s_{jr}})]v^{-2}\, dv. \tag{3.87}$$

As the left-hand side of (3.86) has a finite limit for $\delta \to 0$, the right-hand side must also converge, which implies $|\sin \pi u_{jr}| \to \infty$. The latter requirement entails that $\Im u_{jr} \to \pm\infty$. Assuming $\Im u_{jr} > 0$, balancing the dominant terms yields

$$-e^{2\pi i u_{jr}} \simeq \frac{i\delta}{4\pi}\left(1 + i\frac{\sigma_{jr} - \omega_{j0} - W(\sigma_{jr}/k_j)}{\gamma_{jrL}} + O(\delta)\right) \tag{3.88}$$

where

$$\gamma_{jrL} = \frac{\pi\varepsilon^2 \beta_j^2 N}{2k_j^2} f_t'(\sigma_{jr}/k_j) \simeq \frac{\pi\varepsilon^2 \beta_j^2 N}{2k_j^2} f_t'(\omega_{jr}/k_j) \tag{3.89}$$

would be the Landau damping rate of the wave with coupling constant β_j if its phase velocity were ω_{jr}/k_j. With the parameter

$$\theta_{jr} = \frac{\omega_{jr} - \omega_{j0} - W(\omega_{jr}/k_j)}{\gamma_{jrL}} \tag{3.90}$$

the eigenvalue is given by

$$\gamma_{jr} \simeq -\frac{1}{\tau_{\text{discr}}} \ln\left|\frac{\delta\sqrt{1 + \theta^2}}{4\pi}\right| + \cdots \tag{3.91}$$

$$w_{jr} \simeq \frac{1}{4}\operatorname{sign}\delta + \frac{1}{2\pi}\arctan\theta \tag{3.92}$$

with the angle $-\pi/2 < \arctan\theta < \pi/2$; the recurrence time

$$\tau_{\text{discr}} = \frac{2\pi}{k_j \Delta p} \tag{3.93}$$

characterizes the time-scale on which the approximation of the many beams by a smooth distribution function loses its validity.

The real part of σ_{jr} is thus such that the eigenmode phase velocity is faster than the nearest beam if $\delta > 0$ (i.e. in the Landau unstable case) and slower if $\delta < 0$. The imaginary part of $\sigma_{jr}/(k_j \Delta p)$ does diverge as $\Delta p \to 0$ but only logarithmically, so that eigenvalues do accumulate efficiently towards the real axis. Comparing (3.91), (3.92) with the two-beam case (3.68) shows that the limit of many modes changes the scaling mildly (with the factor $|\ln\delta|$) for γ and not for ω.

Exercise 3.14. Discuss the cross-over case $f_t'(\omega_{j0}/k_j) = 0$, for which the previous analysis is incomplete as $|\ln 0| = \infty$. Hint: expand f_t to higher order.

The time-reversal symmetry of Hamiltonian dynamics is preserved in this analysis, for the previous scaling assumes $\gamma_{jr} > 0$. On assuming $\gamma_{jr} < 0$ one finds the conjugate roots of the dispersion relation. A more accurate estimate for s_{jr} requires estimating the function $W(v)$ explicitly, i.e. the principal-part integral of the continuous limit dispersion relation.

The structure of the eigenvectors $(\delta x, \delta p, Z)$ follows from the structure of the eigenvectors $(c_{jrs}, a_{jrs}, 1)$ of section 3.5, which ensures that an eigenmode with frequency ω_{jr} mainly influences particles in the beams with a velocity near ω_{jr}/k_j. The factor z'_{jr} scaling the eigenvector is given by

$$\frac{1}{z'_{jr}} = 1 + \varepsilon^2 \beta_j^2 N \sum_s \frac{N_s}{N} (\sigma_{jr} - k_j v_s)^{-3} \tag{3.94}$$

so that to dominant order

$$
\begin{aligned}
\frac{k_j \Delta p}{z'_{jr}} &= -\frac{\varepsilon^2 \beta_j^2 N}{k_j^3 \Delta p^2} f_t(\sigma_{jr}/k_j) \frac{\pi^3 \cos\pi u_{rj}}{\sin^3 \pi u_{rj}} + O(\Delta p) \\
&= -\varepsilon^2 \beta_j^2 N k_j^{-2} f_t(\omega_{jr}/k_j) \pi^2 \frac{\delta}{\Delta p} (1 + i\theta) \\
&= -2\pi\gamma_{jrL} - 2\pi i(\omega_{jr} - \omega_{jrL}) \tag{3.95}
\end{aligned}
$$

with the detuned frequency $\omega_{jrL} = \omega_{j0} + W(\omega_{jr}/k_j)$. For a rough estimate, one may approximate the latter by $\omega_{jL} = \omega_{j0} + W(\omega_{j0}/k_j)$, and $f_t'(\sigma_{jr}/k_j)$ by $f_t'(\omega_{j0}/k_j)$ so that $\gamma_{jrL} \simeq \gamma_{jL}$. Note that $W(\omega_{j0}/k_j)$ does not vanish in the continuous limit and in figure 3.7 this correction significantly moves $\omega_{jL} \approx -0.25$ apart from ω_{j0}.

Thus the scaling of the eigenmode complex amplitudes is Lorentz-like,

$$z'_{jr} \simeq (2\pi)^{-1} \frac{k_j \Delta p}{\gamma_{jL} + i(\omega_{jr} - \omega_{jrL})}. \qquad (3.96)$$

Note that σ^*_{jr} is also an eigenvalue of $\mathbf{M}$, and its weight is $z'_{jr}{}^*$. Then the calculations of section 3.4 yield, in the case of a single-wave perturbation $(0, \ldots, 0, \ldots, Z_j(0))$, the contribution

$$Z_j^{\text{non-Landau}}(t) = Z_j(0) \sum_{r \text{ non-Landau}} e^{-i\omega_{jr}t} \Re z'_{jr} \cosh \gamma_{jr} t$$

$$\simeq Z_j(0)\pi^{-1} \int_{-\infty}^{\infty} \frac{-\gamma_{jrL} e^{-i\omega t}}{(\omega_{jL} - \omega)^2 + \gamma^2_{jrL}} \, d\omega$$

$$\simeq Z_j(0) e^{-i\omega_{jL}t - |\gamma_{jL}t|} \operatorname{sign}(-\gamma_{jL}) \qquad (3.97)$$

where the first expression combines the two modes σ_{jr} and σ^*_{jr}, the second expression follows in the continuous limit and the final one is a Fourier transform identity. The sign factor ensures that

- if $\gamma_L < 0$, the van Kampen–Case modes add to reconstruct the initial data $Z_j(0)$, so that

$$Z_j(t) \simeq Z_j(0) e^{-i\omega_{jL}t + \gamma_{jL}|t|} \qquad (3.98)$$

- if $\gamma_{jL} > 0$, the van Kampen–Case modes act subtractively to compensate the extra term in (3.82), so that

$$Z_j(t) \simeq Z_j(0) e^{-i\omega_{jL}t} (e^{\gamma_{jL}t} + e^{-\gamma_{jL}t} - e^{-\gamma_{jL}|t|}) = Z_j(0) e^{-i\omega_{jL}t + \gamma_{jL}|t|}. \qquad (3.99)$$

In this second case the classical Landau theory displays only an unstable eigenmode, which shows that in the kinetic limit the damped eigenmode is cancelled by the van Kampen-like contribution. From the viewpoint of function theory, this result shows how the finite-N system manages to approximate a non-analytic function of time by a sum of analytic functions. In the identification of eigenvalues, the van Kampen-like eigenvalues accumulate to the real axis, to form in the kinetic limit a cut in the complex plane on the support of f_t (this is also the meaning of the ill definition of the dispersion relation due to the poles $v = \omega/k$ in (3.75), (3.76)). Note that (3.98) and (3.99) reflect the time-reversible character of the dynamics.

We reached the striking conclusion that a Langmuir wave defined by its wavenumber is, in fact, the superposition of many beam modes with a broad frequency spectrum. It is interesting to follow numerically the evolution of this Langmuir wave with a given amplitude at $t = 0$. This amplitude must be decomposed on the basis of the many beam modes which leaves a lot of freedom for the initial condition. For a simple example, with mode coefficients

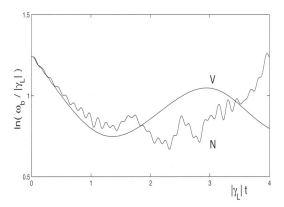

Figure 3.8. Numerical simulation of the evolution of a Langmuir wave for a given $Z_j(0)$ (further discussed in figure 8.13): line N, simulation of $N = 32\,000$ particles and one wave; line V, kinetic limit $N \to \infty$ integrated on a 32×512 (x, p) grid. Here the nonlinear regime starts around $|\gamma_L|t \sim 1$; spontaneous emission is observed from the beginning of the run and competes with nonlinearity. (After Doveil *et al* (2001) and Firpo *et al* (2001).)

defined by (3.96), figure 3.8 gives the result of such a simulation and shows that the amplitude of the Langmuir wave initially decays exponentially with small fluctuations. The kinetic line V also damps initially at the rate predicted from (3.72) if used with a negative slope, in agreement with (3.98).

The statistical approach of chapter 4 will show that this simulation of a single mechanical realization of the plasma simply exhibits a small fluctuation about the Landau damping of the wave due to the phase mixing of the many modes when the average over the initial position of the particles is performed; and it will confirm that the damping rate is indeed given by (3.72). Obviously, non-typical initial conditions can fail to undergo such a damping. This occurs, in particular, if they involve only a few eigenmodes. The following exercise (see also exercise G.3) shows how the finite-N effects drive the departure of the system from the endless exponential Landau damping of kinetic theory: first by generating a thermal type of balance between spontaneous emission and phase mixing; and second by the long-term instability of eigenmodes.

Exercise 3.15. The following results imply that the support of a van Kampen mode is localized over a vanishing range, while the Landau modes involve particles over a finite velocity range.

(i) Consider the eigenvector C_L associated with the Landau eigenvalue $\sigma_L = \omega_{jL} + i\gamma_L$. Show that (3.42)–(3.44) define smooth functions $(a_{jL}(v_s), c_{jL}(v_s), 1)$ in the continuous limit.

(ii) Show that $(a_{jL}^*(v_s), c_{jL}^*(v_s), 1)$ define the eigenvector associated with the anti-Landau eigenmode.

(iii) Consider the eigenvector $(a_{jr}(v_s), c_{jr}(v_s), 1)$ associated with a van Kampen eigenvalue σ_{jr}. Let $v_0 = \lim_{\Delta p \to 0} \sigma_{jr}/k_j$. Given a velocity v 'away' from v_0, show that $\lim_{\Delta p \to 0} a_{jr}(v) = 0$ and $\lim_{\Delta p \to 0} c_{jr}(v) = 0$.
(iv) What happens to (3.56), (3.57) as $\Delta p \to 0$?

Exercise 3.16. Consider a sequence of multibeam reference states, converging for $b \to \infty$ to a smooth distribution function f_t, and choose a simple initial perturbation $(\delta x, \delta p, Z)$ or (C, A, Z). Plot the time evolution of the perturbation over a time interval $-T \le t \le T$. See numerically how (3.98) and (3.99) become sharper as $b \to \infty$ for fixed T and how you may increase T keeping a prescribed quality to the approximation when $b \to \infty$. How does T scale for $\Delta p \to 0$?

3.8.4   Relation with ballistic eigenmodes

Thus in the kinetic limit, the wave–particle system has three types of eigenmode.

(i) For a k_m corresponding to no wave j, the eigenmodes are purely ballistic and evolve in time as $e^{\pm i k_m v_s t}$. They may also be called free-streaming perturbations.

(ii) For a k_j corresponding to a wave j, if $\partial_p f_t > 0$, there exist a Landau eigenmode and an 'anti-Landau' mode, with exponential evolution in time.

(iii) For a k_j corresponding to a wave j, the non-Landau eigenmodes are the van Kampen modes, with a time evolution $e^{\pm i k_j (v_s + u_{jr} \Delta p) t}$ (with $s = s_{jr}$). In the kinetic limit, these van Kampen modes have vanishing growth rates, so that they become analogous to the 'wave-less' ballistic modes. Since van Kampen modes excite particles' velocity locally, causing the denominators in (3.42), (3.43) to vanish, one may rescale the eigenvector by a factor $O(\Delta p^2)$. Then the wave component of the excitation by a single van Kampen mode is small for $\Delta p \to 0$ (in contrast to the ballistic eigenmodes, for which there is no wave excitation at all, and in contrast with the Landau eigenmodes, for which the wave and particles are significantly excited), and the eigenmode appears as a second class of a free-streaming perturbation— as (3.36) and (3.37) imply for the particle motion when $Z_j = 0$.

Exercise 3.17. Relate this classification to the classes 1(a, b, c) and 2 of Case (1959).

Although a single van Kampen mode is a singular object[7], packets of them may be smooth perturbations of the distribution function. Such packets bring two remarkable kinds of collective contribution to the linearized dynamics: a coherent one if a wave is excited initially with an amplitude $Z_j(0)$ (as discussed in section 3.8.3) but also an incoherent one if the wave has a thermal initial amplitude

---

[7] Their lack of smoothness in p is related to the fact that their frequency is just $k_j p$, so that a monochromatic van Kampen mode with wavenumber k_j is associated with a Dirac distribution in p.

(which vanishes in the kinetic limit). The latter case appears in kinetic theory as a zero-field excitation, i.e. a free-streaming one.

A classical physical application of van Kampen modes is the echo phenomenon (Gould *et al* 1967). Consider a plasma with $\partial_p f < 0$, so that two excitations E_1 and E_2 undergo Landau damping. Applying E_1 at time 0 and E_2 at a time $\tau > 0$, one may see each of them first Landau damp but witness a resurgence of a combination of E_1 and E_2 at a time $\tau' > \tau$. This resurgence corresponds to a nonlinear interference between van Kampen or ballistic components of both excitations.

3.8.5 Transition from cold to warm beams

Up to this point, we have considered two extreme cases for the beam, monokinetic or broad in velocity, which are referred to as the hydrodynamic and kinetic limits. Numerically, it is possible to follow the transition between both limits. This was done for the Vlasovian derivation of the dispersion relation (O'Neil and Malmberg 1968). The transition is controlled by the parameter $\rho = (\Delta p_f / v) \eta^{-1/3}$ where v is the average velocity of the beam, Δp_f is the half-width of the beam distribution function and the ratio η of the beam density to the plasma bulk density is small (sections 2.3 and 2.6). If ρ is small, the unstable modes are beam modes but perturbed by the plasma, as we found for $\rho = 0$. As we have just found, if ρ is large, the unstable modes are plasma modes but perturbed by the beam. The transition between these two regimes occurs continuously for ρ of the order of 1. For increasing ρ, the phase velocity of the most unstable mode goes from a value smaller than the smallest beam velocity to a value inside the domain of the beam distribution function with a positive slope after the transition (Self *et al* 1971).

3.9 Synchronization of particles with a wave

3.9.1 Synchronization of particles with a wavelike mode

The previous discussion focuses on the microscopic dynamics of the beam–mode interaction on a short time-scale. In practice, one is interested in this interaction over time-scales much larger than ω_{j0}^{-1}, over which deviations from the nearly periodic behaviour (3.48)–(3.50) are noticed. Over these time-scales the linearized dynamics (3.7)–(3.9) is no longer a satisfactory approximation; and one must take higher orders in the perturbations (δx_l, δp_l, Z_j) into account. The complete nonlinear equations for these perturbations read:

$$\delta \dot{x}_l = \delta p_l \tag{3.100}$$

$$\delta \dot{p}_l = \varepsilon \Re \left(\sum_{j=1}^{M} i \beta_j Z_j e^{ik_j x_l} e^{ik_j \delta x_l} \right) \tag{3.101}$$

$$\dot{Z}_j = -\mathrm{i}\omega_{j0} Z_j + \mathrm{i}\varepsilon\beta_j k_j^{-1} \sum_{l=1}^{N} \mathrm{e}^{-\mathrm{i}k_j x_l}\mathrm{e}^{-\mathrm{i}k_j\delta x_l}. \qquad (3.102)$$

Expanding the exponentials to first order in δx_l yields the second-order (in (C, A, z)) corrections to the reference solution. The acceleration of particle l reads explicitly to second order:

$$\delta\dot{p}_l = \varepsilon\Re\left(\sum_j \mathrm{i}\beta_j Z_j \mathrm{e}^{\mathrm{i}k_j x_l}\right) - \varepsilon\Re\left(\sum_j \beta_j k_j Z_j \mathrm{e}^{\mathrm{i}k_j x_l}\delta x_l\right) + \cdots. \qquad (3.103)$$

Evaluating Z_j, δp_l and δx_l to second order is a simple calculation. The resulting expressions contain terms with a distinct time dependence:

- $\mathrm{e}^{(\alpha_{jr}+\mathrm{i}k_j v_l)t \pm (\alpha_{j'r'}+\mathrm{i}k_{j'}v_{l'})t}$ for terms involving two wavelike eigenmodes r, r'; and
- $\mathrm{e}^{(\alpha_{jr}+\mathrm{i}k_j v_l)t \pm \mathrm{i}k_{j'}v_{l'}t}$ for terms involving one wavelike eigenmode and a ballistic one.

Most of these terms oscillate in time. The simplest case is the one where a single eigenmode jr is excited. Then the acceleration of particle l reads:

$$\mathrm{a}_{jrl} = -\varepsilon^2\beta_j^2 k_j |\zeta_{jr}|^2 \frac{\gamma_{jr}(k_j p_l - \omega_{jr})}{[\gamma_{jr}^2 + (k_j p_l - \omega_{jr})^2]^2}\mathrm{e}^{2\gamma_{jr}t}. \qquad (3.104)$$

As the dependence of this a_{jrl} on p_l is given (except for the sign) by (3.69), line h(p) in figure 3.6 shows that the eigenmode instability corresponds to a mechanism of (average) synchronization of weakly resonating particles with the wave. The conservation of the total momentum (2.5) implies that the wave amplitude grows if $\partial f_\mathrm{t}/\partial p$ is positive, i.e. if there are more fast particles than slow ones. This is consistent with the respective contribution of these particles to the growth rate found in section 3.8.1. Trapped particles have a negligible average acceleration and play little role in the energy exchange. Moreover, if the eigenmode under consideration is the Landau unstable mode (with large γ_L), particles from no beams can strongly resonate with this Landau unstable mode, as $|\alpha_{jr} + \mathrm{i}k_j v_l|^2 = \gamma_{jr}^2 + (\omega_{jr} - k_j p_l)^2 > \gamma_{jr}^2 = \gamma_{j\mathrm{L}}^2$ is large for all beams and this mode. Symmetrically, if the mode under consideration is the anti-Landau (decaying) eigenmode, the same calculation with a negative γ_{jr} implies that the particles are de-synchronized from the mode.

A similar argument applies to van Kampen eigenmodes. Then $\gamma_{jr} = O(k_j\Delta p|\ln\delta|)$, so that the range of beams influenced by a single eigenmode is much narrower than in the unstable Landau case as shown in exercise 3.15. As a result, the particle acceleration by van Kampen modes is more dramatic on the few nearby beams but also limited to them.

3.9.2   Synchronization of particles during Landau damping*

The Landau damping found in the case of a negative slope of the distribution function is the result of a phase mixing of van Kampen modes whose spectrum has a typical width $|\gamma_L|$. Therefore, one cannot consider that $Z_j(t)$ is exponentially decreasing since $t = 0$; and this behaviour sets in at about $t = |\gamma_L|^{-1}$. A particle feels a wave with a given wavelength and a time-decreasing amplitude $A(t)$. Its equation of motion may be written as

$$\ddot{x} = A(t)\sin[k_j x - \omega_L t + \varphi]. \tag{3.105}$$

Here φ is a constant phase as the frequency ω_L is the actual frequency of the mode, taking the detuning (3.87) into account. A first-order calculation in A gives the change in particle position between times 0 and t as

$$\delta x_1(t) = \int_0^t \int_0^{t'} A(t'')\sin[k_j x_0(t'') - \omega_L t'' + \varphi]\, dt''\, dt' \tag{3.106}$$

where $x_0(t) = p_0 t + x_{00}$. To second order in A, the particle acceleration is

$$
\begin{aligned}
\ddot{x}_2(t) &= A(t)[\cos(k_j x_0(t) - \omega_L t + \varphi)]k_j \delta x_1(t)\\
&= \frac{k_j}{2}\sum_{\eta=\pm 1}\int_0^t \int_0^{t'} A(t)A(t'')\sin[k_j(x_0(t'') + \eta x_0(t))\\
&\quad - \omega_L(t'' + \eta t) + (1+\eta)\varphi]\, dt''\, dt'\\
&= \frac{k_j}{2}\sum_{\eta=\pm 1}\int_0^t A(t)A(t'')(t - t'')\sin[k_j(x_0(t'') + \eta x_0(t))\\
&\quad - \omega_L(t'' + \eta t) + (1+\eta)\varphi]\, dt''.
\end{aligned}
\tag{3.107}
$$

When averaging over the initial particle position x_{00} the $\eta = +1$ term vanishes. We denote the average by $\langle \ddot{x}_2 \rangle$. The argument of the sine in the other term is $\Omega(t'' - t)$ where $\Omega = k_j p_0 - \omega_L$. Since $A(t)$ is positive and decays exponentially when t grows, a time-decaying upper estimate for the integral is computed by substituting the sine with the constant 1. However, if $|\Omega|$ is large compared to the decay rate of A, the oscillations in the sine cause the integral to be smaller. Furthermore, the order of magnitude of the integral giving $\langle \ddot{x}_2(t) \rangle$ decreases when $|\Omega|$ grows. For $|\Omega|t \leq \pi$, $\langle \ddot{x}_2(t) \rangle$ has a sign opposite to Ω, which means a synchronization of the particle with the wave. For $t \simeq |\gamma_L|^{-1}$ the synchronization involves particles in a range $|\gamma_L|/k_j$ about the phase velocity, which is much broader than the wave-trapping width.

For larger t the wave may possibly have a desynchronizing effect over some time intervals but because of the decrease of $A(t)$ this effect would be weaker than the initial synchronizing one. Furthermore, conservation of the total wave–particle momentum requires a global synchronization of particles in order to balance the decrease of wave momentum due to the Landau damping. We notice that assuming $A(t)$ to be exponentially decaying from $t = 0$ and applying the argument of section 3.9.1 would have given a desynchronizing effect to the wave[8].

A pure Landau eigenmode was shown by (3.104) to synchronize the particle with the wave but, in general, the unstable wave involves also the decaying anti-Landau and the coherent van Kampen eigenmodes. However the latter contributions compensate since the first one is de-desynchronizing and the second one is synchronizing. This will be made more precise by (4.24).

3.9.3   Fate of particles in the presence of many incoherent modes

Finally, the actual evolution of the beam–wave system seldom involves a single eigenmode. Indeed, exciting a single eigenmode requires tuning the initial data $(\delta x_l, \delta p_l, Z_j)$ to the corresponding eigenvector. More generally, even for the simple Landau instability case, one excites a given Z_j with vanishing particle perturbations or a given family of A_{js}, so that the system evolution involves a superposition of several eigenmodes and these analyses must be refined.

In a typical incoherent regime, each beam is excited by a few nearby van Kampen eigenmodes, whose eigenfrequencies beat at $\omega_{jr} - \omega_{j,r-1} \sim k_j \Delta p$ and, conversely, each van Kampen eigenmode is excited by a few nearby beams. For typical initial data, these modes have independent phases, so that, in contrast to the Landau mode, these excitations are incoherent; and their typical outcome is to apply on the particles a low intensity noisy force, calling for a statistical description. In a naive way one may say that each wavelike eigenmode tries trapping particles from nearby beams, gaining or losing a little momentum in this process. As the number of van Kampen modes interacting with a given beam scales as $|\ln \delta| \to \infty$ (while the velocity range covered by these modes scales as $\Delta p |\ln \delta| \to 0$), it turns out that, in the kinetic limit, their individual cat's eyes are in a strongly overlapping regime, which we discuss further in chapter 6.

Anticipating this analysis, we stress that the particle motion within the velocity range spanned by the resonances is essentially diffusive but not chaotic, so that the major effect of van Kampen modes in the kinetic limit is merely to drive the microscopic particle distribution to a plateau. This diffusive nature for the particle momentum evolution is evidently not a synchronization effect.

3.10   Historical background

The original approach to beam–wave interaction uses kinetic theory, describing both the plasma and the beam through the Vlasov–Poisson integro-differential system of equations. The mathematical treatment of the self-consistent beam–wave system is parallel to this classical one.

---

[8] This is the reason why several textbooks give an intuitive presentation of Landau damping by computing the energy gained by a distribution of particles interacting with a wave with a *constant* amplitude.

Landau (1946) investigates the evolution of the Vlasov–Poisson system with initial conditions leading to the instability (growth) as well as to collisionless damping. His original procedure (see e.g. Lifshitz and Pitaevskiĭ 1981) relies on Laplace transforms in time and an $i\varepsilon$ prescription, tantamount to the principal-part calculation of section 3.8.1. Further insight into the damping was provided by van Kampen (1955, 1957) and Case (1959), who showed that in the unstable case the perturbation of the distribution function f_t can be decomposed into eigenfunctions, whereas in the damped case the 'eigenmodes' involve Dirac distributions, with a continuum of generalized eigenvalues ω/k_j ranging over the domain of velocities with $\partial f_t/\partial p < 0$. Case (1959) further discusses the completeness of the basis of van Kampen eigenmodes and connects van Kampen's and Landau analyses.

Dawson (1960) introduces a many-beam approach to plasma dynamics. Each of his beams is treated as a fluid (rather than as an assembly of particles) and the beams are coupled through the Poisson equation. Our mathematical treatment of the dispersion relation in section 3.8.3 closely follows his analysis of the van Kampen–Case modes. He further discusses the initial value problem in both the finite-N model and the continuous limit, paying attention to the time-reversal symmetry problems. This discussion is relevant not only to the problem of Landau damping and instability in plasmas but also to all instances of N-body systems in a continuum limit. In particular, the approximation of irreversible behaviour by families of analytic functions solving a time-reversible dynamics problem is a core issue of statistical physics.

Equation (3.104) for the force acting on particles was first obtained by Kupersztych (1985) and the synchronization mechanism was suggested in Nicholson's textbook (1983).

Chapter 4

Statistical description of the
small-amplitude wave–particle dynamics

This chapter describes the absorption and both spontaneous and stimulated emissions of Langmuir waves. It continues the description of the self-consistent dynamics in the case of waves with a small amplitude but now in a statistical way. Therefore we must still preclude trapping or chaotic transport; and we keep the same upper bounds in time for the description of the dynamics: the typical trapping time of the particles in the global wave potential or the time of spreading of the position due to chaotic transport. The treatment proceeds with the following steps, corresponding to two different viewpoints. Both viewpoints lead to the same equations but emphasize different aspects thereof.

The first step uses second-order perturbation analysis in ε of the canonical equations of the self-consistent Hamiltonian. The particles are not initially distributed in beams with given velocities but are distributed 'smoothly', so that one may consider their initial positions as independent random variables with uniform distribution. Averaging the second-order equations over the initial positions of the N particles yields an evolution equation for the wave energy. This equation includes both the damping and the spontaneous and stimulated emissions of waves by particles. This induces an exponential relaxation of a stable Langmuir wave to a finite thermal level. A similar treatment for particles shows that a wave tends to synchronize particles with it (as in chapter 3) and enables the derivation of a drag and of a diffusion coefficient related, respectively, to the spontaneous and to the stimulated emission (or absorption) of a broad spectrum of waves.

The second step (section 4.2) consists in obtaining the same evolution equation for the wave intensity through an appropriate averaging in the Floquet approach of chapter 3. This is more involved but gives an interesting insight into the underlying physics by relating the statistical description of the dynamics to that for a single realization of the system. Thanks to the averaging, the calculations in this section are simpler than those of section 3.8.3. A further motivation for introducing this second approach is that, while the first step focuses on the interacting objects, i.e. particles and waves, our second step investigates the dynamics of the eigenmodes identified in the previous chapter.

Finally we show in section 4.3 the link of the present approach with traditional ones in plasma physics.

4.1 Approach using perturbation expansion

This section uses second-order perturbation analysis in ε of the canonical equations of the self-consistent Hamiltonian. The calculation turns out to be simpler and to show its limits of validity better when using the action-angle variables for the electrostatic waves. Therefore we use expression (2.109) of the self-consistent Hamiltonian whose canonical equations are (2.113)–(2.116). Equation (2.115) diverges for I_j small, which restrains the validity of the perturbation expansion in ε in this limit. Indeed at low amplitude the wave dynamics is no longer dominated by the free-wave motion but ruled by the source term due to particles. This is obvious when using (2.114). To perform the perturbation calculation in ε it is necessary to require

$$\frac{I_j}{N} \gg \frac{I_{\min}}{N} \equiv \frac{\varepsilon^2 \beta_j^2 N}{2k_j^2 \omega_{j0}^2} \simeq \frac{\eta}{2k^2 \omega_p} \tag{4.1}$$

where η was defined in (2.106). As η is a small parameter, condition (4.1) is weakly restrictive. Moreover, if one uses the canonical variables X_j, Y_j for the waves, no such singularity appears—but their expansions in powers of ε in the following computations are different (and slightly more difficult).

In the subsequent calculations, we first compute the leading orders in power expansions for the dynamical variables and take advantage from the large number of particles to estimate the evolution of the wave intensities and particle momenta by a statistical average. More precisely, we shall assume that all particles are distributed with initial positions $\{x_{l0}\}$ drawn at random from a uniform distribution over the length L and that any two particles are independent from each other.

4.1.1 Second-order perturbation analysis

We now expand equations (2.111), (2.113), (2.115) and (2.116) in a perturbation series until second order in ε. This is the first order yielding non-zero contributions after averaging over the initial positions of the particles. We write

$$I_j = \sum_{n=0}^{\infty} I_j^{(n)} \tag{4.2}$$

where $I_j^{(n)} = O(\varepsilon^n)$ and $I_j^{(0)} = I_{j0} > 0$ and similarly expand x_l, p_l and θ_j. To remove any arbitrariness in the solution of the evolution equations, we impose, at each order $n \geq 1$, that $I_j^{(n)}(0) = 0$ and similarly for x_l, p_l and θ_j.

The zeroth order in ε corresponds to the uncoupled motion of waves and particles. We define

$$\Omega_{lj} = k_j p_{l0} - \omega_{j0} \tag{4.3}$$

$$\varphi_{lj} = k_j x_{l0} - \theta_{j0} \tag{4.4}$$

where index 0 stands for the initial values of the unperturbed variables x_l, p_l, θ_j. Requiring the perturbed motion to have the same initial conditions, the first-order equations are immediately solved:

$$p_l^{(1)} = \varepsilon \sum_{j=1}^{M} \beta_j \sqrt{2I_{j0}} \Omega_{lj}^{-1} (\cos(\Omega_{lj}t + \varphi_{lj}) - \cos\varphi_{lj}) \tag{4.5}$$

$$I_j^{(1)} = -\varepsilon \sum_{l=1}^{N} \beta_j \sqrt{2I_{j0}} \Omega_{lj}^{-1} (\cos(\Omega_{lj}t + \varphi_{lj}) - \cos\varphi_{lj}). \tag{4.6}$$

Exercise 4.1. Find the first-order perturbations $x_l^{(1)}$ and $\theta_j^{(1)}$.

Now consider the expectation value of (4.6) for a uniform distribution of initial positions of the particles $\{x_{l0}\}$. All the trigonometric functions have a vanishing average, so that

$$\langle I_j^{(1)} \rangle = 0. \tag{4.7}$$

Therefore we turn to the next order in the perturbative expansion. The second-order equation for the wave action is

$$\dot{I}_j^{(2)} = 2\varepsilon^2 \sum_{l=1}^{N} \sum_{j'=1}^{M} \beta_j \beta_{j'} I_{j0}^{1/2} I_{j'0}^{1/2} \Omega_{lj'}^{-2}$$

$$\times \ (\sin(\Omega_{lj'}t + \varphi_{lj'}) - \sin\varphi_{lj'} - \Omega_{lj'}t \cos\varphi_{lj'}) \cos(\Omega_{lj}t + \varphi_{lj})$$

$$- \varepsilon^2 \sum_{l=1}^{N} \sum_{l'=1}^{N} \beta_j^2 k_j^{-2} \Omega_{l'j}^{-1}$$

$$\times \ (\sin((\Omega_{lj} - \Omega_{l'j})t + \varphi_{lj} - \varphi_{l'j}) - \sin(\Omega_{lj}t + \varphi_{lj} - \varphi_{l'j})). \tag{4.8}$$

Perturbation calculation is justified only if the particle and wave orbits do not wander too far away from free orbits. This yields a temporal upper limit for this technique: the divergence time τ_d of nearby orbits for particles, as in chapter 3; this is the trapping time for a regular motion of the particles and, in the case of a diffusive dynamics, the Dupree time (Dupree 1966), $\tau_D = (k^2 D)^{-1/3}$ where k and D are, respectively, the typical wavenumber and diffusion coefficient of the system (see section 6.8.2). For the waves we have already imposed condition (4.1) which is typical of the present perturbative approach, though weakly limiting.

4.1.2   Evolution of waves

Since $I_j^{(0)}$ is constant and $I_j^{(1)}$ has no secularity, the evolution of the amplitude of mode j over long times is given by its second-order expression in ε. Averaging (4.8) over (pairwise) independent uniformly distributed initial positions $\{x_{l0}\}$ and taking into account that $k_j + k_{j'} \neq 0$ for all pairs j, j' (there is no two-wave resonance[1] in the present Hamiltonian dynamics) yields

$$\langle I_j^{(2)} \rangle = \sum_{l=1}^{N} \varepsilon^2 \beta_j^2 I_{l0} \Omega_{lj}^{-2} (\sin \Omega_{lj} t - \Omega_{lj} t \cos \Omega_{lj} t) + \sum_{l=1}^{N} \varepsilon^2 \beta_j^2 k_j^{-2} \Omega_{lj}^{-1} \sin \Omega_{lj} t \tag{4.9}$$

which we may write as

$$\langle I_j^{(2)} \rangle = -\sum_{l=1}^{N} \varepsilon^2 \beta_j^2 I_{l0} \frac{\partial}{\partial \Omega_{lj}} \left(\frac{\sin \Omega_{lj} t}{\Omega_{lj}} \right) + \sum_{l=1}^{N} \varepsilon^2 \beta_j^2 k_j^{-2} \frac{\sin \Omega_{lj} t}{\Omega_{lj}}. \tag{4.10}$$

Now we go to the continuum limit for the particle distribution. Let Δv be the velocity interval supporting the tail distribution function. The continuum limit requires the particle number to be large enough so that the recurrence time $\tau_{\text{discr}} = 2\pi/(k\Delta p)$ defined by (3.93) is much larger than the time over which we follow the system evolution, where $\Delta p \sim N^{-1} \Delta v$ is the characteristic scale of velocity differences[2]. By requiring $t \ll \tau_{\text{discr}}/|\ln \delta|$ we also ensure that the ultimate instability of the van Kampen modes cannot show up in the time range over which our calculations hold.

Let $f(p)$ (dropping the subscript t of f_t in section 3.8) be the coarse-grained velocity distribution function at $t = 0$ defined by

$$N^{-1} \sum_{l=1}^{N} \bullet = \int \bullet f(p) \, \mathrm{d}p. \tag{4.11}$$

Integrating (4.10) by parts yields

$$\langle I_j^{(2)} \rangle = N\varepsilon^2 \beta_j^2 I_{j0} \int \frac{\sin \Omega_j t}{\Omega_j} \frac{\mathrm{d}f}{\mathrm{d}p} \frac{\partial p}{\partial \Omega_j} \frac{\mathrm{d}\Omega_j}{k_j}$$
$$+ N\varepsilon^2 \beta_j^2 k_j^{-2} \int \frac{\sin \Omega_j t}{\Omega_j} f(p) \frac{\mathrm{d}\Omega_j}{k_j} \tag{4.12}$$

where $\Omega_j = k_j p - \omega_{j0}$ is the dummy integration variable relevant to mode j and $p = (\omega_{j0} + \Omega_j)/k_j$ is the corresponding argument of f. Let $\Delta p_f > \Delta p$ be the

---

[1] The two-wave resonance conditions read as $k_j + k_{j'} = 0$, $\omega_{j0} + \omega_{j'0} = 0$.

[2] By analogy with chapter 3, we could consider each particle as a beam here, as we do not discuss the stability of a specific solution of the wave–particle dynamics. But we can also place many particles on a quasi-beam (with velocity in a range of Δp for a given quasi-beam) and let the number of quasi-beams go to infinity more slowly than N, say like $N^{1/2}$, so that $N_s \sim N^{1/2}$ too.

characteristic scale of variation of $f(p)$. If

$$t \gg \frac{2\pi}{k_j \Delta p_f} \tag{4.13}$$

then Δp_f is much larger than $\delta p_t = 2\pi/(k_j t)$, the characteristic scale of variation of $(\sin \Omega_j t)/\Omega_j$ about $p = \omega_{j0}/k_j$. Then the integrals in (4.12) are readily evaluated, and this equation becomes

$$\langle i_j^{(2)} \rangle = 2\gamma_{jL} I_{j0} + S_j \tag{4.14}$$

where

$$\gamma_{jL} = \alpha_j \frac{\mathrm{d}f}{\mathrm{d}p}(\omega_{j0}/k_j) \tag{4.15}$$

$$S_j = \frac{2\alpha_j}{k_j} f(\omega_{j0}/k_j) \tag{4.16}$$

with

$$\alpha_j = \frac{\pi}{2} N\varepsilon^2 \beta_j^2 k_j^{-2}. \tag{4.17}$$

Approximating $\langle \dot{I}_j(t) \rangle$ by $\langle i_j^{(2)}(t) \rangle$, (4.14) becomes

$$\langle \dot{I}_j \rangle = 2\gamma_{jL}\langle I_j \rangle + S_j + \mathrm{O}(N^2\varepsilon^4). \tag{4.18}$$

We notice that, though (2.116) diverges for I_j small, (4.18) is divergence-free in I_j. This is natural since the same calculation could be done (in a more tedious way) using canonical equations (2.114) which have no divergence for I_j small. As long as the linearization of orbits is correct (i.e. for $t \ll \tau_d$), averaging over initial $\{x_{l0}\}$ or over $\{x_l(t)\}$ at time t yields the same second-order result. Hence, as $\tau_{\mathrm{discr}} \to \infty$ for $N \to \infty$ by (3.93) whereas γ_{jL} and τ_d are constant in the continuum limit, (4.18) holds up to times $t \ll \tau_d$. Note that the continuum limit provides the many beams needed to support beam modes with nearby velocities to cause decay by phase mixing.

Exercise 4.2. Show that the second-order expansion in ε of the wave variables Z_j leads to the same evolution of the mode intensities I_j (hint: $2I_j^{(2)} = Z_{j0}Z_j^{(2)*} + 2Z_j^{(1)}Z_j^{(1)*} + Z_{j0}^*Z_j^{(2)}$). Show also that the departure of particles from linear motion, due to interaction with the waves, does not contribute to $\langle i_j^{(2)} \rangle$.

The first term in the right-hand side of (4.18) depends on the wave amplitude and gives an exponential growth or damping of the wave due to the particles. In the case of one mechanical realization of the system, a growth rate was found for a mode j whose phase velocity corresponds to a positive slope of the velocity distribution function. It is given by (3.72) and does not depend on the initial positions of the particles. Furthermore (3.99) recovered the same growth rate for a typical initial condition. It is, therefore, natural to recover the same expression (4.15) for the growth rate in the statistical approach. Equation (3.98) showed the validity of the same formula for a negative slope of the velocity distribution function if spontaneous emission is negligible. However, in the analysis in section 3.8.3 it was difficult to account properly for the intermediate time behaviour of figure 3.8, where wave decay through phase mixing and spontaneous emission balance each other. Thus the statistical approach offers a second viewpoint, complementing (3.98) and (3.99), to explain why the Landau growth and damping are given by the same formula in spite of their different microscopic natures.

In fact, the statistical viewpoint in this section is relevant to the single realization of the particle–wave system: if the average *intensity* of a wave (which is a real positive quantity) goes to zero, this implies that the intensity of the wave does decay in typical realizations.

One may view our averaging over the initial positions of particles as just a technique to estimate the sum over many particles. The simplest viewpoint is to perform averages over quasi-beams: one considers groups of particles with nearby velocities as a perturbation of an equidistributed monokinetic beam following the approach of chapter 3. The condition for this is that $(\delta x_l, \delta p_l, Z_j)$ be small enough, namely that $|k_j \delta x_l| \ll 1$ for all j, l. For the ballistic modes this requires that $t \ll 1/|k_j \delta p_l|$; and for the wavelike modes, this requires that $\varepsilon \beta_j k_j |(\alpha + \mathrm{i} k_j v_s)^{-2} Z_j| \ll 1$. In the limit $\varepsilon \to 0$ of the present section, this condition is fulfilled even for $I_j = O(1)$. Thus the condition $I_j \gg I_{\min}$ (which is equivalent to $\varepsilon \ll 1$) is fully compatible with those required for chapter 3. Therefore, the statistical damping of Langmuir waves can also be interpreted as the consequence of the phase mixing of many van Kampen (beamlike) modes (index jr with many rs in chapter 3). It can, by no means, be the result of the damping of eigenmodes in a Hamiltonian system: indeed, imagine that Landau damping corresponds to some $\gamma_{jr} < 0$; as we further comment in section 5.1, since Hamiltonian dynamics is reversible, there would also be an opposite rate $-\gamma_{jr} > 0$ and, with probability 1, an initial condition would produce exponential growth rather than damping! Let us now discuss the second term in the right-hand side of (4.18), which does not depend on the wave amplitude. It is proportional to the number of particles with velocity ω_{j0}/k_j and may be interpreted as the spontaneous emission of wave j by the particles. It is the equivalent for electrostatic waves of Cherenkov emission for electromagnetic waves. The evolution equation (4.18) has been checked (Doxas and Cary 1997) to see that it correctly reproduces the mode evolution in numerical simulations.

The existence of spontaneous emission implies non-vanishing Langmuir waves in a typical system. Indeed, in the case where $\mathrm{d}f/\mathrm{d}p$ is negative, Langmuir modes cannot damp out but evolve to the limit amplitude such that $\langle \mathrm{d}I_j/\mathrm{d}t \rangle = 0$,

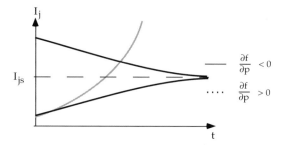

Figure 4.1. Time evolution of the energy of a Langmuir wave. The dotted line corresponds to the exponential growth in the case of a positive slope of f; and the two continuous lines correspond to the relaxation to the 'thermal' level in the case of a negative slope of f.

which is

$$I_{js} = -\frac{f(\omega_{j0}/k_j)}{k_j f'(\omega_{j0}/k_j)}. \qquad (4.19)$$

This corresponds to a 'thermal' level. Indeed, we shall see in section 9.7 that the chaotic motion of particles in the field of a wave (whose intensity and phase vary due to its interaction with particles) favours a non-zero wave intensity. Depending on the total momentum and energy available, the statistical equilibrium intensity may scale as N (in cases corresponding to a Landau instability) or remain O(1) in the limit $N \to \infty$, as is the case in the presence of Landau damping.

Exercise 4.3. Show that $I_{js} = k_B T/(\omega_{j0} - k_j v_0)$ if f is a Maxwell distribution with temperature T and mean velocity v_0.

Imposing I_{js} to be larger than $I_{\min}$ (see (4.1)) yields

$$\frac{k \Delta p_f}{\omega_p} \gg \frac{\eta}{2N} \qquad (4.20)$$

which is verified for N large enough.   Figure 4.1 sketches the temporal dependence of the electrostatic energy of a Langmuir mode as a function of its initial amplitude.

As a result, a damping in (4.18) turns out to be an exponential relaxation to the 'thermal' level. If the intensity of a mode is far above this level, it looks as if it is exponentially damped. This level vanishes in the limit $N \to \infty$. The exponential damping of a Langmuir wave was first discovered by Landau in 1946 and is named after him. However, a mode with an intensity smaller than this level grows until it reaches intensity I_{js}.

When $df/dp > 0$, a mode ends up growing exponentially with the Landau growth rate γ_{jL}, whatever its initial amplitude.

4.1.3 Evolution of particles

The second-order calculation of (2.111), (2.113) yields

$$
\begin{aligned}
\dot{p}_l^{(2)} = & -2\varepsilon^2 \sum_{j=1}^{M} \sum_{j'=1}^{M} k_j \beta_j \beta_{j'} I_{j0}^{1/2} I_{j'0}^{1/2} \\
& \times \Omega_{lj'}^{-2} (\sin(\Omega_{lj'}t + \varphi_{lj'}) - \sin\varphi_{lj'} - \Omega_{lj'}t \cos\varphi_{lj'}) \cos(\Omega_{lj}t + \varphi_{lj}) \\
& + \varepsilon^2 \sum_{l'=1}^{N} \sum_{j=1}^{M} \beta_j^2 k_j^{-1} \Omega_{l'j}^{-1} (\sin((\Omega_{lj} - \Omega_{l'j})t + \varphi_{lj} - \varphi_{l'j}) \\
& - \sin(\Omega_{lj}t + \varphi_{lj} - \varphi_{l'j})).
\end{aligned}
\tag{4.21}
$$

After averaging over initial positions, this implies

$$
\langle \dot{p}_l^{(2)} \rangle = \sum_{j=1}^{M} \varepsilon^2 \beta_j^2 k_j I_{j0} \frac{\partial}{\partial \Omega_{lj}} \frac{\sin \Omega_{lj}t}{\Omega_{lj}} - \sum_{j=1}^{M} \frac{\varepsilon^2 \beta_j^2}{k_j} \frac{\sin \Omega_{lj}t}{\Omega_{lj}}.
\tag{4.22}
$$

By analogy with the derivation for waves, we identify $\langle \dot{p}_l^{(2)} \rangle$ and $\langle \dot{p}_l \rangle$, and we generalize I_{j0} into $I_j(t)$:

$$
\langle \dot{p}_l \rangle = F_{lL} - \sum_{j=1}^{M} \frac{\varepsilon^2 \beta_j^2}{k_j} \frac{\sin \Omega_{lj}t}{\Omega_{lj}}
\tag{4.23}
$$

with $F_{lL} = \sum_{j=1}^{M} F_{ljL}$ and

$$
F_{ljL} = \varepsilon^2 \beta_j^2 k_j I_j \frac{\partial}{\partial \Omega_{lj}} \left(\frac{\sin \Omega_{lj}t}{\Omega_{lj}} \right).
\tag{4.24}
$$

Exercise 4.4. Comparing (4.23) with expression (4.10) of the average evolution of electrostatic energy, check that the average total momentum of the system is conserved, i.e.

$$
\sum_{l=1}^{N} \langle \dot{p}_l \rangle + \sum_{j=1}^{M} k_j \langle \dot{I}_j \rangle = 0
\tag{4.25}
$$

to this order of perturbation theory. Check that the total energy is also conserved (it has to be!).

Comparing relations (4.22) and (4.10) enables us to link the first term of the right-hand side of (4.23) with Landau growth and damping and to link the second term with spontaneous emission.

Figure 4.2 plots the Landau force F_{ljL} acting on particle l due to mode j as a function of $\Omega_{lj}t$. This force vanishes for a particle resonating with the wave,

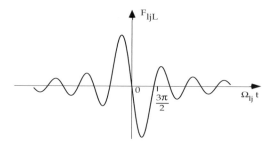

Figure 4.2. Average Landau force acting upon a particle versus $\Omega_{lj}t$.

i.e. for $\Omega_{lj} = 0$, and it takes on mainly positive (respectively negative) values for particles slower (respectively faster) than the wave. Therefore, the Landau effect corresponds to the energy exchange between waves and particles related to a synchronization mechanism of the latter with the former. It mainly involves particles whose velocity is close to the phase velocity of the wave. As the typical width of F_{ljL} as a function of $p_l - \omega_{j0}/k_j$ is of the order of $1/(k_j t)$ and as the time related to the Landau effect is $1/|\gamma_{jL}|$, synchronization is efficient for velocities in a range $|\gamma_{jL}|/k_j$ about ω_{j0}/k_j. This defines a resonance interval about the phase velocity of a given wave. This interval is much larger than the wave-trapping width. Therefore the corresponding resonance is a weak one as we found in section 3.9.

4.1.4 Fokker–Planck equation for the particles

We assume orbits to be weakly perturbed, k to be the typical wavenumber and Δv to be the typical width of the domain with a positive slope for the velocity distribution function. Then a broad spectrum of modes is Landau unstable, and $\tau_c = 1/(k\Delta v)$ is the correlation time of the electric field as seen by the particles. When $t \gg \tau_c$ waves act incoherently on the particle, and a Markovian description is possible, then the velocity distribution function $f(p, t)$ satisfies an equation of the type

$$f(p, t + \Delta t) = \int P(p - \delta p, \delta p, \Delta t) f(p - \delta p, t) \, d\delta p \qquad (4.26)$$

where $P(p, \delta p, \Delta t)$ is the probability density of a transition from p to $p + \delta p$ during time lapse Δt. This Markovian property enables us to write a so-called Fokker–Planck equation for the distribution function $f(p, t)$ (see section E.3)

$$\frac{\partial f}{\partial t} = \frac{1}{2} \frac{\partial^2}{\partial p^2} \left(\frac{\langle \Delta p \Delta p \rangle}{\Delta t} f \right) - \frac{\partial}{\partial p} \left(\frac{\langle \Delta p \rangle}{\Delta t} f \right). \qquad (4.27)$$

We now estimate the coefficients of this equation with the previous perturbative technique. Then

$$\Delta p = p_l(\Delta t) - p_{l0}. \tag{4.28}$$

The expression for $\langle \Delta p^2 \rangle$ to second order in ε is obtained from (4.5). For $\Delta t > \tau_c$ small enough

$$\frac{\langle \Delta p^2(\Delta t) \rangle}{\Delta t} = \frac{\langle \Delta p^2(\Delta t) \rangle - \langle \Delta p^2(0) \rangle}{\Delta t} \simeq \frac{\partial}{\partial t} \langle \Delta p^2 \rangle = 2 \sum_{j=1}^{M} \varepsilon^2 \beta_j^2 I_{j0} \frac{\sin \Omega_{lj} \Delta t}{\Omega_{lj}} \tag{4.29}$$

and the second-order expansion (4.22) in ε yields

$$\frac{\langle \Delta p \rangle}{\Delta t} \simeq \langle \dot{p} \rangle = \sum_{j=1}^{M} \varepsilon^2 \beta_j^2 k_j I_{j0} \frac{\partial}{\partial \Omega_{lj}} \frac{\sin \Omega_{lj} \Delta t}{\Omega_{lj}} - \sum_{j=1}^{M} \varepsilon^2 \beta_j^2 k_j^{-1} \frac{\sin \Omega_{lj} \Delta t}{\Omega_{lj}}. \tag{4.30}$$

Now we go to the limit of a continuous wave-spectrum. Given the wavenumber spectrum (3.1), the intensity spectrum defines an interpolating function $J(k)$ such that for the allowed values k_j

$$J(k_j) = I_j \frac{L}{2\pi}. \tag{4.31}$$

However, as the drive for wave–particle resonance is sensitive to velocities, it will prove useful to consider the intensity spectrum $I(v)$ interpolating with respect to phase velocities $v_j = \omega_{j0}/k_j$, i.e.

$$I(v_j) = \frac{I_j}{\Delta v_j} \tag{4.32}$$

where $\Delta v_j \equiv v_j - v_{j-1}$. Then the continuous spectrum limit reads:

$$\sum_j I_{j0} \bullet \rightarrow \int \bullet I_j \frac{L}{2\pi} \, dk = \int \bullet J(k) \, dk = \int \bullet I(v) \, dv. \tag{4.33}$$

We define a momentum diffusion coefficient $D_{QL}(p)$ and a force $F_c(p)$ by

$$\frac{\langle \Delta p^2 \rangle}{2 \Delta t} = \frac{L \varepsilon^2 \beta^2}{2p} I(p) = \frac{L}{p} |e E_j|^2 = D_{QL}(p) \tag{4.34}$$

$$\frac{\langle \Delta p \rangle}{\Delta t} = \frac{d D_{QL}}{dp} + F_c(p) + O(M N \varepsilon^4) \tag{4.35}$$

where $\beta(p)$, $k(p)$ and $\omega(p)$ interpolate respectively for β_j, k_j and ω_{j0} (with j chosen such that $|p - \omega_{j0}/k_j|$ is minimum) and

$$F_c(p) = -\frac{L \varepsilon^2 \beta^2}{2\omega(p)} = -\frac{D_{QL}}{kI}. \tag{4.36}$$

In this calculation we assumed Δt verifies

$$\Delta t > \frac{2\pi}{k_j \Delta v}.$$

(4.37)

Rewriting (4.27) as

$$\frac{\partial f}{\partial t} = \frac{1}{2} \frac{\partial}{\partial p} \left(f \frac{\partial}{\partial p} \frac{\langle \Delta p^2 \rangle}{\Delta t} + \frac{\langle \Delta p^2 \rangle}{\Delta t} \frac{\partial f}{\partial p} \right) - \frac{\partial}{\partial p} \left(\frac{\langle \Delta p \rangle}{\Delta t} f \right)$$

(4.38)

yields

$$\frac{\partial f}{\partial t} = \frac{\partial}{\partial p} \left(D_{QL}(p) \frac{\partial f}{\partial p} \right) - \frac{\partial}{\partial p} (F_c f)$$

(4.39)

where the Fokker–Planck equation displays two terms on the right-hand side. The first one involves only the wave amplitude through the diffusion coefficient; and the second one is a friction term due to spontaneous, or Cherenkov, emission. The first term in the right-hand side of (4.39) is the so-called quasilinear diffusive term (Vedenov *et al* 1962, Drummond and Pines 1962), associated with the particle's spreading over a time $\tau_c \ll \Delta t \ll \tau_d$. The second term in (4.39) accounts for the dynamical friction or drag due to the spontaneous emission responsible for S_j in (4.18).

Exercise 4.5. Check that (4.18) and (4.39) ensure that spontaneous processes conserve momentum ($\int k S L \, dk / (2\pi) + \int F_c f N \, dp = 0$). Note that this balance does not follow directly from the conservation of total momentum (which also includes non-spontaneous processes).

When particles diffuse, the local momentum conservation underlying the Landau effect no longer corresponds to synchronization with a single wave: this effect can correspond to a coherent mechanism or to a (quasilinear) diffusive one.

The coupled equations (4.18) and (4.39) may be used to describe the nonlinear evolution of the kinetic beam–plasma instability, as long as the previous perturbation theory remains correct. They are called the quasilinear equations in the literature when the Cherenkov term is omitted. As will be discussed in section 7.1 they are a good approximation of the dynamics even when the particle orbits are strongly perturbed due to chaos. As will be shown in section 7.6, these equations without the Cherenkov term drive the initial bump (corresponding to the beam) in the distribution function to flatten into a plateau (so that $\partial f / \partial v = 0$ in a given velocity interval). The presence of the Cherenkov term implies a relaxation toward a distribution function with a negative slope on a longer time-scale. This is the prelude to the relaxation toward a new thermal equilibrium for the whole plasma[3].

[3] This further relaxation cannot be described by the self-consistent Hamiltonian since the latter corresponds to a given bulk.

Exercise 4.6. Find the relation between $I(p)$ and $f(p)$ at such an equilibrium.

Assumptions on time-scales similar to those used for deriving the Fokker–Planck equation enable the derivation of further results with the same perturbation theory. In particular, the shift in frequency of modes due to tail particles reads:

$$\langle \delta\omega_{j0} \rangle = \langle \dot{\theta}_j^{(2)} \rangle = \frac{\varepsilon^2}{2} \sum_{l=1}^{N} \beta_j^2 \Omega_{lj}^{-2} (1 - \cos\Omega_{lj}t - \Omega_{lj}t \sin\Omega_{lj}t) \qquad (4.40)$$

yielding in the continuous limit

$$\langle \delta\omega_{j0} \rangle = \frac{\varepsilon^2 \beta_j^2 N}{2k_j} \mathrm{PP} \int \frac{\partial f}{\partial p} \frac{1}{\Omega_j} \, dp \qquad (4.41)$$

which generalizes (3.75), (3.76) to the case of damping, in agreement with (3.87). Computing the variance of the phase and amplitude of mode j yields

$$\langle \Delta\theta_j^2 \rangle = \frac{\pi}{4} \varepsilon^2 \beta_j^2 k_j^{-3} I_j^{-1} f(\omega_{j0}/k_j)t = \langle \Delta I_j^2 \rangle / (4I_j^2) \qquad (4.42)$$

$$\langle \Delta I_j^2 \rangle = \pi \varepsilon^2 \beta_j^2 k_j^{-3} I_j f(\omega_{j0}/k_j)t = N^{-1} I_j S_j t \qquad (4.43)$$

whose secular, linear growth with t is the signature of a diffusive process[4]. The diffusion coefficients so obtained depend on the value of the distribution function at the phase velocity of the mode of interest. They become small when $\varepsilon \to 0$. In the macroscopic limit, for a fixed $\eta = N_t/N_0$ defined by (2.106), the limit $N \to \infty$ implies that $\varepsilon \to 0$. Then in this statistical approach the phase and amplitude of a Langmuir wave, as characterized by its wavenumber, do not diffuse and are well defined, macroscopically relevant, physical quantities.

4.2 Approach using Floquet equation*

We now compute the average temporal evolution of the electrostatic energy of a given mode in a set of plasmas with a single (non-coarse-grained) velocity distribution function of the multibeam type. In contrast to section 4.1, momenta are fixed here but the positions of particles of a given beam are independent random variables uniformly distributed in space. In order to take advantage of the Floquet approach of chapter 3, one could think about writing the position of a particle of a given beam as that of one of the particles on the beam array, plus a shift. However, averaging directly over this shift would mean allowing it to take on values too large for the linearization of chapter 3 to hold (any $|k_j \delta x_l(0)|$ may be close to π). As a result, for each sample of the set of random positions of the particles of a beam, we redefine the numbering of the particle of the

---

[4] The phase variance may grow beyond π^2 because the phase is followed here continuously in time, so that one does not consider it on the circle modulo 2π.

corresponding array particles in agreement with the spatial order of the particles of the sample. Then the corresponding $\delta x_l(0)$ remain typically of order L/N_s. Under these conditions, when the number of particles per beam is high enough, the linearization is valid for most samples and the average may be done by using the linearized equations of motion.

Equations (3.7) and (3.8) yield a formal expression of the particle motion

$$\delta x_l(t) = \delta x_l(0) + \delta p_l(0)t + \frac{\varepsilon}{2}\sum_{j=1}^{M} i\beta_j \int_0^t \int_0^{t'} (Z_j e^{ik_j x_l} - Z_j^* e^{-ik_j x_l})\,dt''\,dt'.$$

(4.44)

We define the envelope of wave j, a slow component of the amplitude $Z_j(t)$ of mode j

$$R_j(t) = Z_j(t)e^{i\omega_{jL}t}$$

(4.45)

where ω_{jL} is the pulsation (3.76) of Langmuir waves. Setting (4.44) into the evolution equation (3.9) of mode j, and taking advantage of spatial equidistribution of particles of a given beam array, yields the evolution equation

$$2\frac{dI_j}{dt} = \varepsilon^2 \beta_j^2 \sum_{s=1}^{b} N_s \int_0^t \int_0^{t'} \Re[iR_j^*(t)R_j(t'')e^{-i\Omega_{js}(t-t'')}]\,dt''\,dt'$$

$$+ \varepsilon\beta_j \sum_{l=1}^{N}(\delta x_l(0) + \delta p_l(0)t)(Z_j e^{ik_j x_l} + Z_j^* e^{-ik_j x_l})$$

(4.46)

where $\Omega_{js} = k_j p_s - \omega_{jL}$ (note that this Ω_{js} differs from the Ω_{lj} defined in section 4.1). Averaging over the set of initial particle positions yields

$$\frac{d\langle I_j \rangle}{dt} = S_j' + \frac{\varepsilon^2 \beta_j^2}{2}\int_0^t \int_0^{t'} \Re\left(i\langle R_j^*(t)R_j(t'')\rangle \sum_{s=1}^{b} N_s e^{-i\Omega_{js}(t-t'')}\right) dt''\,dt'$$

(4.47)

where

$$S_j' = \frac{\varepsilon\beta_j}{2}\left\langle \sum_{l=1}^{N}(\delta x_l(0) + \delta p_l(0)t)(Z_j e^{ik_j x_l} + Z_j^* e^{-ik_j x_l})\right\rangle.$$

(4.48)

To estimate the double time integral, we first consider some properties of the functions inside the integrand. The s-sum in (4.47) is $F_j(t - t'')$ where the function

$$F_j(t) = e^{i\omega_{jL}t}\sum_{s=1}^{b} N_s e^{-ik_j p_s t}$$

(4.49)

has a typical width of the order of $1/(k_j \Delta v)$ and is periodic with period $1/(k_j \Delta p)$. Here Δv is again the width of the velocity distribution of particles, and Δp is the velocity mismatch between neighbouring beams ($\Delta p \ll \Delta v$).

We consider the case where $df/dp < 0$ and we leave the (simpler) opposite case as an exercise. As we need only times $t > 0$, $t'' > 0$, (3.50) implies

$$R_j(t) = H(t \geq 0) \sum_{r=1}^{2b+1} \zeta_{jr} e^{\delta\alpha_{jr}t} \tag{4.50}$$

where $H(t \geq 0)$ is the Heaviside function (section B.3) and $\delta\alpha_{jr} = \alpha_{jr} + i\omega_{jL}$. The key quantity in (4.47) is the covariance of R_j, which we now show to be almost constant for the values of t and t'' under consideration. To this end, we first consider its Laplace transform[5]

$$\tilde{R}_j(\sigma) = \int_0^\infty R_j(t)e^{-\sigma t}\, dt = \sum_j (\sigma - \delta\alpha_{jr})^{-1}\zeta_{jr} \tag{4.51}$$

and its formal Fourier transform $\hat{R}_j(\omega) = \tilde{R}_j(i\omega)$.

Exercise 4.7. Show that $\langle R_j(t) \rangle = 0$.

For a given k_j, beam s supports two modes with same frequency ω_{jr} and opposite growth rates γ_{jr}. In the continuum limit, if $df/dv(\omega_{j0}/k_j) < 0$, both γ_{jr} and $\omega_{jr} - k_j v_s$ go to zero with Δp (at nearly the same rate). Therefore, in (3.54) and in $\delta p_{ls}(t)$, the dominant contributions among beam modes come from $O(|\gamma_{jr}|/\Delta p)$ modes around beam s.

The amplitudes ζ_{jr} are related to initial conditions through (3.59) and (3.16), (3.17). The Lorentzian-like factors $(\alpha_{jr} + ik_j v_s)^{-2}$ in (3.54) dominate the near-diagonal terms of the mapping between the ζ_{jr} and initial data $(Z_j(0), \delta x_{ls}(0), \delta p_{ls}(0))$. For a beam s denote by $r_\pm(s)$ the modes jr for which $|\alpha_{jr} + ik_j v_s|$ is minimum, i.e. the modes with which it is closest to resonance. Then let us denote $\alpha_{jr_\pm} = \pm\gamma - ik_j v_s - i\varphi$ (where we omit subscripts for simplicity), with $\varphi = wk_j\Delta p = O(\Delta p)$ as in section 3.8.3. These modes dominate the sum to the right-hand side of (3.54), so that

$$C_{js}(0) = \frac{\varepsilon\beta_j}{2}\frac{\zeta_+}{(\gamma - i\varphi)^2} + \frac{\varepsilon\beta_j}{2}\frac{\zeta_-}{(-\gamma - i\varphi)^2} + \frac{\varepsilon\beta_j}{2}\sum_{r=1}^{2b+1}{}' \frac{\zeta_{jr}}{(\alpha_{jr} + ik_j v_s)^2} \tag{4.52}$$

where the first two terms are $O(z/\Delta p^2)$ while in the sum (with the prime denoting that the selected $r_\pm$ are excluded from the sum) all terms have a larger denominator. In the continuous limit (where the number of terms scales like

---

[5] We must use a Laplace transform because of weakly unstable van Kampen modes. The integral defining $\tilde{R}_j(\sigma)$ diverges if $\Re\sigma \leq \max_{jr} \gamma_{jr}$. However, the rational function so defined can be continued analytically to a unique function of σ over the whole complex plane, except for its poles. This gives a definite meaning to the Fourier transform $\hat{R}_j$. Moreover, one must use a single-sided Laplace transform, i.e. introduce the Heaviside function, as (3.50) diverges for $t \to -\infty$ because of damped van Kampen modes.

Δp^{-1}), one estimates $\Delta p^{-1} \int_{\Delta p}^{\infty} u^{-2} \, du = 1/\Delta p^2$, so the sum in (4.52) is also $O(z/\Delta p^2)$. Moreover, in (3.53) a similar estimate yields

$$A_{js}(0) = \frac{i}{2}\varepsilon\beta_j \frac{\zeta_+}{\gamma - i\varphi} + \frac{i}{2}\varepsilon\beta_j \frac{\zeta_-}{-\gamma - i\varphi} + \frac{i}{2}\varepsilon\beta_j \sum_{r=1}^{2b+1}{}' \frac{\zeta_{jr}}{\alpha_{jr} + ik_j v_s} \qquad (4.53)$$

with a similar analysis. Finally, in (3.55) no coefficient becomes singular.

One may expect the most singular terms to dominate in the continuous limit, so that we approximate the solution to (4.52), (4.53) and (3.55) by the solution to the reduced system of two equations (4.52), (4.53) omitting the primed sum, for the two unknowns $\zeta_\pm$. This leads to

$$\varepsilon\beta_j \zeta_\pm^{(0)} = i \frac{(\varphi \pm i\gamma)(\gamma^2 + \varphi^2)}{\gamma} C_{js}(0) \pm i \frac{(\pm\gamma - i\varphi)^2}{\gamma} A_{js}(0) \qquad (4.54)$$

which determines $2b$ amplitudes ζ_{jr} among the $2b + 1$ unknowns. The remaining one may be deduced from them either by (3.55) or by setting it equal to $Z_j(0)$—as we compute first an approximate solution. If the initial perturbations $\delta x_l(0)$, $\delta p_l(0)$ are independent random variables $O(L)$, $O(\Delta p)$ respectively[6], the coefficients $A_{js}(0)$ and $C_{js}(0)$ are random variables $O(L/N_s^{1/2})$ and $O(\Delta p/N_s^{1/2})$. This determines the scaling for $\zeta_{jr}^{(0)}$. As each beam s is initialized independently of the others, the coefficients $\zeta_{jr}^{(0)}$ with different r are independent from each other.

Substituting this approximate solution $\zeta_{jr}^{(0)}$ into the neglected terms of the full system (4.52), (4.53) and (3.55) yields a solution by iteration:

$$\zeta_{jr+(s)} = \zeta_{jr+(s)}^{(0)} + \frac{(\gamma - i\varphi)^2}{2\gamma} \sum_{r'=1}^{2b+1}{}' \frac{\zeta_{jr'}^{(0)}}{\alpha_{jr'} + ik_j v_s}$$

$$+ \frac{(\gamma - i\varphi)(\gamma^2 + \varphi^2)}{2\gamma} \sum_{r'=1}^{2b+1}{}' \frac{\zeta_{jr'}^{(0)}}{(\alpha_{jr'} + ik_j v_s)^2} + \cdots \qquad (4.55)$$

and similarly for $\zeta_{jr-(s)}$. Both sums on the right-hand side have diverging coefficients in the limit $\Delta p \to 0$ (with the worst case $r' = r_\pm(s)$ duly excluded) but these divergences are controlled by the prefactors. Specifically, the first sum has terms $O(1/\Delta p)$ but its prefactor is $O(\Delta p)$ (to a possible $|\ln \delta|$ correction); the second sum has faster decaying terms, going at worst like $O(1/\Delta p^2)$ with a prefactor $O(\Delta p^2)$. Therefore, the most important correction to (4.54) comes from the first sum (whose terms decay more slowly), with coefficients decaying algebraically.

---

[6] The estimate $\delta x_l(0) = O(L)$ is most pessimistic and could be improved by a factor N_s^{-1} as we saw earlier.

Now we estimate the covariance of $R_j(t)$. This is easy using the power spectrum of $R_j(t)$,

$$\langle \hat{R}_j(\omega)\hat{R}_j^*(\omega)\rangle = \sum_r \sum_{r'} \frac{\langle \zeta_{jr}\zeta_{jr'}^*\rangle}{(i\omega - \delta\alpha_{jr})(-i\omega - \delta\alpha_{jr'}^*)}. \qquad (4.56)$$

The preceding discussion leads to estimating the amplitudes by $\zeta_{jr}^{(0)}$; these coefficients have expectation $\langle \zeta_{jr}^{(0)}\rangle = 0$ (as $\langle A_{js}(0)\rangle = 0$ and $\langle C_{js}(0)\rangle = 0$) and covariances $\langle \zeta_{jr}^{(0)}\zeta_{jr'}^{(0)*}\rangle = 0$ for $r' \neq r$ since they relate to independent beams. Thus

$$\langle \hat{R}_j(\omega)\hat{R}_j^*(\omega)\rangle \simeq \sum_r \frac{\langle |\zeta_{jr}^{(0)}|^2\rangle}{|\omega + i\delta\alpha_{jr}|^2}. \qquad (4.57)$$

Finally, the variances $\langle |\zeta_{jr}^{(0)}|^2\rangle$ have all the same order of magnitude, with contributions $O(|A_{js}(0)|^2\varepsilon^{-2}\beta_j^{-2}k_j^2\Delta p^2) = O(\varepsilon^{-2}\beta_j^{-2}k_j^2\Delta p^4/N_s)$ from the random $\delta p_l(0)$ and $O(|C_{js}(0)|^2\varepsilon^{-2}\beta_j^{-2}k_j^4\Delta p^4) = O(\varepsilon^{-2}\beta_j^{-2}k_j^4\Delta p^4 L^2/N_s)$ from the random $\delta x_l(0)$.

Thus the power spectrum of R_j is like a Lorentzian $|\omega - \omega_{jrL} + \omega_{jr} + i\gamma_{jr}|^{-2}$ and its Fourier transform is the correlation function $\langle R_j^*(t)R_j(t'')\rangle \simeq 2\langle I_j(t)\rangle e^{-|\gamma_{jr}(t-t'')|}$, which does not vary much on the characteristic scale $1/(k_j\Delta p)$ of $F_j(t - t'')$. Hence we may approximate the correlation function of R_j by $2\langle I_j(t)\rangle$. Straightforward calculations then reduce (4.47) to

$$\frac{d\langle I_j\rangle}{dt} = \varepsilon^2\beta_j^2 \sum_{s=1}^b N_s(\sin\Omega_{js}t - \Omega_{js}t\cos\Omega_{js}t)\Omega_{js}^{-2}\langle I_j\rangle + S_j'. \qquad (4.58)$$

By going to the continuous limit for the velocity distribution, condition $t \ll 1/(k_j\Delta p)$ is satisfied and the mean electrostatic energy $2\omega_{j0}\langle I_j\rangle$ verifies

$$\frac{d\langle I_j\rangle}{dt} = 2\gamma_{jL}\langle I_j\rangle + S_j' \qquad (4.59)$$

where γ_{jL} is given by (4.15). By comparison with (4.18) we deduce that $S_j' = S_j$, where S_j is given by (4.16).

Exercise 4.8. Using (3.23), (3.24) and (3.53), (3.54) and the estimates on the $\langle \zeta_{jr}\zeta_{jl}^*\rangle$, check that $S_j' = S_j$. Calculations are easier if you assume that all $\delta p_l(0) = 0$ to avoid secular terms.

Exercise 4.9. Discuss the case where $df/dp > 0$, using a similar argument. As $\gamma_{jL} > 0$ replaces $\gamma_{jr} \to 0$, note that the sums are less singular and the Lorentzian distributions have finite width.

To describe the particle evolution in this averaged Floquet approach, we insert the formal expression of $\delta x_l(t)$ into the second-order expression for $\delta \dot{p}_l$. The statistical distribution of the initial data for beam particles implies that the resulting $\delta p_l(t)$ are random variables, which we must characterize by their moments $\langle \delta p_l \rangle$, $\langle \delta p_l^2 \rangle$ etc. The calculations are similar to those for the average intensity and lead to the estimates (4.29), (4.30). The physical argument of section 4.1.4 for the Markovian description still holds, so that the statistical Floquet description leads to the Fokker–Planck equation (4.39) for the particle momentum distribution function.

Finally, we must recall that, as for the approach through perturbation theory, the time limit for these calculations is given by τ_d.

4.3 Link with traditional descriptions

4.3.1 Landau effect from a Vlasovian point of view

The traditional derivation of the Landau effect is done by linearizing the Vlasov equation in the vicinity of some spatially uniform distribution function and solving the linearized Vlasov–Poisson system by a Fourier–Laplace transform. If the velocity distribution function has a tail with a positive slope, the Langmuir modes are found to be unstable eigenmodes if their phase velocity lies in the interval of the tail with positive slope. Their growth rate is given by the Landau formula (3.72) or (4.15). There is a perfect match between the mechanical and Vlasovian conclusions in this case: the existence of unstable eigenmodes with the Landau growth rate. The mechanical approach adds yet the information that this is a property of a special realization of the system, while a typical initial condition also involves the damped 'anti-Landau' mode and the van Kampen-like eigenmodes as shown in (3.99).

In the domain of velocities where the distribution function has a negative slope, the Vlasovian approach does not exhibit eigenmodes with eigenvalue given by Landau's formula but an asymptotically damped solution is found by an integral (Laplace) representation. A single realization of the mechanical model provides a satisfactory analogue of this behaviour over a time interval limited by $\tau_{discr}/|\ln \delta|$ but ultimately the unstable van Kampen modes grow. However, before the most unstable ones emerge, their destructive interference has first caused the exponential decay of the wave intensity and later generated a transient of weak intensity. This transient is sensitive to the specific values of the ζ_{jr} and appears as noise on the wave intensity. The source of this noise is the spontaneous emission.

The statistical version of the mechanical approach sharpens this statement: I_j is exponentially damped for times verifying (4.13) as the thermal level vanishes in the Vlasovian limit. The thermal noisy regime is now accurately described by the balance between spontaneous emission and wave decay by van Kampen phase mixing. The deterministic version of the mechanical approach yields the explanation for the absence of a damped eigenmode: the reversibility of Hamiltonian dynamics would then provide an unstable eigenmode with the opposite growth rate as well, which would be dominant with probability 1. Landau damping turns out to be a consequence of the phase mixing of the many-beam modes with the same k_j which are found in the deterministic approach. This confirms the result found by van Kampen (1957, 1959) and Case (1959) when solving the Landau problem by allowing Dirac distributions (cold beams) to be added to the perturbed velocity distribution function. They found a continuum of beam modes which is the continuous limit of the discrete beam modes of the mechanical approach (Dawson 1960) and interpreted Landau damping as a phase mixing of the beam modes (Balescu 1963).

The mechanical approach has the advantage of clearly showing why Landau damping cannot be the signature of an eigenmode; and how the phase mixing occurs when going to the statistical description when N is finite. It proves the Landau effect is not a resonant mechanism in the sense of section 1.2 but a weakly resonant one in the sense that the strict nonlinear resonance is broadened by the time dependence of the wave: it involves velocities in a range $|\gamma_{jL}|/k_j$ about ω_{j0}/k_j. The synchronization mechanism of particles with the wave found in the mechanical approach was anticipated by Nicholson (1983).

Note that the accumulation of eigenmodes indexed by $1 \leq r \leq 2b+1$ occurs even if $M = 1$, i.e. if one restricts the analysis to a single Langmuir wave interacting with a family of monokinetic beams with a broad velocity spread and with a negative $\partial f_t/\partial p$. Thus the continuum of modes involved in Landau damping is not the continuum of Langmuir waves but is directly associated with the continuum of beam modes[7].

Even for specialists, the reality of collisionless Landau damping was fully recognized only after its experimental observation (Malmberg and Wharton 1964), almost two decades after its prediction (Landau 1946). See e.g. Ryutov (1999) for a historical sketch. Indeed the derivation of this effect in the Vlasovian framework does not make it intuitive. As a result many textbooks tried to derive it with approximate models whose physics is simpler or proposed intuitive pictures to explain it. The present mechanical approach shows that some of these models and pictures should be used with caution. In particular, since trapping is ruled out as an explanation of the Landau effect, the surfer model (Chen 1984) is to be understood in a loose sense. Furthermore, a single-mode model $(Z_j \sim \exp(i\omega + \gamma)t)$ is appropriate for describing the Landau instability but not for the damping as it is due to the phase mixing among many eigenmodes.

The following exercise may be considered as a synthesis of the themes of this and the previous chapter.

Exercise 4.10.   (i) Compute by first-order perturbation theory in E_0 the

---

[7] Actually, Landau damping may also occur in non-plasma systems involving a continuum of subsystems, such as a collisionless neutral gas (Stubbe and Sukhorukov 1999), diphasic fluids with bubbles (Ryutov 1975) or dusty plasmas (Bliokh *et al* 1995).

motion of an electron (charge q) in the field of a small longitudinal wave $E(x, t) = \Re(E_0 \exp[\gamma t + i(kx - \omega t)])$; let $x = x_0 + vt$ be the unperturbed motion. Show that the mean work $w = \langle q\dot{x}E(x, t)\rangle$ done by the field on the electron by unit time is proportional to

$$\frac{d}{dv} \frac{v\gamma}{\gamma^2 + (kv - \omega)^2}.$$

(ii) Use this result to compute the mean work per unit time W on a distribution of electrons defined by $f(v)$. Integrating by parts and using (B.17), show that $W \sim df(\omega/k)/dv$ and that $W\gamma(df(\omega/k)/dv) < 0$ if γ/k is small with respect to the scale of variation of $df(v)/dv$.

(iii) We now consider the wave of interest to be a Langmuir wave whose growth/damping rate γ is due to an energy exchange with the electron distribution. Show that energy conservation forbids $df(\omega/k)/dv < 0$ whatever the sign of γ. In terms of the eigenvalue analysis of chapter 3, to what do $\gamma > 0$ and $\gamma < 0$ correspond? Why is the present model unable to reproduce Landau damping? Recalling that the electrostatic wave energy density (in volume) is $|E(x, t)|^2\epsilon_0/2$, use energy conservation to compute the Landau growth rate if $df(\omega/k)/dv > 0$. This calculation takes into account wave–particle self-consistency in a partial way. What is missing? For a given k, what is the value of ω to be taken?

(iv) We now identify $\pm\gamma$ as the Landau growth rate. What is the range of velocities of the particles exchanging the largest amount of energy with the wave? For a given time, compute the bouncing frequency ω_b of the particles in the wave troughs and the trapping width of the wave. Show that $\omega_b \ll |\gamma|$ is the necessary and sufficient condition of validity for the previous perturbation calculation. Deduce from this that particles most involved in exchanging energy with the wave are far from being trapped in it. Interpret the growth or damping of the wave for $df(\omega/k)/dv > 0$ as an energy exchange between the wave and particles slower or faster than it. Show that trapped particles make a small contribution to this energy exchange.

4.3.2 Spontaneous emission

As the Vlasov equation applies in the limit of an infinite number of particles in the Debye sphere, it corresponds to a noiseless limit and cannot describe spontaneous emission by itself. Textbooks recover this emission by adding a test particle to a Vlasovian plasma. The absence of spontaneous emission of a Vlasovian plasma means that the thermal level of Langmuir waves is zero (the finite value of I_{js} in (4.19) corresponds to a vanishing wave amplitude in the limit of large N because of the ε prefactor). This explains why Landau predicts an exponential damping of Langmuir waves, when experiments and the mechanical approach show a relaxation to a finite thermal level. Due to spontaneous emission, the exponential (Landau) damping or growth of a wave occurs only for a large enough amplitude: only special initial conditions corresponding to eigenmodes enable the exponential growth of a wave without spontaneous emission.

The mechanical approach has the advantage of unifying the Landau effect and spontaneous emission. Altogether, the mechanical derivation is much simpler and intuitive when starting from first principles than the one going through the Liouville equation, the BBGKY hierarchy[8] and the Landau treatment of the Vlasov equation. It fully accounts for the Hamiltonian character of the microscopic dynamics. In addition to the previous discussion, note that equation (4.18) is germane to that obtained by Harris (1969) through a quantum mechanical treatment of the three-dimensional problem.

This chapter ends our discussion of linear dynamics. In the following chapters, we turn to the nonlinear evolution of the wave–particle system, for the understanding of which the self-consistent model was originally constructed. A prerequisite is the introduction of concepts and tools for dealing with Hamiltonian chaos, which is being covered in the next two chapters.

---

[8] The Vlasov equation can be derived in a much shorter way by using the mean-field limit (Spohn 1991) sketched in appendix G.

Chapter 5

Hamiltonian chaos

This chapter introduces basic concepts of Hamiltonian chaos necessary to understand the nature of the nonlinear wave–particle dynamics defined by the self-consistent Hamiltonian[1]. As these concepts are also useful in other contexts, this chapter is written in a self-contained way and some applications to other Hamiltonian systems are indicated. Chaotic diffusion is extensively studied in the next chapter.

The central model in this chapter is the motion of an electron in the field of prescribed longitudinal waves. Wave–particle resonance, which was introduced in section 1.2, is a key concept in this problem. Chaos occurs when at least two waves with different phase velocities are present. It has several signatures, the simplest one being the exponential divergence of nearby orbits, a property which is not specific to Hamiltonian chaos. This feature rules out an analytical calculation of chaotic orbits over times long with respect to the divergence time τ_d. Fortunately, for most physical applications the precise knowledge of individual orbits is useless and statistical descriptions are appropriate. The next chapters give several instances of such descriptions of chaotic dynamics.

Another consequence of the exponential divergence of nearby orbits is that the numerical calculation of a chaotic orbit over times long with respect to τ_d is impossible, since it would require computers with exponentially large number of bits. Yet, somewhat paradoxically, numerical calculations of chaotic orbits played a central role in the development of the science of chaos! To some extent, numerical chaotic orbits have features which are consistent with the mathematical knowledge about true chaotic orbits[2], provided that some care is taken in choosing

---

[1] Computing some macroscopic features of chaos can be done without studying this chapter. The reader interested in deriving a diffusion equation for the chaotic motion may skip this chapter and directly go to section 6.7.

[2] For completely hyperbolic systems the shadowing lemma ensures that numerical chaotic orbits have features which are consistent with the mathematical knowledge about true chaotic orbits (Guckenheimer and Holmes 1983, Alligood *et al* 1996). Typical Hamiltonian systems are not hyperbolic. However, several types of orbit with specific properties can also be identified rigorously from numerical approximate dynamics, possibly using integer arithmetic (Earn and Tremaine 1992) or interval arithmetic, and symplectic integration schemes reproduce the orbits with a good overall accuracy (Benettin and Giorgilli 1994, Benettin and Fassò 1999).

the integration scheme. For simulating Hamiltonian dynamics, one must use a so-called symplectic integration scheme, which means that the integration map must be area-preserving for the simple cases considered in this chapter (appendix D). The error incurred at each integration step with respect to the true orbit looks like having lenient consequences.

Numerical calculations provide another signature of chaos. This is obtained in a graphical way through the so-called Poincaré map which is a stroboscopic map of the dynamics in the simplest cases. When the wave amplitudes are small, each one appears in the map by its own chain of islands (or cat's eyes) in phase space, but most phase space is still covered by the so-called Kolmogorov–Arnold–Moser (KAM) tori which are continuous sets of regular orbits with a given mean velocity. When wave amplitudes increase, KAM tori break up, the edge of the islands is smeared out by chaos and, eventually, they merge to generate a connected chaotic sea. Higher-order resonances bring in chains of islands arbitrarily close to a KAM torus. These features are typical of Hamiltonian chaos. We introduce geometrical tools for this chapter in section 5.1. Some of them are used in the numerical observations of section 5.2 which provides a concrete introduction to Hamiltonian chaos.

Magnifying the phase space close to a KAM torus reveals that the corresponding cat's eyes close down in the vicinity of the torus. There is a tool to magnify phase space analytically: the renormalization transformation, which we introduce heuristically in section 5.3.1 and investigate more systematically in section 5.4. It may be used to compute the thresholds for break-up of a KAM torus and for the occurrence of global chaos in a domain of phase space or to find the critical exponent for the opening of holes in a former KAM torus.

Unstable periodic orbits play a crucial role to distinguish order from chaos. Indeed the sets of points whose images converge toward such orbits in the Poincaré map when time goes toward $\pm\infty$ have a very different structure in each case. For the ordered motion these sets are simple curves connecting the traces of the periodic orbits (or cycles). For the chaotic motion they take on a wiggling structure which creates a tight (so-called homoclinic) trellis covering the chaotic sea as discussed in section 5.5. In the large-wave-amplitude limit, where any wave would individually trap particles in the same domain of phase space, the extension of this trellis and of the chaotic sea can be defined geometrically in a simple way. As a result, the large- and small-amplitude limits of the waves provide complementary viewpoints where the geometrical aspects of chaos and order are simple.

A motto throughout this chapter is that, in spite of the complexity of chaotic behaviours, many of their characteristics can be understood with elementary constructions. In particular, we insist on perturbative estimates (i.e. start from sketchy descriptions and refine them) and on the notion of resonance, with both its physical and its geometrical connotations.

In this introductory book, we shall not discuss the geometrical problems intrinsically related to the coupling of several degrees of freedom.

5.1   Geometrical tools for Hamiltonian chaos

As seen in chapter 1, Hamiltonian dynamics rest on the existence of canonical variables. A time-independent Hamiltonian system with N degrees of freedom is described by its $2N$-dimensional phase space $\mathcal{M}$, spanned by a coordinate system $(q_1, \ldots q_N, p_1, \ldots p_N)$ where (q_i, p_i) are pairs of conjugate variables, and its Hamiltonian function $H : \mathcal{M} \to \mathbb{R}$, such that for all $1 \leq i \leq N$

$$\dot{q}_i = \frac{\partial H}{\partial p_i} \tag{5.1}$$

$$\dot{p}_i = -\frac{\partial H}{\partial q_i}. \tag{5.2}$$

Such a system is said to be autonomous. For a non-autonomous Hamiltonian system, H depends also on time and (5.1), (5.2) still hold.

A classical theorem (see, e.g., Hirsch and Smale 1974) shows that orbits do not cross for reasonable dynamics. For the two-dimensional phase space $\mathcal{M}$ (with the topology of a plane, cylinder or sphere) of a one-degree-of-freedom Hamiltonian, the orbits' mutual avoidance also ensures some order. Indeed any closed orbit separates $\mathcal{M}$ into two unconnected domains, say $\mathcal{D}_1$ and $\mathcal{D}_2$. Then any orbit with initial condition in $\mathcal{D}_1$ remains in $\mathcal{D}_1$ for all time. The mutual avoidance argument applies to both the pendulum and the electron in a wave. More complicated dynamical systems may have topologically more complicated phase spaces. In a higher-dimensional phase space $\mathcal{M}$, a single orbit is still a line, which cannot separate $\mathcal{M}$ in unconnected domains. Then orbits can braid around each other, in more or less complicated patterns, and chaos can occur. More information about the constraints of mutual avoidance is given in appendix C.

5.1.1   Poincaré surface of section

Before looking at chaotic dynamics we must introduce a tool enabling one to reduce the dimension of the space used to visualize the orbits: the Poincaré section method. Imagine that instead of looking at the dynamics of an electron in the presence of a wave by going in the rest frame of this wave, as we did in section 1.2, we look at it in the laboratory frame. The corresponding Hamiltonian becomes

$$H_w(p, q, t) = \frac{p^2}{2} - A \cos(q - t) \tag{5.3}$$

by choosing the time unit such that the phase velocity of the wave is 1. If we look at the motion in (q, p) space (and draw it continuously with time as a parameter), a unit translation velocity is added to that of all orbits in the rest frame of the wave. In particular, trapped motions become cycloid-like; and the picture drawn by several orbits is bound to be rather messy. A way to solve this problem is to look at the motion with a stroboscope at the wave frequency. Then the potential well of the wave and the cat's eye corresponding to the trapping domain are at the same position at each stroboscopic flash if q is plotted modulo 2π.

In general, the time necessary to shift q by 2π for a passing orbit, or to make a full oscillation for trapped ones, is not commensurate with the period 2π of the stroboscope (almost all real numbers are irrational). Therefore the successive positions of an orbit on its continuous trace $p_1(q)$ of figure 1.1 should progressively cover this trace densely. By waiting long enough, the shape of most orbits should become visible by this technique. The existence of the constant of the motion $H_1(p, q - t)$ is visible because most orbits correspond to a set of stroboscopic points drawing a continuous curve in (p, q). The former one-dimensional trace $p_1(q)$ in (q, p)-space now corresponds to a two-dimensional surface $p_2(q, t) = p_1(q - t)$ in (q, p, t)-space.

Exercise 5.1. Draw some orbits generated by (5.3) in the three-dimensional (q, p, t)-space, and draw their projections on two-dimensional spaces with coordinates (q, p), (q, t) and (p, t).

We now embed the time-dependent dynamics of Hamiltonian (5.3) into the time-independent dynamics of the two-degree-of-freedom Hamiltonian

$$H_e(p, w, q, \tau) = \frac{p^2}{2} + w - A \cos(q - \tau) \tag{5.4}$$

where (p, q) and (w, τ) are the pairs of conjugate variables. As $\dot{\tau} = \partial H / \partial w = 1$, one may retain the orbit such $\tau = t$.

Exercise 5.2. Show that the equations of motion for p and q generated by (5.4) are identical to those generated by (5.3), so that H_e is indeed the two-degrees-of-freedom embedding of H_w. Generalize your proof and associate a time-independent Hamiltonian $H_2(p, w, q, \tau) = w + H_{1.5}(p, q, \tau)$ with any time-dependent Hamiltonian $H_{1.5}(p, q, t)$, irrespectively of whether it be integrable. This justifies calling $H_{1.5}(p, q, t)$ a 1.5-degrees-of-freedom Hamiltonian, since its dynamics has some features of a two-degrees-of-freedom one, yet with a strong constraint on its structure.

For H_e the previous surface $p = p_2(q, t)$ in (p, q, t)-space becomes a surface $p = p_2(q, \tau)$ in (p, w, q, τ)-space. When p, q and τ are known, w is immediately found by using the constancy of H_e along the motion, so that $p = p_2(q, \tau)$ generates $w = w_2(q, \tau)$.

Exercise 5.3. Show that, for any c, given an orbit $(q(\tau), p(\tau), w(\tau))$ emerging from initial data (q_0, p_0, w_0), the orbit emerging from $(q_0, p_0, w_0 + c)$ is $(q(\tau), p(\tau), w(\tau) + c)$. Therefore w is of little interest for the dynamics and one may characterize the (p, q, τ) evolution without referring to w.

Let us now discuss orbits on these surfaces. Untrapped orbits stay on one of the two branches of p_2 related to their energy, and on each of them

$z = q - \tau$ increases or decreases monotonically; and τ increases steadily. As the Hamiltonian (5.4) is a 2π-periodic function of both τ and z, these two angles parametrize each branch of the surfaces $p = p_2(q, \tau)$.

However, trapped orbits successively visit both branches of their surface $p = p_2(q, \tau)$. Then it is convenient to refer a point on the orbit using the angle ϕ of (p, q) with the q-axis and to consider that p_2 and w_2 are functions of ϕ and τ.

As a result, for given $(H_e, H_1(p - 1, q - \tau))$, and for a given sign of p for untrapped orbits, a good system of coordinates on a surface (p, w) is provided by a pair of angles[3], so that, mathematically, the manifold $\mathcal{T}$ defined by these (p, w) (a single branch for the untrapped case) is topologically a two-dimensional torus.

The previous stroboscopic plot for an orbit defined by H_w corresponds to cutting the corresponding orbit of H_e by the hyperplane $\mathcal{P}$ defined by $\tau = 0 \bmod (2\pi)$. This is the definition of the Poincaré surface of section which carries over to more general two-degrees-of-freedom Hamiltonian systems. The successive points of the stroboscopic plot correspond to successive iterates of the so-called Poincaré map.

By extension, this name is also used for the sequence of points of each orbit in the Poincaré surface of section, which lie on the section $\mathcal{S}$ of $\mathcal{T}$ by $\mathcal{P}$. Let u be the time-average value of $p(t)$ for an orbit belonging to $\mathcal{T}$. If u is irrational, the successive points of the orbit in the Poincaré map tend to fill $\mathcal{S}$ densely. If u is a rational m/n, after n iterates the orbit is back to its initial position: the orbit has period n.

Finally, the stability properties of orbits of the continuous-time dynamics appear on the Poincaré section as well.

Exercise 5.4. Consider the $M = N = 1$ case for Hamiltonian (2.110). Let $\omega_0 = 0$ by a good choice of the reference frame and choose units so that $k = 1$ and $\varepsilon\beta = 1$. The crucial difference with (1.12) and (5.4) is that the wave–particle coupling is not given by the constant A but depends on the variable $w = I$, so that exercise 5.2 does not apply. Let the configuration space be bi-periodic (i.e. the torus $\mathbb{T}^2 = S_{2\pi} \times S_{2\pi}$), with both θ and x defined on the circle modulo 2π.

(i) What is the phase space $\mathcal{M}$?

(ii) Show that this phase space is foliated by two-dimensional invariant surfaces $(E, P) = (E_0, P_0)$, where the value of the constants of the motion $E = H$, $P = p + I$ is determined by initial conditions. Are all choices $(E, P) \in \mathbb{R}^2$ physically permitted?

(iii) To describe the motion on each invariant two-dimensional surface, let $z = \theta - x$ and show that the evolution equations read:

$$\dot{z} = I - P - \frac{\beta}{\sqrt{2I}} \cos z \qquad (5.5)$$

$$\dot{I} = -\beta\sqrt{2I} \sin z \qquad (5.6)$$

---

[3] These are not the angles conjugate to the actions of section 5.1.3.

on the cylinder. Are variables z and I canonically conjugate? Does system (5.5), (5.6) derive from a one-degree-of-freedom Hamiltonian?

(iv) For various choices of (E, P), find fixed points and separatrices (if there exist any) of system (5.5), (5.6). Plot typical orbits in coordinates (z, I). The resulting family of pictures forms a 'phase portrait' of the dynamics. Do they undergo qualitative changes (i.e. bifurcations) for some values of (E, P)?

(v) The $I = 0$ line is singular, which reflects the singularity at the origin for polar coordinates $\sqrt{2I}\,e^{-i\theta} = X + iY$. In a loose sense, one may identify $(I, \theta + \pi)$ with the continuation of (I, θ) for 'negative' $\sqrt{2I}$. This 'd'Alembert singularity' disappears if one uses variables (u, v) with $u + iv = \sqrt{2I}\,e^{-iz}$. How do (5.5), (5.6) read in these variables? What do the phase portraits look like? How do these variables relate to the wave–particle variables (x, p, X, Y)?

(vi) Discuss your (z, I)-space fixed points and orbits in terms of original particle motion and wave evolution.

5.1.2   Conservation of areas, symplectic dynamics and flux*

Hamiltonian dynamics preserves the volume in phase space. For more than one degree of freedom other invariants are also preserved (appendix D). The conservation of volume in the phase space implies that the orbits of a Hamiltonian system cannot converge in volume to an attractor, in strong contrast to orbits of dissipative dynamics. However, orbits may well approach each other along some 'locally stable' directions and diverge from each other along other 'locally unstable' directions.

In particular, the Lyapunov exponents[4] λ_i, $1 \le i \le 2N$, of a periodic orbit of an autonomous Hamiltonian dynamics with N degrees of freedom fulfil a pairing rule $\lambda_i + \lambda_{2N+1-i} = 0$. Moreover, thanks to the conservation of energy and the arbitrariness in the choice of initial time, $\lambda_N = \lambda_{N+1} = 0$; and, more generally, each pair of zero Lyapunov exponents hints at the existence of a constant of the motion (Arnold *et al* 1988, chapter 6).

---

[4] Lyapunov exponents measure the way orbits diverge from or converge to each other asymptotically in time. Given an initial condition $x(0) \in \mathcal{M}$, consider $2N$ linearly independent small perturbations $\delta x^i(0) = (\delta q_1^i(0), \dots \delta q_N^i(0), \delta p_1^i(0), \dots \delta p_N^i(0)) \in \mathbb{R}^{2N}$, indexed by $1 \le i \le 2N$. Then consider the orbit $x(t)$ and the evolution $\delta x^i(t)$ of each small perturbation. If it exists, the largest Lyapunov exponent of the orbit x is almost surely

$$\lambda_1 = \lim_{t \to \infty} (1/t) \ln(\|\delta x^i(t)\| / \|\delta x^i(0)\|) \tag{5.7}$$

for any i, i.e. the average exponential growth rate of the separation of nearby orbits. The second exponent λ_2 is given by the sum $\lambda_1 + \lambda_2 = \lim_{t \to \infty} (1/t) \ln(\|\delta x^i(t) \wedge \delta x^j(t)\| / \|\delta x^i(0) \wedge \delta x^j(0)\|)$ for $i \ne j$, i.e. the average exponential growth rate of areas defined by small vectors, and so on (see section D.1 for the definition of the wedge product $\wedge$). Associated with the exponents are directions defined in the neighbourhood of each orbit, providing a natural system of coordinates related with the stable and unstable manifolds discussed in section 5.5.1. The existence of Lyapunov exponents is generally guaranteed for periodic orbits. This is also the case for non-periodic orbits in hyperbolic dynamics (see, e.g., Guckenheimer and Holmes 1983).

For discrete time, one uses multipliers $\mu_i = e^{\lambda_i}$, so that $\mu_i\mu_{2N+1-i} = 1$. For the Poincaré map of a 1.5-degrees-of-freedom system, only the two non-trivial exponents matter, as we have already eliminated the zero exponent corresponding to the choice of the origin for time.

Area preservation is so central to Hamiltonian dynamics[5] that it also provides a simple invariant (i.e. coordinate-system independent) concept for transport: transport may be measured by the flux from an initial domain $\mathcal{D}_0$ into a target domain $\mathcal{D}_1$ (Meiss 1992).

Given an N-degree-of-freedom autonomous Hamiltonian dynamics with conjugate variables $(\boldsymbol{q}, \boldsymbol{p})$ and Hamiltonian H, consider a new set of generalized coordinates $\boldsymbol{\theta}$ and generalized momenta $\boldsymbol{J}$. Assume that there exists a function $\Phi(\boldsymbol{q}, \boldsymbol{J})$, such that, for all $1 \leq i \leq N$,

$$p_i = \frac{\partial \Phi}{\partial q_i} \tag{5.8}$$

$$\theta_i = \frac{\partial \Phi}{\partial J_i}. \tag{5.9}$$

Then Φ is called the generating function of the canonical transformation $(\boldsymbol{q}, \boldsymbol{p}) \mapsto (\boldsymbol{\theta}, \boldsymbol{J})$ and this mapping is symplectic. Moreover, the equations of motion in the new variables $(\boldsymbol{\theta}, \boldsymbol{J})$ derive from the Hamiltonian $H'(\boldsymbol{\theta}, \boldsymbol{J}) \equiv H(\boldsymbol{q}, \boldsymbol{p})$ on solving (5.8), (5.9) for the arguments.

Exercise 5.5. For Hamiltonian (5.4), show that the mapping $(q, \tau, p, w) \mapsto (x, y, u, v)$ generated by $\Phi(q, \tau, u, v) = (u - cv)\tau + (c + v)q$ is a Galileo transformation with velocity c. What is the Galileo transformation for the self-consistent Hamiltonian in the forms (2.110) and (2.109)?

Exercise 5.6. Show that the generating function

$$F = \sum_{s=1}^{b} \sum_{n=1}^{N_s} \delta x_{ns} \left[A_{-\mu,s} + 2 \sum_{m \in \mu_s^+} [A'_{ms} \cos k_m x_{ns}(t) - A''_{ms} \sin k_m x_{ns}(t)] \right] \tag{5.10}$$

generates the canonical form of the small perturbation dynamics of the multibeam system of exercise 3.4.

There exist other kinds of generating function, enabling one, e.g., to interchange the roles of generalized coordinates and momenta (Percival and

---

[5] The fact that area preservation is more stringent than mere volume conservation in phase space is emphasized in the 'principle of the symplectic camel' (Gromov 1985, see Gosson 2001): a time-dependent volume-preserving transformation can squeeze a camel through the eye of a needle, by first stretching the camel so that its cross section on the needle plane is small enough—whereas a time-dependent symplectic transformation cannot reduce the camel cross section below its original value if the coordinates in the needle plane are conjugate. From a quantum mechanical viewpoint, the fact that each pair of conjugate variables fulfils the Heisenberg inequality may be understood as a quantization of the symplectic area; one cannot reduce a product $\Delta q_x \Delta p_x$ below the indeterminacy limit, even if one is willing to make $\Delta q_y \Delta p_y$ arbitrarily large.

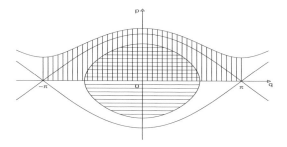

Figure 5.1. Sketch of areas used in the definition of action.

Richards 1982). By considering families of generating functions Φ_λ, depending continuously on a parameter λ, one also introduces the fruitful concept of infinitesimal transformations; in particular, the flow (see appendix D) defined by Hamilton equations is a continuous family of canonical transformations.

5.1.3   Action-angle variables

An important class of canonical transformations is used to simplify Hamiltonian dynamics, and even to solve the dynamics in the integrable case, by changing variables to the so-called action-angle variables. These variables are an essential tool for investigating the structure of phase space. We illustrate their construction by considering the pendulum or the wave–particle interaction as defined by Hamiltonian (1.12).

For the trapped motion, the action variable I is defined as the area bounded by the orbit in the (p, q)-plane divided by 2π (figure 5.1). For the untrapped motion, the same definition applies with the relevant area defined between the orbit and the q-axis for q running from $-\pi$ to π (figure 5.1). Both definitions may be written as

$$I = \frac{1}{2\pi} \oint p \, dq \qquad (5.11)$$

where points differing by 2π in q are identified for a passing orbit. This definition[6] shows that I is a constant of the motion; and that there is a one-to-one relation between I and the energy E of the orbit in the trapped as well as the untrapped domains: $E = H_0(I)$. For each domain, H_0 can be computed through elliptic integrals (see, e.g., Lichtenberg and Lieberman 1983). Moreover, in each domain one can define a second variable θ such that each point (p, q) is precisely represented by data (I, θ) and this angle θ is the generalized coordinate conjugate to the action I. Explicitly, let $p(I, q)$ be the value of p on an orbit with action I

---

[6] The need for a reference line (here the q-axis) with which to associate a closed domain with passing orbits suggests that the difference in action between two orbits may be more important than the value of I for a single orbit (recall the Bohr–Sommerfeld quantization rule). It also makes it difficult to compare actions related to orbits with different types of motion.

at position q. This definition is complete when the sign of p is known at q_0; for a trapped orbit, the functions $q(p)$ and $p(q)$ are two valued, and the right definition of $p(I, q)$ must be selected by continuity. Then θ is defined as

$$\theta = \frac{\partial \Phi}{\partial I} \tag{5.12}$$

where

$$\Phi(I, q) = \int_{q_0}^{q} p(I, q')\, dq'. \tag{5.13}$$

With these definitions[7] θ varies by 2π when the orbit makes one period, which justifies calling it an angle.

As the change of variables $(q, p) \mapsto (\theta, I)$ is canonical, it preserves the phase space area: $\det[\partial(I, \theta)/\partial(p, q)] = 1$. This ensures that, in the new conjugate variables (I, θ), the equations of motion are generated by the new Hamiltonian H_0. Now, as H_0 is independent of the angle variable θ, these equations read:

$$\dot{I} = \frac{\partial H_0}{\partial \theta} = 0 \tag{5.14}$$

$$\dot{\theta} = \frac{\partial H_0}{\partial I} \equiv \Omega(I) \tag{5.15}$$

so that the period of an orbit with energy E is just $2\pi/\Omega(I)$, and motions within each domain are constant-velocity orbits for the angle θ, with fixed action I.

In addition to this 'trivialization' of the equations of motion, action-angle variables also enable one to identify separatrices between domains in phase space (q, p): the change of variable $(q, p) \mapsto (\theta, I)$ is ill defined on these lines, because they correspond to orbits with 'infinite period'. Actually, these mappings are defined separately for each domain.

Exercise 5.7. For the harmonic oscillator $H(X, Y) = \omega_0(X^2 + Y^2)/2$, with generalized coordinate X and conjugate momentum Y, find the action I and the angle θ. Express the Hamiltonian in these variables and find the angular velocity function $\Omega(I)$. Are there separatrices, on which $\Omega = 0$, in the phase space?

Exercise 5.8. Consider a Hamiltonian in action-angle variables (I, θ). If its orbits are isochronous, i.e. $\Omega(I) = \omega_0$ is independent of I, show that it reduces to the harmonic oscillator.

Exercise 5.9. Consider a Hamiltonian $H(q, p) = \frac{1}{2}p^2 + V(q)$, where the potential V is smooth and has a maximum at $q = 0$ and a minimum at $q = a > 0$. Assume there is $b > a > 0$ such that $V(b) = V(0) = 0 > V(a)$. Let

---

[7] $\Phi(I, q)$ is called the generating function of the canonical transformation from (p, q) to (I, θ) as introduced in section 5.1.2.

$c = -V''(0) > 0$. For simplicity, assume there is no other extremum of V in $]0, b[$ and that $V'(b) > 0$.

(i) Sketch the graph of $V(q)$ and constant-energy lines in the (q, p)-plane.
(ii) Find the equilibrium points of H in the phase space and discuss their stability.
(iii) Find the separatrix associated with the unstable equilibrium.
(iv) Show that all orbits with energy $E < 0$ are periodic. Denote by $T(E)$ their period, and show that $T(E) = \oint (2E - 2V(q))^{-1/2}\, dq$.
(v) Sketch the projection of a few orbits in (q, t) space and in (p, t) space.
(vi) Show that $T(E) \simeq \alpha \ln |E/\beta| + \cdots$ as $E \to 0^-$, with constants α, β independent of E. How does α depend on V?
(vii) Denote $I(0)$ by I^*. How does $I(E)$ behave for $E \to 0^-$?

5.2   Motion of one particle in the presence of two waves

The machinery of the previous section is not essential to study a simple system like one particle in one prescribed wave. Indeed the wave modifies the structure of phase space with the occurrence of trapped orbits but does not generate chaos. Chaos occurs as soon as two waves with different phase velocities are present. We now turn to this case.

The motion of an electron in the presence of two electrostatic waves with different phase velocities is generated by the Hamiltonian

$$H_2(p, q, t) = \frac{p^2}{2} - A_1 \cos q - A_2 \cos k(q - vt) \qquad (5.16)$$

by choosing the reference frame of the first wave and a length unit such that the wavenumber of the first wave is 1. The time unit may be chosen such that the phase velocity of the second wave is $v = 1$. The role of the two waves may be exchanged by defining a new position $q' = k(q - vt)$, a new time $t' = -kt$ and a new momentum $p' = v - p$. This yields a Hamiltonian $H_2'(p', q', t')$, named the *equivalent Hamiltonian*, which is H_2 with (k, A_1, A_2) substituted with $(1/k, A_2, A_1)$.

Exercise 5.10. Show that the equations of motion generated by $H_2'(p', q', t')$ with respect to its conjugate variables (p', q') are equivalent to the equations of motion generated by $H_2(p, q, t)$ with respect to its own conjugate variables (p, q). Note that the algebraic value of $H_2'(p', q', t')$ for a given (p', q', t') is not the algebraic value of $H_2(p, q, t)$ for the corresponding values of p, q, t: the Hamiltonian's major role is to define the dynamics, not to determine the energy of the system (which is not even conserved if H is a function of t).

The case where one of the A_i vanishes was treated in section 1.2 (up to a rescaling of q and t if $k \neq 1$). Define the (dimensionless) stochasticity parameter

$$s = \frac{2\sqrt{A_1} + 2\sqrt{A_2}}{|v|} \qquad (5.17)$$

as the sum of the individual wave trapping widths, normalized to the relative velocity of the waves. In the rest of this chapter, we let $v = 1$ in (5.16) with no loss of generality.

5.2.1 Small resonance overlap and cantori

If the wave amplitudes are small, the analysis in section 1.2 is still expected to be useful due to the property of locality in p for the influence of one resonance in phase space. Indeed if $s \ll 1$, we may expect the Poincaré map of the system to display separated trapping domains, with the passing orbits in between being slightly squeezed due to the presence of the cat's eyes. This naive picture is supported by the KAM theorem (section 5.3.2) which states that, if the velocity u of a torus $\mathcal{T}$ is a typical irrational for $s = 0$, then for s small enough this torus persists and the total measure of such preserved tori is positive.

Numerical calculations of orbits enable the trace of these KAM tori to be visualized. We consider the case where $k = A_1/A_2 = 1$. Then H_2 is its own equivalent Hamiltonian and the Poincaré map is symmetrical with respect to $p = 1/2$. In the following we avoid plotting all symmetrical orbits in order to avoid overcrowding the figures. Figure 5.2 displays the Poincaré map[8] of the orbits (stroboscopic map for $t/(2\pi)$ integer) for $s = 0.5$. We recognize KAM tori which are slightly distorted with respect to those corresponding to $s = 0$. Above a certain threshold for s, a KAM torus breaks up into a different object called a cantorus (Percival 1979, Aubry 1978, Meiss 1992). In the coordinates of Hamiltonian (5.4), the position of a point on the cantorus is still defined by coordinates (q, τ) but q belongs to a Cantor set[9], while τ still belongs to a circle. Orbits with mean velocity u no longer belong to a continuum and holes which never close (in the limit of an infinite number of Poincaré iterates) appear in the Poincaré map of these orbits. The higher s, the larger are the holes in the cantorus, and the larger is the flux of orbits through them.

In figure 5.2 we also see orbits which seem to be trapped in each wave. They take on the grossly elliptic shape of the trapped tori of the pendulum case. For these tori the analogue of u is the average number of turns ρ an orbit makes about the stable equilibrium point[10]. The KAM theorem states that, at fixed A_i, a torus trapped in resonance i for $A_j = 0$, $j \neq i$, whose ρ is a typical irrational, is

---

[8] If $k = r/r'$ where r and r' are mutually prime, the periodicity in q of the Poincaré map is $2\pi r'$.

[9] A Cantor set is a set which has neither internal points nor isolated points. A classical example of such a set is constructed as the limit of the following process: take the closed unit interval [0, 1] and take out the open interval centred at 0.5 of length 1/3, take out in a similar way the central one-third part of the remaining closed intervals, and repeat the same procedure on the successive closed subintervals so produced. See also exercise 5.14.

[10] Both u and ρ are called the rotation numbers for the corresponding orbits.

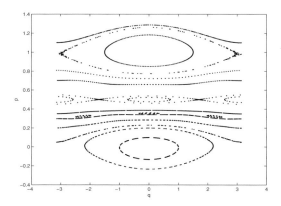

Figure 5.2. Poincaré map of the dynamics of Hamiltonian (5.16) for $s = 0.5$, $A_1 = A_2$, $k = v = 1$.

preserved for A_j small enough. Therefore the trapped orbits of figure 5.2 sketch the trace of preserved trapped KAM tori.

Close to the centre of the cat's eye, the motion is harmonic-like as for the single wave or the pendulum. The pulsation of the harmonic bounce motion close to the bottom of one of the potential wells can be computed by retaining in Hamiltonian (5.16) only the corresponding ith resonant term and by setting $A = A_i$ in (1.15). The almost harmonic character of the motion in the bottom of the potential well makes this zone non-KAM. This may generate chaos for very small amplitudes A_2, in particular for k large and appropriately chosen values of A_1 (Chernikov *et al* 1987, 1988). Physically, conditions favourable to chaos are the resonance conditions, which are more easily met over a 'large' range of u near u_* (or of ρ near ρ_*) if $d\Omega/dI$ vanishes at the I_* for which $\Omega = u_*$. Therefore, chaos is tamed by the so-called twist condition $d\Omega/dI \neq 0$.

The X-point related to each wave may be followed numerically when increasing s, and be seen to persist for these values (the implicit function theorem[11] ensures that such a point is structurally stable under a small enough perturbation). The existence of regular trapped and untrapped orbits is an incentive to extend the naive picture for $s \ll 1$: there could be a separatrix for each wave which would define the border of the corresponding trapped and untrapped motions. This extension will prove to be wrong when increasing s

---

[11] The implicit function theorem, which may be called 'the fundamental theorem of perturbation theory', is also the starting point of the qualitative theory of dynamical systems, including bifurcation theory (Guckenheimer and Holmes 1983, Iooss and Joseph 1990). Indeed, a perturbation method amounts to solving an equation $F(x, \epsilon) = 0$ in the form $x = y(\epsilon)$, finding an asymptotic expansion for $y(\epsilon)$ in the vicinity of a parameter value $\epsilon = 0$ for which a solution x_0 is known. Subtleties of perturbation theory are related to the fact that x may be a function in some specific space (rather than just a number) and F a functional (rather than a function), i.e. the problem may belong in functional analysis rather than elementary calculus. Besides, perturbative solutions may be merely asymptotic expansions rather than convergent series (see Bender and Orszag 1978).

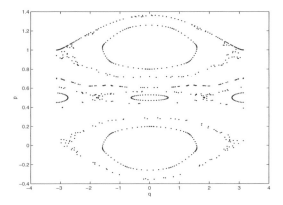

Figure 5.3. Poincaré map of the dynamics of Hamiltonian (5.16) for $s = 0.68$, $A_1 = A_2$, $k = v = 1$.

further. In fact the neighbourhood of the upper X-point is slightly blurred.

Together with these elements of continuity of the Poincaré map with respect to the previous integrable cases, figure 5.2 exhibits new qualitative features: chains of islands passing domains. The centres of these islands correspond to periodic orbits which are remnants of tori with rational u. This is shown by the Poincaré–Birkhoff theorem, as explained in the next section. The perturbation of the free motion by one wave is non-typical as all tori persist whatever the amplitude of the wave. However, the perturbation by two (or more) waves induces the typical break-up of rational tori. Each island may be viewed as a small cat's eye whose centre is a point of one of the Poincaré–Birkhoff stable periodic orbits and whose corners correspond to two points of the related unstable periodic orbit. This picture relies upon the existence of nonlinear resonances which will be proved in section 5.3.1. However, when the perturbation increases, stable periodic orbits may become unstable. Therefore we cannot expect the existence of a dense set of chains of islands on the basis of the Poincaré–Birkhoff theorem. There are many more than shown in figure 5.2 but they are small, though.

5.2.2 Moderate resonance overlap and stochastic layers

By increasing s to $s = 0.68$, a new qualitative behaviour shows up at the edge of the trapping regions where a separatrix would be expected to exist. The corresponding numerical Poincaré map (figure 5.3) displays orbits whose points do not look as if they belong to a curve but rather seem to fill a layer. This is especially visible close to the X-point. Furthermore, when using a display following the successive points of the orbit, one sees an intriguing feature: sometimes the orbit behaves like a passing one with $|u|$ small, which keeps a definite sign to p and crosses the abscissa of the X-point; and sometimes it behaves like a trapped orbit with $|\rho|$ small, which rotates about the O-point and crosses the ordinate of the X-point. It switches apparently unpredictably from one behaviour to the other: the orbit is chaotic. Another orbit started close enough to this orbit displays the same behaviour. It first diverges exponentially from the first orbit, when two regular orbits would diverge linearly from each other. At some close encounter with the X-point the two orbits separate in a more radical way: one adopts the passing behaviour and the other the trapped behaviour. Together they provide a more precise definition of a layer where orbits are typically chaotic (or stochastic). For this reason this layer is called a stochastic layer (for some time 'stochasticity' was a challenger to 'chaos'). A close inspection of other chains of islands shows that their apparent separatrix is, in fact, also a thin stochastic layer.

A stochastic layer is bounded by KAM tori: one inside the trapping domain and one in each of the two passing domains. The stochastic layer can be split into the contributions located between the former separatrix and each of the three tori. As these tori are slightly distorted versions of tori of the pendulum corresponding to $A_2 = 0$, they can be characterized by the related energy E of the pendulum. The width of each of the three domains can, therefore, be characterized in terms of the relative energy $w = |E - A_1|/A_1$ since A_1 is the separatrix energy. One can estimate w using a technique sketched in section 5.4.2 (Escande 1982). If $A_1 \ll 1$, one finds

$$w \sim \frac{A_2}{A_1 \mu^{2k+1}} e^{-\alpha/\mu} \tag{5.18}$$

where $\mu = 2\sqrt{A_1}/(\pi k)$ and $\alpha = 1$ for the inner domain and the outer upper domain, while $\alpha = 3$ for the outer lower domain of the stochastic layer. For $A_1 \gg 1$, the reference motions for the perturbative estimates are different and this technique yields instead

$$w \sim \frac{A_2}{A_1^{k+1}} \tag{5.19}$$

for $k = 1$ or $1/2$. These formulae apply to the stochastic layer of resonance 2 by using them for the equivalent Hamiltonian.

When increasing s from small values, one should use (5.18). One notes that $w(\mu)$ suddenly blows up when increasing μ and has an inflection point later on. For the most symmetric case $k = A_1/A_2 = 1$, the tangent at the inflection point cuts the μ-axis at $\mu = 1/9$, so that one can expect the width of the stochastic layer to suddenly increase at $s \simeq 2\pi/9$, which turns out to be correct. This value of s corresponds to the 'two-third' rule ($s = 2/3$), originally proposed by Lichtenberg and Lieberman (1983) to improve upon the resonance overlap criterion ($s = 1$) which will be introduced later in this section. Using an explicit construction of KAM surfaces, Celletti and Chierchia (1988) found the accurate estimate $s \geq 0.48$ for the break-up of the golden mean torus which appears to be the most robust barrier betweeen the two resonances.

More information about the stochastic layer is provided by Nekhoroshev's theorem which asserts that, for A_2 small, the chaotic orbit cannot wander away from the separatrix of wave 1 by more than a quantity of order A_2^α with $0 < \alpha < 1$

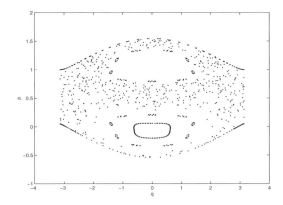

Figure 5.4. Poincaré map of the dynamics of Hamiltonian (5.16) for $s = 1$, $A_1 = A_2$, $k = v = 1$.

over an exponentially long time $\sim \exp(A_2^{-\beta})$ with $0 < \beta < 1$ (Lochak and Meunier 1988, p 163). Further insight results from the geometrical viewpoint of section 5.5.1.

A constant of the motion $C(p, q, t) = C_0$ for a given orbit is also a constant of the motion for this orbit shifted in space (respectively time) by an integer number of spatial (respectively temporal) periods of H_2. Therefore C must have the same periods as H_2. As our Poincaré map corresponds to $t = 0$, the existence of a constant of the motion for this orbit means that in the Poincaré map it must belong to the curve defined by $C(p, q, 0) = C_0$ and not to a layer with a finite width. Thus the existence of stochastic layers shows the non-integrability[12] of the dynamics defined by H_2.

When $s = 1$ (figure 5.4), the passing KAM tori between resonances 1 and 2 are no longer present. Large-scale chaos now exists: a chaotic orbit connects the neighbourhoods of $p = 0$ and $p = 1$. When following the successive points of the orbit, it looks like being trapped sometimes in wave 1 and sometimes in wave 2, switching from one behaviour to the other unpredictably. These intermittent trappings seem for the particle like being in resonance with the two waves and for two waves like being in conflict about the particle control.

This substantiates the basic intuition underlying Chirikov's resonance overlap criterion (Chirikov 1979) which rests on the simple idea that large-scale chaos should occur when an orbit may be trapped in both waves if considered separately. This corresponds to the overlap of the two unperturbed separatrices, i.e. to $s = 1$. For a general Hamiltonian system, Chirikov defines the resonance

---

[12] We saw in sections 1.2 and 5.1 that, if the dynamics is independent of t, it is integrable. Therefore the dependence on time is a minimum condition for chaos to occur. Systems with such a dependence are said to have 1.5 Hamiltonian degrees of freedom. Time dependence is not a sufficient condition for chaos though, as shown by the dynamics defined by H_2 for $A_1 = 0$.

overlap parameter s as the sum of the half-widths of the two resonances divided by the difference in their phase velocities, the exact generalization of definition (5.17). The resonance overlap criterion is intuitive and easy to implement. However, it must be used with caution: it yields a correct order of magnitude estimate for k and A_1/A_2 of the order of 1 (see section 5.4.4) but it is obvious that no chaos occurs if $s = 1$ and $A_2 = 0$. More generally, a strong resonance overlap due to one very large resonance yields a dynamics where chaos is weak, as can be seen from (5.19). For more general Hamiltonian systems, the large amplitude of a resonance may even correspond to the vanishing of its separatrix—which also makes chaos weak (Escande *et al* 2000).

KAM tori are visible in the expected trapping domains of waves 1 and 2 (by symmetry), along with chains of islands whose centres are periodic orbits with rational ρ. Increasing s further induces the break-up of the trapped tori as well.

5.2.3 Physical summary

These observations may be summarized in a weaver's tale. As sketched in figure 5.5, a free-particle phase space is like a single-patch well woven fabric, with ordered threads following orbits. Each resonance opens a cut in this phase-space fabric by tearing it open and seaming some new fabric to both sides of the cut. A typical integrable phase space is a fabric with several patches, seamed along the separatrices. Non-integrability means that threads connect different sides of the separatrices. According to (5.18) and (5.19), tearing the phase space to squeeze a new patch into it devastates the fabric more dramatically near the pre-existing seams than just in the middle of a single patch.

The dramatic consequences of resonance overlap for Hamiltonian (5.16) are seen, e.g., in a plasma column where a standing Langmuir wave is present (Doveil 1981). Indeed a standing Langmuir wave corresponds to two counter-propagating waves. If their amplitude is small, the plasma bulk is weakly distorted by the waves. However, if resonance overlap occurs, the bulk particle motion becomes stochastic over a width in velocity much larger than the initial thermal velocity of the plasma, and a strong plasma heating occurs.

Exercise 5.11. Examine more closely this experiment in terms of the present section, focusing on the general physical picture and order-of-magnitude estimates.

The dramatic consequences of resonance overlap are also visible in a very elementary experiment using the simplest instance of a synchronous motor. The device is made of a compass located in an oscillating magnetic field perpendicular to its axis and can be modelled by Hamiltonian (5.16) if friction is neglected. In some range of values of the amplitude and frequency of the field, the compass can be locked to one of the rotating torques to which it is subjected and rotates with the imposed frequency. However, when being locked, if the frequency is lowered enough, or if the amplitude is increased enough, the compass enters a chaotic

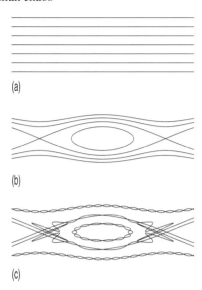

Figure 5.5. (*a*) Free-particle phase space: trajectories are parallel. (*b*) Integrable cat's eye: an X-point with separatrices and periodic trajectories on either side of them in disconnected action-angle patches. (*c*) Chaotic cat's eye: the stable manifold of an X-point intersects transversally the unstable manifold of an X-point, generating a heteroclinic tangle; trajectories starting in the tangle wander.

motion where it alternates in an unpredictable way from one sense of rotation to the other (Croquette and Poitou 1981).

Revisiting the Poincaré maps more critically reveals that it is not certain that orbits which seem to belong to a KAM torus actually do. Indeed, as stated in the introduction to this chapter, numerical orbits are not computed with an infinite precision and the orbit cannot exactly follow a given torus. As all KAM tori are not preserved for a finite amplitude of the perturbation, there are stochastic layers and higher-order island chains between two nearby tori. As yet, no theorem proves that for small enough numerical errors the numerical orbit stays close to a real orbit. However, the qualitative and quantitative agreements between numerical calculations and mathematical expectations for many Hamiltonian systems support such a conjecture.

One should not take the difference between the regimes $0 < s \ll 1$ and $s \sim 1$ too dramatically. There is, indeed, a qualitative difference between the existence of KAM barriers in between the main resonances and their absence but the existence of cantori and of stochastic layers are actually related. Locally, parts of phase space have a similar geometry for $0 < s \ll 1$ and $s \sim 1$. This observation will be further discussed in the renormalization approach (section 5.4).

The qualitative behaviour described in this section also appears in the dynamics of a particle in many waves. This is a consequence of the locality principle. For the same reason, more general Hamiltonian systems can be reduced locally to a many-wave Hamiltonian (Escande 1985). In all cases, KAM tori are barriers for chaos for two-degrees-of-freedom systems.

5.3   Construction of orbits

5.3.1   Origin of higher-order resonances

The numerical calculations of section 5.2 exhibit several chains of islands which hint at the existence of corresponding resonances in the dynamics. We now prove the existence of such resonances by a perturbation calculation. Consider the amplitudes A_i in Hamiltonian (5.16) with $v = 1$ to be of order ε and compute perturbatively the dynamics of an orbit $(q(t), p(t))$ with average velocity u. At zeroth order in ε we get the free motion

$$q^{(0)}(t) = \alpha + ut. \tag{5.20}$$

At first order we have

$$\ddot{q}^{(1)}(t) = -A_1 \sin q^{(0)}(t) - k A_2 \sin \varphi_0(t) \tag{5.21}$$

where $\varphi_0(t) = \omega_{01} t + k\alpha$ with

$$\omega_{ln} = (l + nk)u - nk. \tag{5.22}$$

If u is 0 or 1, the perturbative solution has secularities: $q^{(1)}(t)$ can grow linearly with time, which limits the temporal validity of the solution.

Exercise 5.12. For $u = 0$ and a prescribed accuracy $\varepsilon' > 0$, estimate the first time τ_1 at which $|q^{(1)}(\tau_1)| = \varepsilon'$, assuming the initial condition $q^{(1)}(0) = \dot{q}^{(1)}(0) = 0$. How does τ_1 depend on α?

The corresponding orbits are trapped in the first or second resonance of Hamiltonian (5.16). It is not surprising that perturbation theory is unable to describe such orbits since they are generally cycloid-like in (q, p)-space (helix-like in (q, p, t)-space), which is qualitatively different from free orbits. For each resonance, two special orbits are free from secularities: for resonance 1 they correspond to $\alpha = 0$ or π (mod 2π) and for resonance 2 they correspond to $k\alpha = 0$ or π (mod 2π). These orbits are the stable one at the centre of the cat's eye, and the unstable one at the corner. This agrees with the Poincaré–Birkhoff theorem (Arnold and Avez 1968, section 20) which states that a rational torus (i.e. one with periodic orbits) is typically destroyed by an arbitrary small perturbation but that at least one pair of stable and unstable orbits of this torus are preserved. This does not depend on the non-integrability of the dynamics and already holds for the pendulum ($A_2 = 0$) considered as a perturbation of free motion.

If u is neither 0 nor 1, (5.21) yields

$$q^{(1)}(t) = \frac{A_1}{\omega_{10}^2} \sin q^{(0)}(t) + \frac{kA_2}{\omega_{01}^2} \sin \varphi_0(t) + \alpha_1 + u_1 t \qquad (5.23)$$

where α_1 and u_1 are determined by initial conditions $(q^{(1)}(0), \dot{q}^{(1)}(0))$. As we search for orbits $(q(t), p(t))$ with average velocity u, we require $u_1 = 0$. A proper choice of α in terms of physical initial data $(q(0), p(0))$ enables one to set also $\alpha_1 = 0$.

If u comes close to 0 or 1, a denominator in (5.23) becomes small and perturbation theory is correct for smaller and smaller ε only, in agreement with the secularities found earlier. This calculation shows that, in the Poincaré map, the straight line related to the torus with velocity u (if it exists) is rippled due to resonances 1 and 2 of Hamiltonian (5.16).

Exercise 5.13. For $u \neq 0$ and $u \neq 1$, estimate the range of values of (A_1, A_2) such that $|q^{(1)}(t)| < \varepsilon'$ for any time t, given a prescribed accuracy $\varepsilon' > 0$. Conversely, for given amplitudes (A_1, A_2), what are the values of u for which this accuracy ε' is respected for all times? Display your results on a plot in the (ε', u)-plane.

At second order we must solve

$$\ddot{q}^{(2)}(t) = -\frac{A_1^2}{2\omega_{10}^2} \sin 2q^{(0)}(t) - \frac{k^3 A_2^2}{2\omega_{01}^2} \sin 2\varphi_0(t)$$
$$- \frac{kA_1 A_2}{2} \left(\left(\frac{k}{\omega_{10}^2} + \frac{1}{\omega_{01}^2} \right) \sin(q^{(0)}(t) + \varphi_0(t)) \right.$$
$$\left. + \left(\frac{k}{\omega_{10}^2} - \frac{1}{\omega_{01}^2} \right) \sin(-q^{(0)}(t) + \varphi_0(t)) \right). \qquad (5.24)$$

The first two terms yield secularities for the same values of u as at order 1. The next two terms have pulsations $\omega_{\pm 1,1}$ which vanish for $u = k/(k \pm 1)$ and yield two new secularities.

At first order the secularities occur for $\omega_{ln} = 0$ with $|l| + n = 1$. At second order this condition becomes $|l| + n = 2$. In fact, one shows by induction that, at order r, secularities occur for $\omega_{ln} = 0$ with $n \geq 0$ and $|l| + n = r$. As pulsation ω_{ln} is produced by terms of the type $\left(A_1 \exp[iq^{(0)}(t)] \right)^{|l|} \left(A_2 \exp[i\varphi_0(t)] \right)^n$, the amplitude of the corresponding terms in the perturbation expansion scales like $A_1^{|l|} A_2^n \sim \varepsilon^r$.

We introduce the *zoning number*

$$z = k/u - k + 1 \qquad (5.25)$$

such that $u = k/(k + z - 1)$. In the following, an invariant torus is said to be rational or irrational according to the corresponding nature of z.

When r increases, more and more frequencies occur which must be large enough to avoid small denominators in the perturbation expansions. Clearly our naive approach cannot face this problem but the KAM theorem does for z irrational enough and ε small enough. The proof uses canonical transformations and the fact that it is possible to take away a finite interval about each rational without removing all real numbers.

Exercise 5.14. Consider the velocities between the two waves, i.e. $0 < u < 1$. To each rational number $0 < r'/r < 1$ associate a 'forbidden interval' $r'/r - a^r b < u < r'/r + a^r b$, with some $a > 0$, $b > 0$. Show that the total length of forbidden intervals is, at most, $2ab + \sum_{r=2}^{\infty} \sum_{r'=1}^{r-1} 2ba^r < 2ab + 2ba^2(1-a)^{-2}$, which tends to zero for $a \to 0$.

Following the previous arguments, it suffices to take $a \sim \varepsilon^{1/2}$ in exercise 5.14 to estimate the total range of velocities u capable of causing secularities. The remaining velocities u form a Cantor set with positive measure, also called a 'fat fractal'. Note that a similar discussion of velocities $u > 1$ and $u < 0$ would also exclude the neighbourhood of all rationals but the scaling of a with ε depends on u in a 'favourable' way, thanks to the locality in velocity of a wave influence on the particle. When worked out more carefully, the resulting plot in the (ε, u)-plane yields the so-called Arnol'd tongues (Iooss and Joseph 1990) emerging from each rational $u = r'/r$.

The existence of singularities at first order is related to the existence of primary resonances in Hamiltonian (5.16) with $v = 1$. In a wider sense, for ω_{ln} vanishing, we say that this is the signature of a resonance, named resonance (l, n). Some of these resonances show up in figure 5.2: the chains of v islands in between resonances 1 and 2 correspond to $n = 1$ and $l = v - 1$; their mean velocity is $u = 1/v$.

For a KAM torus $\mathcal{T}$ the previous perturbation expansion converges and its truncation at order M may be written formally as

$$q_M(t) = q^{(0)}(t) - \sum_{r=1}^{M} \sum_{n=0}^{r} \sum_{l=\pm(r-n)} W_{ln} \sin((l+nk)q^{(0)}(t) - nkt) \quad (5.26)$$

where $W_{ln} \sim A_1^{|l|} A_2^n$. This shows that $q_M(t)$ fluctuates about the unperturbed motion $q^{(0)}(t)$ due to the action of resonances of orders 1 up to M (i.e. from ε to ε^M). Define a new coordinate q' implicitly by

$$q = q' - \sum_{r=1}^{M} \sum_{n=0}^{r} \sum_{l=\pm(r-n)} W_{ln} \sin((l+nk)q' - nkt). \quad (5.27)$$

The unperturbed motion at order M, $q(t) = q_M(t)$, corresponds to $q'(t) = q^{(0)}(t)$. Therefore all the fluctuations due to the resonances up to order M vanish in the coordinate q'. The unperturbed motion at order $M + 1$, $q(t) = q_{M+1}(t)$, corresponds to

$$q'(t) = q^{(0)}(t) - \sum_{n=0}^{M+1} \sum_{l=\pm(M+1-n)} W_{ln} \sin((l + nk)q^{(0)}(t) - nkt) + O(\varepsilon^{M+2}).$$

(5.28)

As this equation has the same structure as (5.23), it is tempting to consider it as the calculation to first order in $\eta = \varepsilon^{M+1}$ of an orbit with mean velocity u of the Hamiltonian

$$H(p', q', t) = \frac{p'^2}{2} - \sum_{n=0}^{M+1} \sum_{l=\pm(M+1-n)} U_{ln} \sin((l + nk)q' - nkt) \qquad (5.29)$$

where the U_{ln} are appropriate constants. A similar Hamiltonian is provided by KAM theory[13] and (5.29) is its approximation in the vicinity of $\mathcal{T}$.

5.3.2 Poincaré and KAM theorems*

After the just described physical picture, we state the problem of regular motion from the mathematical viewpoint of Poincaré (see, e.g., Arnold *et al* 1988, chapter 6). Consider an N-degree-of-freedom Hamiltonian

$$H(q, I) = H_0(I) + \varepsilon H_1(q, I) \qquad (5.30)$$

so that H_0 is integrable, with actions $I \in \mathcal{D} \subset \mathbb{R}^N$ and angles $q \in \mathbb{T}^N$.

The velocities are $\Omega_i(I) \equiv \partial H_0 / \partial I_i$, and we Fourier-expand the perturbation as

$$H_1 = \sum_{m \in \mathbb{Z}^N} a_m(I) e^{i m \cdot q} \qquad (5.31)$$

where $m \cdot q \equiv \sum_{i=1}^N m_i q_i$. The Poincaré set is the set of values $I \in \mathcal{D}$ for which the angular velocities are relevant ($a_m(I) \neq 0$) and resonant (there exist $N - 1$ linearly independent vectors $k_1, \ldots k_{N-1} \in \mathbb{Z}^N$ such that $k_s \cdot \Omega(I) = 0$).

Considering the class $\mathcal{A}(V)$ of functions which are analytic in a domain $V \subset \mathbb{R}^N$, a subset $B \subset V$ is called a set of uniqueness for $\mathcal{A}(V)$ if any analytic function that vanishes on B vanishes identically on V. For instance, a set of points of an interval $[a, b] \subset \mathbb{R}$ is a set of uniqueness if and only if it has a limit point in $]a, b[$ (hence a set of measure zero can be a set of uniqueness).

Poincaré's theorem reads:

Nonexistence of uniform constants of the motion:
For Hamiltonian (5.30), assume that $\det(\partial \Omega_i / \partial I_j) \neq 0$ in $\mathcal{D}$. Consider $I_0 \in \mathcal{D}$ such that $\Omega(I_0) \neq 0$. Assume that in every neighbourhood U

---

[13] Note that a similar Hamiltonian, with large coefficients U_{ln}, will be discussed in the next chapter.

of I_0 the Poincaré set is a set of uniqueness for $\mathcal{A}(U)$.

Then the equations of motion for (5.30) do not admit a formal constant of the motion F, independent of H, representable as a formal power series $\sum_{n \geq 0} F_n(I, q) \varepsilon^n$ with coefficients F_n analytic in $\mathcal{D} \times \mathbb{T}^N$.

A corollary of Poincaré's work relates non-integrability with the existence of a dense family of periodic orbits such that, in a small neighbourhood of one such orbit with period T, there is no periodic orbit with a nearby period.

Kolmogorov, Arnold and Moser have stated a family of theorems, differing by the regularity of the Hamiltonian H_0, which may be analytic or simply sufficiently differentiable (Arnold and Avez 1968, section 21, Benettin *et al* 1984 and references therein).

KAM theorem:

For $\varepsilon \to 0$, for almost every (in the sense of Lebesgue measure) $\mathbf{\Omega}_*$ there exists a torus $\mathcal{T}(\mathbf{\Omega}_*)$, invariant under the dynamics of (5.30) and $\mathcal{T}(\mathbf{\Omega}_*)$ is close to the invariant torus $\mathcal{T}_0(\mathbf{\Omega}_*)$ of H_0.

More precisely, let I_* be the action for $\mathcal{T}_0(\mathbf{\Omega}_*)$. Then for any $\eta > 0$ there exist $\varepsilon' > 0$ and a mapping $Q \mapsto (I, q)$ from a torus $\mathbb{T}^N$ to $\mathcal{D} \times \mathbb{T}^N$ such that $\dot{Q} = \mathbf{\Omega}_*$ and that

$$|I(Q) - I_*| < \eta \tag{5.32}$$

$$|q(Q) - Q| < \eta \tag{5.33}$$

provided that $0 \leq \varepsilon < \varepsilon'$.

Moreover, the tori $\mathcal{T}(\mathbf{\Omega}_*)$ form a positive measure set in $\mathcal{D} \times \mathbb{T}^N$ and the measure of their complement tends to zero in the limit $\varepsilon \to 0$.

For area-preserving maps, Moser established a similar 'twist theorem' (Guckenheimer and Holmes 1983).

5.3.3 Higher-order resonances from action-angle variables

Another way to exhibit resonances related to the chains of islands of section 5.2 is to use canonical changes of variables. In the new variables, the dynamics is still described by a Hamiltonian which can be computed from the previous one in a definite way—whereas the calculations of section 5.3.1 did not warrant that equations resulting from our approximations would preserve the Hamiltonian structure.

By identifying A_1 to A, Hamiltonian (5.16) with $v = 1$ reads as Hamiltonian (1.12) plus a resonance. In the action-angle variables of (1.12), H_2 becomes

$$H_2'(I, \theta, t) = H_0(I) - A_2 \sum_{n=-\infty}^{\infty} a_n(I) \cos(k_n \theta - kt) \tag{5.34}$$

where k was assumed to be a positive rational, where the Fourier expansion of $\cos k(q(I, \theta) - t)$ was used, and $k_n = k + n$ (respectively $k_n = n$) for (p, q) in the rotation (respectively libration) domain of the pendulum, i.e. for passing (respectively trapped) motion of the electron. Coefficient a_n is of order A_1^n for A_1 small and $n \geq 0$ (Escande and Doveil 1981b, Escande 1985). If the resonance condition

$$\Omega(I) = \frac{k}{k_n} \tag{5.35}$$

is met for some $I = I_n$, the nth term in Hamiltonian H_2' is resonant. The reasoning in section 5.2 for the Poincaré map of Hamiltonian (5.3) can be repeated here and we may expect to see chains of islands if A_2 is small enough.

Exercise 5.15. Show that $n < 0$ corresponds to orbits above resonance 2 or below resonance 1.

For passing orbits, and by writing $k = r/r'$, where r and r' are mutually prime, $\Omega(I_n) = r/(r + nr')$. Thus resonance n should correspond to a chain of $r + nr'$ islands. For $A_1 \to 0$, (I, θ) converges to (p, q) and $\Omega(I)$ to p. Hence, for moderate values of A_1, the velocities in the islands of the chain should be about $r/(r + nr')$. This enabled us to identify in figures 5.2 and 5.3 a passing chain of ν islands with velocities fluctuating about $1/\nu$ as related to resonance $n = \nu - 1$. In particular the single-island chain $\nu = 1$ corresponds to wave 2 of Hamiltonian (5.16). The harmonic trapping pulsation close to the O-point of resonance n can be computed by retaining only this resonance in Hamiltonian (5.34) and identifying the resulting Hamiltonian with a pendulum Hamiltonian.

Exercise 5.16. Estimate the bounce frequency for resonance n, expanding $H_0(I)$ to second order near I_n and identifying $a_n(I)$ with $a_n(I_n)$. This is the *centred resonance approximation*. A change of origin in I removes the linear term in I, which yields the requested pendulum Hamiltonian.

For trapped orbits, resonance n corresponds to a chain of n islands. In figure 5.4 the chain of nine islands trapped in wave 1 corresponds to $n = 9$. In the trapped domain, $\Omega(I)$ is maximum for $I = 0$, and $\Omega(0) = \sqrt{A_1}$. This sets a lower bound to the index n of the resonant terms in (5.34). When A_1 grows, chains of n islands with decreasing n bifurcate out of the centre of the trapping domain of wave 1. Harmonic bounce frequencies for these trapped chains of islands may be computed as for the passing ones.

Action-angle variables enable us to identify some chains of islands visible in the Poincaré maps of Hamiltonian (5.16) for various values of s. This identification works by interpreting some chains as the imprint of a resonant term in Hamiltonian (5.34).

5.4 Renormalization for KAM tori

In this section, we first discuss the calculation of a single orbit with pedestrian perturbation theory. Then we turn to a more elaborate perturbation method, adapted to Hamiltonian dynamics. Having properly formulated the renormalization procedure for a given torus, we discuss it first for the study of specific tori, specially relevant to the transition to large-scale chaos. Finally, we discuss quantitative implications of the renormalization for typical Hamiltonians.

5.4.1 Simple approach to renormalization

Until now we considered resonances according to their order in ε, which is natural from the viewpoint of perturbation theory. However, (5.21) and (5.24) showed that the contribution of resonance (l, n) to the perturbation expansion involves a denominator ω_{ln}^2. If this denominator is small, the corresponding term is large. Therefore all terms of the same order in ε are not of the same size. Furthermore, when ε is close to the threshold for break-up of the corresponding KAM torus $\mathcal{T}$, the perturbation expansion converges slowly, which means that terms of different order may have about the same size. Then small denominators bring a dominant contribution to the perturbation expansion. We are thus left with the problem of finding for a given u the family of resonances providing the smallest ω_{ln}. This will prompt us to re-organize the perturbation series and compute some re-summation for it, which is a procedure usual in physics.

Number theory (Khinchin 1964) provides the tool for this reorganization. Indeed $\omega_{ln} = kn(l/n + 1 - z)/(k + z - 1)$ where z is the irrational zoning number of $\mathcal{T}$ defined by (5.25). Writing $z = [a_0, a_1, a_2, \ldots]$, where the a_i are the positive integers of the continued fraction expansion of z

$$z = a_0 + \cfrac{1}{a_1 + \cfrac{1}{a_2 + \cfrac{1}{\cdots}}} \tag{5.36}$$

the best approximants to z are obtained by truncating the continued fraction at successive orders N which yields the rationals z_N. Approximating z by a rational n'/n usually yields an error $|z - n'/n| \sim 1/n$. In contrast, for k and z of order 1 and $n'/n = z_N$, where n' and n are mutually prime, this error is $\sim 1/n^2$, so that $\omega_{ln} \sim k/[n(k + z - 1)] \sim 1/n$.

Estimating in a single step the size of the resonance related to z_N for N large would be a formidable task. It is thus desirable to estimate this size by successive steps. This gradual approach is provided by the so-called renormalization transformation. Taking $A_2 = O(1)$, it focuses on resonances with $n = 1$ closest to $\mathcal{T}$ (if $\mathcal{T}$ is between the chains of two and three islands in figure 5.2, the renormalization focuses on resonances related to these chains). They correspond to $z = a_0$ and $z = a_0 + 1$ (to alleviate the notation we write m for a_0 in the following). Explicit contributions to $q(t)$ from these resonances are obtained by taking $M = m+1$ in (5.26). The W_{l1} entail denominators ω_{l1} taking their smallest value for $l = l_1 \equiv m - 1$ and $l = l_2 \equiv m$. Considering again that $\mathcal{T}$ is close to break-up, the size of the terms in the perturbation expansion is weakly related to their order and the $W_{l_i 1}$, $i = 1, 2$, take prominent values. For this reason, we single out these two terms and introduce the change of variable removing the fluctuations due to all other terms. We implicitly define a new position Q by

$$q = Q - \sum_{r=1}^{m+1} \sum_{n=0}^{r} \sum_{l=\pm(r-n)}^{*} W_{ln} \sin((l+nk)Q - nkt) \qquad (5.37)$$

where the star excludes the cases $(r, l, n) = (m + 1, m, 1)$ and $(r, l, n) = (m, m - 1, 1)$. At lowest order in $\eta \equiv q - Q$, the value of $q_{m+1}(t)$, as defined by (5.26), corresponds to

$$Q(t) = q^{(0)}(t) - \sum_{l=m-1}^{m} W_{l1} \sin(k_l q^{(0)}(t) - kt) \qquad (5.38)$$

where $k_l = k + l$ and the W_{l1} are small. As with (5.28) for $q'(t)$, it is tempting to consider $Q(t)$ as the calculation to first order in the $W_{l_i 1}$ of an orbit with mean velocity u for the Hamiltonian

$$H(P, Q, t) = \frac{P^2}{2} - \sum_{l=m-1}^{m} A_2 U_l \sin(k_l Q - kt) \qquad (5.39)$$

where the U_l are appropriate constants of order A_1^l. This Hamiltonian generates the second-order differential equation

$$\ddot{Q} = -A_2 \sum_{l=m-1}^{m} k_l U_l \sin(k_l Q - kt) \qquad (5.40)$$

after eliminating P in the canonical equations of motion. We now define new time and position variables by

$$t' = -kt/k_m \qquad (5.41)$$
$$q' = k_{m-1}Q - kt. \qquad (5.42)$$

With these new variables, (5.40) becomes

$$\frac{d^2 q'}{dt'^2} = -A_1' \sin q' - k' A_2' \sin k'(q' - t') \qquad (5.43)$$

where

$$A_i' = \frac{A_2 U_{m-2+i}}{\Delta u_m^2} \qquad i = 1, 2 \qquad (5.44)$$

$$k' = \frac{k + m}{k + m - 1} \qquad (5.45)$$

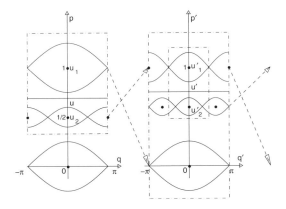

Figure 5.6. Principle of renormalization.

and

$$\Delta u_m = u_{m-1} - u_m \tag{5.46}$$

with

$$u_l = k/k_l. \tag{5.47}$$

Equation (5.43) is the equation of motion for q' which can be deduced from the canonical equations for a Hamiltonian H_2' of the type (5.16) with $v = 1$ where all quantities have been primed. In these new variables, the mean velocity u' is rescaled linearly with respect to u so that $u' = 0$ for $u = u_{m-1}$ and $u' = 1$ for $u = u_m$,

$$u' = \frac{(k+m)[k - (k+m-1)u]}{k}. \tag{5.48}$$

The mapping $\mathcal{R}$ of (u, H_2) to (u', H_2') is called a *renormalization transformation*. It involves a rescaling as shown in figure 5.6 but also a straightening of orbits which results from using coordinate Q. The rescaling corresponds to magnifying a box in a phase space including only two of the resonances of Hamiltonian (5.34). When magnified, this box reveals other subchains of islands which enable $\mathcal{R}$ to be iterated. These successive sets of resonances are nested like Russian dolls. The first instance of $\mathcal{R}$ was derived by Escande and Doveil (1981a, b) with a technique explained in the next subsection.

5.4.2 More explicit derivation*

The KAM torus $\mathcal{T}$ with a mean velocity $0 < u < 1$ may be spotted by locating it first between resonances $m - 1$ and m of Hamiltonian (5.34) for some value of m, i.e.

$$u_{m-1} < u < u_m. \tag{5.49}$$

The one-resonance approximation made for chains of islands would be arbitrary for studying a KAM torus trapped between two such chains: no resonance may be considered *a priori* as dominant. The next reasonable approximation is a two-resonance approximation in Hamiltonian (5.34) which retains the two resonant terms with index $m - 1$ and m. This yields a two-resonance Hamiltonian[14] similar to that in (5.16) but the amplitude of the resonances is momentum dependent and $H_0(I)$ is not quadratic in I.

For A_2 small, the value of I on $\mathcal{T}$ stays close to the value I_u defined by $\Omega(I_u) = u$ corresponding to $A_2 = 0$. Approximations similar to those made for computing the harmonic bounce frequency in resonance n may be made for the two-resonance Hamiltonian: $H_0(I)$ is expanded to second order close to I_u, and the $a_l(I)$ are substituted with $a_l(I_u)$. In fact the centred resonance approximation $a_l(I) \simeq a_l(I_l)$ used for computing the amplitude of the isolated resonances is also better for the case with two resonances (Escande *et al* 1984, Escande 1985).

Exercise 5.17. Consider a Hamiltonian

$$H(p, q, t) = \frac{p^2}{2} + \epsilon \sum_{m=1}^{M} V_m(p) \cos(k_m q - \omega_m t) \qquad (5.50)$$

where ϵ is a small quantity, and where the dV_m/dp are slowly varying functions. Derive with respect to t the canonical equation for q, which brings factors $\dot{p}$ in the expression. Substitute $\dot{p}$ with its canonical expression. Check that terms coming from the product of two $\cos(k_m q - \omega_m t)$ are negligible. Then check that it is a good approximation to replace the two $V_m(p)$ of interest by the $V_m(\omega_m/k_m)$, when making the two-resonance approximation for an orbit with mean velocity u. Check that the same reasoning holds for a Hamiltonian like (5.50) where $\frac{1}{2}p^2$ is substituted with a function whose second derivative is slowly varying. This justifies the centred-resonance approximation.

This yields a Hamiltonian

$$H_2''(I, \theta, t) = H_0(I_u) + u(I - I_u) + \frac{\sigma_u(I - I_u)^2}{2} - A_2 \sum_{l=m-1}^{m} a_l^* \cos(k_l \theta - kt) \qquad (5.51)$$

writing $a_l^* \equiv a_l(I_l)$, which generates a second-order differential equation for θ by eliminating I in the canonical equations of motion

$$\ddot{\theta} = -\sigma_u A_2 \sum_{l=m-1}^{m} k_l a_l^* \sin(k_l \theta - kt). \qquad (5.52)$$

[14] The validity of this two-resonance approximation when dealing with chaotic thresholds may be understood from (5.18). Indeed a resonance with the same amplitude as resonance 2 but twice as far from resonance 1 yields a coefficient μ which is half the size in (5.18), which yields a later blow-up of chaos when increasing the wave amplitudes.

We now use the new time and position defined by (5.41) and (5.42). With these new variables, (5.52) becomes (5.43) with

$$A_i' = \sigma_u A_2 a_{m-2+i}^* / \Delta u_m^2 \qquad \text{for } i = 1, 2 \tag{5.53}$$

where k' is defined by (5.45) and Δu_m by (5.46).

As yet the renormalization of the A_i is not explicit because $a_n(I)$ is not. For the sake of simplicity, as $\mathcal{R}$ is approximate, when m is small we prefer to work with the lowest-order expression in A_1 of (5.52). This corresponds to substituting the action with its approximation given by first-order perturbation theory in A_1. The corresponding generating function is

$$\Phi'(I, q) = Iq + \frac{A_1 \sin q}{I} \tag{5.54}$$

instead of (5.13), so that

$$p = I + \frac{A_1 \cos q}{I}. \tag{5.55}$$

Then the lowest-order expression in A_1 of the A_i' becomes

$$A_i' = c_{l_i}(k) A_1^{l_i} A_2 \qquad \text{for } i = 1, 2 \tag{5.56}$$

with

$$c_l(k) = \frac{(k+l)^{3l-1}(k+m)^2(k+m-1)^2}{l! 2^l k^{2l+1}}. \tag{5.57}$$

As a result $\mathcal{R}$ is monomial in the A_i. Taking the logarithm of equation (5.56) transforms it into an inhomogeneous linear equation, expressing $(\ln A_1', \ln A_2')$ in terms of $(\ln A_1, \ln A_2)$ using matrix algebra (Mehr and Escande 1984, MacKay 1988).

Exercise 5.18. Show that the transformation $(\ln A_1, \ln A_2) \mapsto (\ln A_1', \ln A_2')$ is area preserving, which implies that its fixed points or those of iterates (with different successive m, k) have either two real eigenvalues λ and $1/\lambda$ with $|\lambda| > 1$ or only unit-modulus eigenvalues. How many fixed points does a single mapping $\mathcal{R}$ admit with $|\ln A_i| < \infty$? Let $\mathcal{R}_1$ and $\mathcal{R}_2$ be $\mathcal{R}$ computed for two different values of m. How many fixed points can a chained map $\mathcal{R}_1 \circ \mathcal{R}_2$ admit?

If m is large, another approach must be used (Escande 1982) which keeps the definition of the a_l in terms of action variables. It is easily checked (Escande 1982) that the larger m, the larger the range Δm over which $k_{m \pm \Delta m}/k_m$ and $a_{m \pm \Delta m}^*/a_m^*$ may be approximated by 1. Defining a new time $t' = kt$, changing the origin of I and rescaling it then shows that the dynamics of $\mathcal{T}$ is close to being defined by Hamiltonian

$$H(p, q, t) = \frac{p^2}{2} + A \sum_{m=-\infty}^{\infty} \cos(q - mt) \tag{5.58}$$

where $A = \sigma_u A_2 a_m^* k_m^4 / k^2$.

We define the area-preserving (actually, symplectic—see appendix D) map

$$q_{n+1} - q_n = J_n \tag{5.59}$$

$$J_{n+1} - J_n = K \sin q_{n+1} \tag{5.60}$$

where $q_n = q(t_n)$, $J_n = 2\pi p(t_n + 0)$ and $K = 4\pi^2 A$ with $t_n = 2n\pi$. The jump $(J_{n+1} - J_n)/(2\pi)$ is the kick impulsed on the particle. This map is called the standard map. Historically, the standard map played a major role as it was used as the paradigm for chaotic area-preserving (Hamiltonian) maps, in particular since its numerical calculation is easy and fast. Its threshold for global chaos is[15] $K_c \simeq 0.9716$ (Greene 1979) and plays a key role in the study of stochastic layers (section 5.5.1).

Exercise 5.19. Using the formal identity (B.23), show that the equations of motion for (5.58) yield a periodically impulsive force on the particle. Integrate them to find the standard map.

The use of an infinite sum in (5.58), or of Dirac distributions, heralds a subtle limit. Though the equations of motion for (5.58) formally lead to the discrete-time map (5.59), (5.60), the kicks imply that the actual equations of motion are not smooth with respect to time. Yet smoothness is important to the conservation of invariants and construction of quasi-invariants; and to the statement and proof of the KAM and related theorems. Taking advantage of analytic properties of the right-hand side of (5.59), (5.60) with respect to (q_n, J_n), a careful study of the fixed points of the standard map and of their stable and unstable manifolds (see section 5.5.1) shows, for instance, that the actual width of the stochastic layer for $K \ll 1$ is about twice the estimate resulting from the simple resonance overlap argument. It enables the derivation of formulae (5.18) and (5.19). For a KAM torus with a zoning number whose $a_i = m$ is large for all i, the renormalization mapping is simpler as it maps K to a new value K' (Escande 1982).

Other approximate renormalization schemes and the accuracy of approximations in various schemes are discussed by Escande (1985). A renormalization scheme was developed for area-preserving maps, in particular Hamiltonian Poincaré maps (MacKay 1983). It is well suited to the precise description of the critical fixed points defined in the next section. The stochastic layer close to the X-point of resonance 1 has a rescaling invariance (see Abdullaev (2000) and references therein): in particular, the structure of the phase space stays invariant there if A_2 is rescaled by an appropriate coefficient λ, the positions close to the X-point by $\sqrt{\lambda}$ and the initial phase of the perturbation is shifted by π.

The break-up of tori where $d\Omega/dI = 0$ at I_u, i.e. the non-twist case, has also been investigated with a specific renormalization method (del-Castillo-Negrete *et al* 1997).

[15] For the two-resonance approximation of the standard map Hamiltonian, the threshold would be $K_c \simeq 1.7$.

Exercise 5.20. Investigate (numerically and using the centred resonance approximation) the $M = 2$, $N = 1$ self-consistent wave–particle system, for various choices of its total energy and momentum and of its parameters $(\varepsilon\beta_j, k_j, \omega_{j0})$.

Exercise 5.21. Same questions for $N = 2$, $M = 1$ (Firpo and Doveil 2002). These observations will be used in section 9.7.

5.4.3 Study of the renormalization mapping

The rescaling (5.48) of u corresponds to a simple arithmetic rule when interpreted in terms of the zoning number. Then the correct value of m in (5.49) is[16] a_0, the integer part of z ($a_0 > 0$ as $0 < u < 1$), and (5.45), (5.48) and (5.25) imply that the renormalized value of z is

$$z' = \frac{1}{z - a_0} = [a_1, a_2, a_3, \ldots]. \tag{5.61}$$

Let the b_i be the coefficients of the continued fraction expansion of k, i.e. $k = [b_0, b_1, b_2, \ldots]$. We extend the renormalization mapping to irrational k by continuity, as it is defined for all rational k. Then (5.45) implies that $k' = [1, a_0 - 1 + b_0, b_1, b_2, \ldots]$. The next renormalization step maps k' to $k'' = [1, a_1, a_0 - 1 + b_0, b_1, \ldots]$. The renormalization $\mathcal{R}$ shifts the indices of the continued fraction expansion of z by -1 and discards the first digit, which becomes (from the second iterate on) the second digit of the continued fraction expansion for the new iterate of k, whose successive digits are deduced from those of the preceding one by shifting indices by $+1$.

Thus renormalizing is like peeling off an onion for z and piling up the successive peels to build k. The higher the order of a coefficient in the continued fraction expansion of a number is, the weaker is its influence on the value of this number. Therefore, when $\mathcal{R}$ is iterated l times, $k^{(l)}$ becomes independent of the initial value of k when l grows but is increasingly determined by the details of the initial value of z. Note that $\mathcal{R}$ acts on the zoning number z and its 'conjugate variable' k in different ways: the evolution of k is slaved to the autonomous evolution of z. However, given an arbitrary initial z, the iterates by $\mathcal{R}$ do not converge: actually the continued-fraction map $z \mapsto z'$ is locally expanding, and one proves that it is ergodic (see, e.g., Khinchin 1964, p 71, or an introduction to ergodic theory). To some extent, this ergodicity suggests that, at small scales in u, chaos develops similarly near many values of $u \in \mathbb{Q}$—a property called a universal behaviour in physics.

Moreover, the mapping $(A_1, A_2) \mapsto (A_1', A_2')$ is also slaved[17] to both (z, k). The mapping for the amplitudes is important, since the initial (A_1, A_2) convey the physical importance of both resonances.

---

[16] No confusion should arise with the coefficients $a_n(I)$.

[17] In mathematics, $\mathcal{R}$ is said to have a skew-product structure.

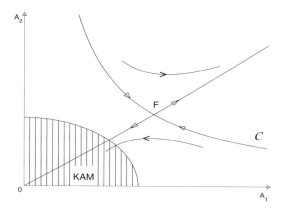

Figure 5.7. One fixed point of the renormalization.

If the a_i are periodic with a period L (equivalently z is a quadratic irrational), $z^{(l)}$ is periodic with the same period. When l increases, $k^{(l)}$ becomes also periodic with this period, and $k^{(mL)}$ converges toward an irrational value independent of the initial k.

The behaviour of $\mathcal{R}' = \mathcal{R}^L$ is easy to describe for a quadratic irrational z_* with a continued fraction expansion of period L. Indeed $z^{(mL)}$ is fixed and $k^{(mL)}$ converges toward a quadratic irrational k_* whose continued fraction expansion is the mirror image of the z_*. Iterating $\mathcal{R}'$ with the same z_* and $k = k_*$ reduces the renormalization dynamics to that of $(A_1^{(mL)}, A_2^{(mL)})$. It is easy to check that this dynamics has two fixed points: a trivial sink at $(A_1, A_2) = (0, 0)$, and a non-trivial hyperbolic one F_{z_*} with finite values of the A_i. In the vicinity of the sink lies the domain where the KAM theorem applies. As sketched in figure 5.7, the hyperbolic fixed point has an unstable manifold of dimension 1 which goes through $(0, 0)$, and a stable manifold of dimension 1.

When the k dynamics is included in the renormalization, the unstable manifold remains of dimension 1 but the stable manifold has dimension 2. There are good reasons (Chandre and Jauslin 2000) to believe that the present approximate renormalization scheme is the approximation of an exact renormalization scheme acting in an infinite-dimension space of Hamiltonians instead of the three-dimensional space of Hamiltonians (5.16). In this direction, Koch (2002) provided a computer-assisted proof of the existence of a non-trivial renormalization fixed point associated with the break-up of golden invariant tori. For such an exact scheme, the unstable manifold remains of dimension 1 and the stable one has codimension 1, i.e. it separates the space of Hamiltonians into two domains, with one of them containing the trivial fixed point.

The latter domain is called the KAM domain. Indeed, if the large-scale dynamics is chosen in this domain $((A_1, A_2)$ below the stable manifold in figure 5.7), its iterates converge toward $(0, 0)$ and enter the domain where KAM theorem applies. Therefore, within our approximate renormalization scheme, the previous KAM torus with zoning number z_* is preserved at a small scale and thus at any scale in the phase space. At small scales, the orbit looks like that of the free-particle motion but the bundles of straight lines of the integrable case are lost.

As verified in numerical simulations, the non-KAM domain corresponds to the break-up of the KAM torus into a cantorus. The amplitudes of the two waves grow to infinity when renormalizing. The two-resonance approximation made for deriving $\mathcal{R}$ cannot hold over the whole domain. However, this suggests investigating the limit of infinite overlap of two resonances, which might be to chaos what the vanishing overlap limit is to order. This proves interesting indeed as shown in section 5.5.

For more general values of z, the renormalization dynamics is more intricate as the map for z is ergodic and the fixed point is replaced by a strange attractor (MacKay 1988, Stark 1988, Chandre and Jauslin 2000). What is left is a manifold separating the KAM and non-KAM domains. The renormalization approach reveals that the more one enters small scales the more one finds chains of islands if KAM tori are preserved. The existence of trapped KAM tori reinforces this fact. Indeed, another renormalization transformation may be defined which enables us to deal with trapped tori (Escande 1985). An approximate version of it can be derived with a three-resonance Hamiltonian which is Hamiltonian (5.16) with $v = 1$ supplemented with a third resonance $A_3 \cos(k'q + kt)$. In order to look inside resonance 1, one uses the action-angle variables for the trapped motion instead of those for the passing motion. This yields a Hamiltonian similar to (5.51). Either one considers a trapped KAM torus, and one deals with this new Hamiltonian as with (5.51), or one considers a trapped chain of islands corresponding to one of the resonant terms. In the latter case one keeps this term and the two neighbouring ones and one performs rescalings and approximations similar to those for $\mathcal{R}$, which again yield a three-resonance Hamiltonian like the initial one (if $k' > 0$, resonance 1 lies between resonances 2 and 3). Combining both renormalization transformations enables one to focus on a large variety of regions in phase space: for instance, on a KAM torus which is in a chain of islands inside the chain of two islands of figure 5.2.

5.4.4 Thresholds and exponents

The simplest instance of renormalization enables us to show its predictive efficiency and, in particular, what is meant by universality. Indeed the case with $L = a_0 = 1$, which corresponds to z equal to the golden mean $g = \frac{1}{2}(\sqrt{5} + 1)$, is of practical importance, as numerical calculations (MacKay 1983) reveal that the most robust tori belong to the so-called noble class where all a_i are 1 after a certain rank: their zoning number becomes g after a finite number of iterates of $\mathcal{R}$. For $z = g$, $c_0(k) = (k + 1)^2$, $c_1(k) = (k + 1)^4/(2k)$, $k_* = g$. The non-trivial fixed point of $\mathcal{R}$ corresponds to $(A_1, A_2) = (2/g^7, 2/g^{11})$. The stable manifold

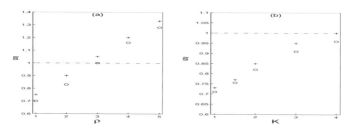

Figure 5.8. Comparison of theoretical (dots) and numerical (crosses) estimates of the threshold $\tilde{s}$ for global chaos for (a) $k = 1$, $\rho \equiv (A_1/A_2)^{1/2}$ varying, and (b) $A_1 = A_2$, k varying; the line $\tilde{s} = 1$ corresponds to the resonance overlap criterion. (After Escande and Doveil (1981b).)

of $\mathcal{R}$ in (k, A_1, A_2)-space is defined by

$$A_1 A_2^g = R(k) \tag{5.62}$$

where $R(k)$ can be estimated numerically. A more precise approximate renormalization scheme can be derived by using the generating function suppressing both waves to first order instead of (5.54) and yields an estimate of $R(k)$ (Escande 1985).

When applying $\mathcal{R}$ to the noble tori of type $z = l + g$ (where l is an integer such that $l \leq l_0$ with l_0 a given integer) and to the tori of this type for the equivalent Hamiltonian, the renormalized torus is a golden mean torus; and (5.62) can be used for (A_1', A_2') to compute its break-up threshold s_l for a given ratio A_2/A_1. As another KAM torus might be more robust than the noble ones of interest, the highest s_l yields an approximate lower bound to the threshold for global chaos between waves 1 and 2. Figure 5.8 displays these lower bounds together with numerical upper bounds for this threshold (Escande and Doveil 1981a, b) for various values of $\rho = (A_2/A_1)^{1/2}$ and k when $l_0 = 4$. The agreement is within 4% for the values and much better for the slopes. For the case $k = A_1/A_2 = 1$, the actual large-scale chaos threshold is $s \simeq 0.7$ (as mentioned earlier, Celletti and Chiercia (1988) found the lower bound to be 0.48). This turns out to be the minimum value for the threshold. The dotted line $s = 1$ corresponds to Chirikov's resonance overlap criterion.

The analytical calculation of large-scale chaos thresholds for general Hamiltonians is a tedious task. If they are written as an integrable part with resonant perturbations, the resonance overlap criterion often yields the right order of magnitude. Better estimates can be obtained with numerical calculations. Then upper bounds for the threshold can be obtained by monitoring the transition we observe for H_2 for $0.68 < s < 1$.

The renormalization ideas can be used to obtain good numerical approximations of the threshold for break-up of the most robust KAM torus in a given domain. This uses the fact that when s grows the stable orbit in the centre of

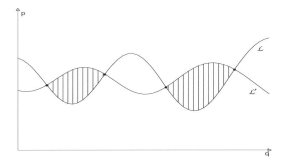

Figure 5.9. Flux through a cantorus.

any island becomes unstable. We illustrate this on H_2. Imagine we want to know the break-up threshold of the most robust KAM torus for $1 < z < 5$. As a first step the destabilization threshold of periodic orbits with $z = [a_0]$, $a_0 = 1 \ldots 5$ is computed numerically. Empirically, for regular enough Hamiltonians, the highest two thresholds correspond to nearby values m_0 and $m_0 + 1$ of a_0, and the most robust torus has a zoning number between these two values. As a second step one computes the destabilization threshold of periodic orbits with $z = [m_0, a_1]$, a_1 from 1 up to some number (say 5), and one proceeds as in the first step. This defines a number m_1, and the third step considers periodic orbits with $z = [m_0, m_1, a_2]$, and so on. This sequence of highest thresholds converges quickly towards the threshold of the most robust KAM torus (Greene 1979, Schmidt 1980, Schmidt and Bialek 1982, Doveil and Escande 1982) with $1 < z < 5$. The reason for this fast convergence is given at the end of this section.

Though the Poincaré map looks fuzzy in a chaotic domain, there is a hidden order in chaos. This order will be described in the next section but it is already visible during the passage of orbits through the holes of a cantorus $\mathcal{C}$. A continuous line $\mathcal{L}$ going through each point of $\mathcal{C}$ becomes a continuous line $\mathcal{L}'$ after one iteration of the Poincaré map. As shown in figure 5.9, $\mathcal{L}'$ is rippled with respect to $\mathcal{L}$. Let $\mathcal{A}_{\mathcal{L}\mathcal{L}'}$ be the total area of the ripples of $\mathcal{L}'$ above $\mathcal{L}$. Conservation of area in Hamiltonian dynamics implies that $\mathcal{A}_{\mathcal{L}\mathcal{L}'}$ is equal to the total area of the ripples of $\mathcal{L}'$ below $\mathcal{L}$. The minimum value of $\mathcal{A}_{\mathcal{L}\mathcal{L}'}$ over the set of $\mathcal{L}$ is $\mathcal{A}_\mathcal{C} = fT$ where f is the flux of the area through $\mathcal{C}$ and T the period of the Poincaré map. By $\mathcal{R}$, in the vicinity of the fixed point F_{z_*}, area is rescaled by the factor

$$\xi = (k_* + a_0 - 1)^2 (k_* + a_0)/k_*, \tag{5.63}$$

time is rescaled by the factor

$$\tau = -k_*/(k_* + a_0) \tag{5.64}$$

and q is rescaled by the factor

$$d = k_* + a_0 - 1. \tag{5.65}$$

Thus $\mathcal{R}$ rescales the flux f by the factor

$$\chi = \frac{\xi}{|\tau|d} \tag{5.66}$$

which is always larger than 1.

Imagine that the point corresponding to the large-scale system is close to the unstable manifold of F_{z_*}. Let $x \ll 1$ be the distance from this point to F_{z_*} and let δ be the unstable eigenvalue of F_{z_*}. Then $\mathcal{R}$ rescales x by δ. Imagine after M rescalings x is equal to some $x_0 \ll 1$. For the corresponding Hamiltonian system, f is f_0. Then the original f is $\chi^{-M} f_0$ and the original x is $\delta^{-M} x_0$. We now focus on the H_2 such that $\mathcal{A}_C$ equals some small value ε. Such a set corresponds to a manifold in (k, A_1, A_2) space whose trace in figure 5.7 would be a line slightly above the stable manifold, intersecting the unstable manifold at a distance from the fixed point which is the x_0 of our previous reasoning. When M grows, the logarithms of both f and x are linear functions of M. This implies $f \sim x^\nu$ with

$$\nu = \frac{\ln|\chi|}{\ln \delta} \tag{5.67}$$

which is always positive. The hyperbolic nature of the fixed point implies that the rescaling of x also holds for the distance to the stable manifold $s - s_c$ for a one-parameter family of Hamiltonians with (k, ρ) fixed (Escande 1985). Therefore

$$f \sim (s - s_c)^\nu \tag{5.68}$$

for such a family. This means that the flux through C is governed by the positive exponent ν called a critical exponent. This exponent is confirmed numerically (Peyrard and Aubry 1983, MacKay 1982). Other critical exponents can be defined for cantori.

The break-up of a KAM torus into a cantorus may be viewed as a continuous process. The existence of a renormalization group and of critical exponents makes this break-up analogous to a second-order phase transition. For $s - s_c$ small, f is small and a cantorus acts as a (nearly impermeable) barrier to chaotic transport in phase space. If $z_* = g$, $\chi = g^3$ and $\delta = g$ which yields $\nu = 3$ which is close to the exact value $\nu_* \simeq 3.011\,722$. Other critical exponents can be computed by the same method, in particular for the Lyapunov exponent of cantori.

This hyperbolicity argument can be used for other issues and, for instance, for the fast convergence with M of the destabilization thresholds of the nearby periodic orbits $z_M = [a_0, \ldots, a_M]$ toward the threshold for break-up of the KAM torus with $z = [a_0, \ldots, a_M, \ldots]$. Consider, for simplicity, the golden case where all a_i are 1. Instead of the previous set of H_2 with A having a small value, consider the set such that the stable periodic orbit with $z = 1$ becomes unstable. The hyperbolicity argument shows that the thresholds of destabilization found after M renormalization steps are only $O(\delta^{-M})$ away from the threshold for break-up of the golden mean torus. A similar argument can be used to show how any (reasonable) criterion for torus break-up can give estimates with $O(\delta^{-M})$ accuracy if the M first renormalization steps are performed exactly (Escande 1985). If $z_M = l/n$ where l and n are mutually prime, this means applying perturbation theory up to order ε^{l+n-1} using a canonical transformation and making appropriate rescalings.

5.5   Order in chaos

5.5.1   Wiggling arms of the X-point

As shown in section 5.2, KAM tori are the essential geometrical objects underlying regular motion in non-integrable Hamiltonian systems. Are there essential geometrical objects underlying chaotic motion? Yes, indeed. A hint as to their nature was given in section 5.2 when describing the stochastic layer. This layer develops in the vicinity of the separatrix of the unperturbed pendulum and its chaotic orbits switch in an apparently erratic way from trapped to passing behaviour and back, a fact which is evidenced when they come close to the X-point of the unperturbed separatrix. For the pendulum, the separatrix was found in section 1.2 to be the set of points of orbits asymptotic to the X_i for $t \to \pm\infty$.

The stable (respectively unstable) manifold of an X-point is defined as the set of points whose image through the dynamics converges toward the X-point for $t \to +\infty$ (respectively $t \to -\infty$). *A priori* there is no reason for the stable and unstable manifolds to merge. For the pendulum, the low dimensionality of the dynamics provides such a reason: the phase space is two dimensional, and the separatrix is the set of points having the same constant of the motion (the energy) as the X-point; both the stable and the unstable manifolds contain orbits, which are one-dimensional lines, corresponding to the same constant of the motion, and merge into the separatrix. Such an argument no longer holds for a non-integrable Hamiltonian, for which the phase space is higher dimensional[18].

Consider a time-periodic 1.5-degrees-of-freedom Hamiltonian $H(p, q, t) = H(p, q, t + T)$. In general, there is no constant of the motion, and the role of an X-point is played by a periodic orbit $X(t) = X(t + nT)$ for some integer n (which gives a fixed point in the period-nT Poincaré map). Its stable manifold is a two-dimensional surface (tracing a line in the Poincaré section), and the unstable manifold is also a two-dimensional surface. As phase space is three dimensional, there is no reason for the stable and unstable manifolds to coincide.

For the two-wave Hamiltonian, the two manifolds split when both waves have non-vanishing amplitudes. This may be seen by numerical calculation of the two manifolds and may be proved analytically in some limits.

---

[18] However, for an N-degrees-of-freedom autonomous Hamiltonian system, the energy hypersurface of a fixed point X has dimension $2N - 1$. If no eigenvalue of the linearized dynamics around X has a zero real part, the stable and unstable manifolds of X both have dimension N. Then they still typically intersect along a one-dimensional line.

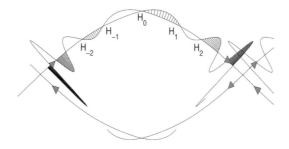

Figure 5.10. Homoclinic intersections.

One of the cases where this splitting may be proved is a perturbing wave with a small amplitude. This is done using a technique introduced by Melnikov (1963) which consists in calculating orbits asymptotic to the X-point by perturbation calculus. In agreement with intuition, such a calculation reveals that when starting from the X-point the two manifolds stay close to the separatrix. This is true up to the time when they would again come close to the X-point (see figure 5.10). This first part of the manifolds is called their first branch. The invariance of the equations of motion of Hamiltonian (5.16) under the transformation $(x, p, t) \mapsto (-x, p, -t)$ shows that the first branches of the two manifolds have in common the point with $x = 0$, $t = 0$. Therefore the orbit passing through this point at $t = 0$ belongs to both manifolds for all times and the Poincaré map displays an infinity of intersections, called homoclinic points. As a result, the two manifolds wiggle with respect to each other (as the Poincaré map is orientation preserving, an odd number of intersections—here 1—must be present between those linked through the mapping). These properties extend to the case where the perturbing wave has an O(1) amplitude.

The closed domain defined by the arcs of the two manifolds joining nearby intersections is called a homoclinic lobe. When the intersections are asymptotic to the X-point, one of the arcs of the lobe shrinks to vanishing length but the preservation of the lobe area through the mapping forces the lobe to increase its length in the other direction (see figure 5.10). A stable or unstable manifold cannot self-intersect, for this would imply a sequence of similar self-intersections converging to the X-point, which is forbidden in the (asymptotically valid) linear dynamics near this point. The stable (or unstable) manifolds of two distinct X-points cannot intersect for similar reasons.

Exercise 5.22. Show more explicitly that these intersections are forbidden by trying to draw successive images of the intersections if they are assumed to exist (Abraham and Shaw 1992, chapter 14).

For Hamiltonian dynamics, a second reason why a given manifold may not self-intersect is that the points of a loop due to self-intersection cannot all converge to the X-point, because of the conservation of the loop area. This imposes a strong constraint on the wiggles of the manifolds and the homoclinic lobes are forced to bend when stretching in order to avoid self-intersection of the manifolds (see figure 5.10). In turn this generates new homoclinic intersections and the two manifolds (the 'arms' of the X-point) create an intricate pattern called a homoclinic tangle or trellis which already challenged Poincaré at the end of the 19th century. The two manifolds are well defined but make many folds—an infinity!

As the successive iterates of a given homoclinic lobe must become increasingly thin while becoming increasingly long, chaos generates small scales in the phase space, which may be described, among other means, by complex symbolic dynamics (positive topological entropy). Moreover, the successive homoclinic and heteroclinic intersections generate topological structures called Birkhoff signatures (Abraham and Shaw 1992) enabling one to characterize them further.

An orbit belonging to one of the manifolds must follow all its folds and looks rather erratic though deterministic. The stochastic layers found numerically show that this behaviour is quite general, though the mathematical knowledge about the orbits in chaotic layers is still limited. The Smale–Birkhoff homoclinic theorem (Guckenheimer and Holmes 1983) shows that the existence of homoclinic intersections implies that of a zero-measure set (Smale horseshoe) where the Poincaré map is topologically chaotic.

The area of homoclinic lobes enables one to quantify the number of orbits which go from the trapped to the untrapped state (and vice versa). As the first branches of the manifolds are close to the unperturbed separatrix, it is natural to define the boundary of the trapping domain by the arc $X'H_0$ of the unstable manifold and by the arc H_0X of the stable one. In figure 5.10 consider the lobe corresponding to the nearby homoclinic points H_{-1} and H_0. Its close enough antecedents are certainly in the trapped domain but its successor and its close followers are not. Therefore the lobe area gives the 'number' of orbits which transit from being trapped to being untrapped during a Poincaré time. This area is computed in the small perturbation limit by the Melnikov integral[19] (Melnikov 1963, Guckenheimer and Holmes 1983).

Note that the Melnikov technique is not correct if we consider the homoclinic tangle replacing the separatrix of a small wave when it is perturbed by another small wave (as for Hamiltonian H_2 with $A_1 = A_2$ and s small). Lazutkin (1984, see Gelfreich and Lazutkin 2001) computed a correct estimate for the energy of KAM tori bounding the stochastic layer. This estimate yields a value about 2.1 times larger than the naive estimate obtained through the Melnikov integral as well as through the renormalization technique of section 5.4.2 for m large. The mechanism for splitting the separatrix into stable and unstable manifolds and (5.18) explain why chaos first appears close to the unperturbed separatrices when s grows.

---

[19] The Melnikov integral for Hamiltonian H_2 is given by Chirikov (1979).

Until now we have focused on the X-point related to the unperturbed pendulum (the static wave). Similar tangles may be defined for the X-points of the other wave or the higher-order resonances described in section 5.3.1. Therefore we must envision the coexistence of (un)stable manifolds related to different X-points. As two (un)stable manifolds may not intersect, numerical calculations reveal that in chaotic domains two (un)stable manifolds make close parallel wiggles far from their corresponding X-points. The homoclinic lobes must interpenetrate and the flux through cantori can be eventually traced back to the area of homoclinic lobes of nearby resonances.

Exercise 5.23. Show that neither the stable manifold of the X-point nor its unstable manifold can intersect a KAM torus.

For small perturbations of a typical integrable Hamiltonian, such as H_2 with $A_1 = O(1)$ and $A_2 = \varepsilon \ll 1$, denote the total area of the domain between the nearest KAM tori on all sides of the X-point by $\mathcal{A}_{\text{chaos}}$. Let $\mathcal{A}_{\text{lobe}}$ be the area of a homoclinic lobe and denote by λ the Lyapunov exponent of the X-point for the Poincaré map ($\lambda \sim A_1^{1/2} 2\pi/k$ for H_2). Treschev (1998) shows that

$$\mathcal{A}_{\text{chaos}} \sim \frac{\mathcal{A}_{\text{lobe}} \ln \mathcal{A}_{\text{lobe}}^{-1}}{\lambda^2}. \tag{5.69}$$

Thus the non-KAM domain has an area which scales to zero with ε almost like the area of a single lobe. This means that the whole homoclinic tangle of X cannot extend 'very' much farther than the first lobes, in this limit. Estimate (5.69) does not mean that the ratio $\mathcal{A}_{\text{chaos}}\lambda^2/(\mathcal{A}_{\text{lobe}} \ln \mathcal{A}_{\text{lobe}}^{-1})$ tends to a definite limit as $\varepsilon \to 0$; in general it oscillates within a bounded range.

In the case where $\lambda \to 0$ with the perturbation size, as, e.g., for $A_1 \sim \varepsilon$, $A_2 \sim \varepsilon$, the estimate (5.69) does not apply. Rather, one expects

$$\lim_{\varepsilon \to 0} \frac{\mathcal{A}_{\text{chaos}}\lambda^2}{\mathcal{A}_{\text{lobe}} \ln \mathcal{A}_{\text{lobe}}^{-1}} = \frac{8\pi^2}{K_c} \tag{5.70}$$

provided some symmetry conditions are met (Treschev 1998). Here $K_c = 0.971\,635\ldots$ is the threshold for large-scale chaos in the standard map.

Finally, the parallel wiggles of manifolds of the same kind naturally also create a new kind of intersection, called heteroclinic intersections, between the stable manifold of one X-point and the unstable manifold of another X-point. The existence of sizable chaotic domains was related in section 5.2 to the break-up of KAM tori. Now we may relate it to the heteroclinic intersection of manifolds coming from the X-points of the two waves of Hamiltonian (5.16). After the break-up of a KAM torus, these manifolds go through the small holes of the corresponding cantorus. In this respect, the Chirikov resonance overlap criterion may be viewed as an approximate heteroclinic intersection criterion, computed by

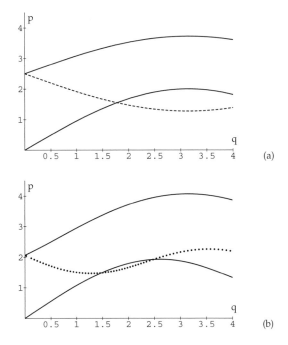

Figure 5.11. (*a*) Chirikov's resonance overlap picture and (*b*) the actual beginning of the stable and unstable manifolds for the three-wave Hamiltonian (5.71) in the Poincaré section at $\lambda = 0$ for $\varepsilon = 2.5$, $\mu = 0.75$, $\mu' = 1$ (stochasticity parameter $s \simeq 1.29$). Light full (dotted) lines are branches of unstable (stable) manifolds. (After Elskens and Escande (1991, 1993).)

approximating the manifolds of the X-points of the two waves by branches of the corresponding separatrices (figure 5.11).

The stable and unstable manifolds of X-points provide the looked-for geometrical objects underlying chaos. They are to chaos what KAM tori are to order.

5.5.2   Large resonance overlap: numerical results

We now turn to the limit of strong resonance overlap where the size and shape of the chaotic domain can be simply defined with geometrical arguments based on the structure of the homoclinic tangle. The argument is easier to introduce (Elskens and Escande 1993) when considering the dynamics of an electron in three electrostatic waves defined by Hamiltonian

$$H_3(p, q, t) = \frac{p^2}{2} + A \cos q - \frac{\mu}{2}[\cos(q+\epsilon t)+\cos(q-\epsilon t)]-(A-\mu \cos \epsilon t) \quad (5.71)$$

where $0 < \epsilon \ll 1$, $0 < \mu < A$, μ is a quantity of order 1; the last term is added for our later convenience and plays no role in the equations of motion. The first wave is static and the two propagating waves have phase velocities $\pm\epsilon$. The resonance overlap parameter of these two waves is

$$s = \frac{\sqrt{\mu/2}}{\epsilon} \tag{5.72}$$

and that of either of these waves with the static one is

$$s' = \frac{2\sqrt{A} + \sqrt{2\mu}}{\epsilon}. \tag{5.73}$$

The small ϵ limit of interest is thus the limit of strong overlap of these three waves.

Inspecting the equations of motion shows that the point $X = (0, 0)$ is a fixed point of the dynamics. Linearizing the dynamics close to this fixed point yields for the small displacement δq

$$\delta\ddot{q} = (A - \mu\cos\epsilon t)\delta q \tag{5.74}$$

which is a Mathieu equation. Floquet theory (see, e.g., Whittaker and Watson (1980) or Iooss and Joseph (1990)—or Bloch theorem in condensed matter theory) implies that this equation admits two linearly independent solutions of the type $f(\pm t)e^{\pm\nu t}$, where f is periodic with period $2\pi/\epsilon$ and ν is either real or pure imaginary. Classical results on the Mathieu equation show that ν is real for $A > \mu > 0$, so that X is a hyperbolic fixed point whatever ϵ may be. Thus in Hamiltonian (5.71) the effects of the two travelling waves compensate and let the X-point of the static wave remain a fixed point for any value of μ; furthermore, this fixed point remains an X-point.

Hamiltonian (5.71) may be rewritten in the *pulsating separatrix* form[20]

$$H_{ps}(p, q, \lambda) = \frac{p^2}{2} + g(\lambda)(\cos q - 1) \tag{5.75}$$

where $\lambda = \epsilon t$ and

$$g(\lambda) = A - \mu\cos\lambda. \tag{5.76}$$

Then Hamiltonian (5.71) may be interpreted as the Hamiltonian of a pendulum in a slowly modulated gravity field. When λ is frozen at a given value, X is the X-point of the corresponding pendulum. The frozen separatrix is defined by $H = 0$ for all λ, and the corresponding cat's eye has an area which is minimal ($\mathcal{A}_m$) for $\lambda = 0$ and maximal ($\mathcal{A}_M$) for $\lambda = \pi$. Given λ, a frozen action may be defined by (5.11) for any point (p, q) in the phase space. For any λ we consider the trace $\mathcal{T}_0$ of the trapped orbits with $I = \mathcal{A}_m/(2\pi)$ and the traces $\mathcal{T}_\pm$ of the passing orbits with $I = \mathcal{A}_M/(4\pi)$ and $\pm p > 0$. The domain $\mathcal{S}$, outside $\mathcal{T}_0$ and in between $\mathcal{T}_+$ and $\mathcal{T}_-$, is called the domain swept by the slowly pulsating separatrix corresponding to the continuous sequence of frozen separatrices obtained when $\lambda = \epsilon t$.

---

[20] An effective dynamics of a similar kind will be encountered in chapter 8.

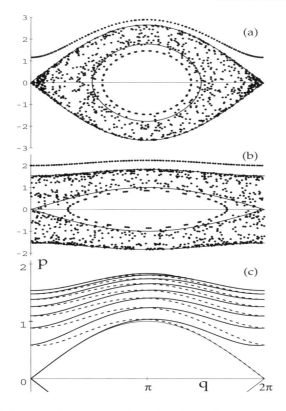

Figure 5.12. Poincaré sections of the modulated pendulum for $\epsilon = 0.1$, $\mu = 0.75$, $\mu' = 1$ (stochasticity parameter $s' \simeq 32$). Light full lines indicate the boundaries of the swept region. At (a) $\lambda = \pi$ and (b) $\lambda = 0$, dots mark three typical orbits: 1000 points for the orbit in the swept domain, 100 points for each other orbit. (c) Beginning of the stable (dotted line) and unstable (full line) manifolds of the X-point at $\lambda = 0$. (After Elskens and Escande (1991, 1993).)

Figure 5.12 displays two Poincaré surfaces of section for the dynamics of Hamiltonian (5.71) at $\epsilon t = 0 \mod 2\pi$ and $\epsilon t = \pi \mod 2\pi$, for $A = 1$, $\mu = 0.75$ and $\epsilon = 0.1$ which corresponds to $s' \simeq 32$. Three typical orbits are displayed, as well as the traces $\mathcal{T}_0$ and $\mathcal{T}_\pm$ materializing the boundaries of $\mathcal{S}$. The two orbits outside $\mathcal{S}$ look regular but that inside $\mathcal{S}$ looks chaotic and seems to fill in this domain in a fairly uniform way. As explained later, for ϵ small there is a tight homoclinic trellis which covers $\mathcal{S}$ and comes quite close to its boundaries. In figure 5.12, ϵ is not very small but the chaotic sea related to the trellis fits well within $\mathcal{S}$. We note that no large-size island is visible in the chaotic sea.

Figure 5.12(c) displays the beginning of the stable (unstable) manifold of X at $\epsilon t = 0$ computed numerically as the pre-image (image) of the part of the manifold with length 1.4×10^{-13} close to X. The plotted branches of the manifolds look as if they touch $\mathcal{T}_+$. They correspond to seven turns of the pendulum around its fixation point. They intersect transversally which shows the absence of a separatrix and the presence of a homoclinic trellis. The meshes of the trellis are a kind of parallelogram with sides which are pieces of stable and unstable manifolds.

Exercise 5.24. Show that the eigenvalues $\sigma_\pm$ of the Poincaré map linearized at X scale as $\epsilon \to 0$ like $\sigma_\pm \simeq e^{\pm c/\epsilon}$ with $c = O(A^{1/2}) = O(1)$. Hint: first show that $\sigma_+\sigma_- = 1$, and find $0 < c_1 \le c_2$ such that on the unstable manifold of X one has $c_1 t \le \ln |\delta q(t)|/|\delta q(0)| \le c_2 t$ for $t > 0$, so that $c_1 \le \epsilon \ln \sigma_+ \le c_2$. Can you estimate c more precisely? Beware of factors 2π in numerical estimates.

5.5.3 Large resonance overlap: analytical results*

These numerical facts about the stable and unstable manifolds of X can be substantiated analytically. A classical approach to Hamiltonian dynamics with a slow time dependence is adiabatic theory (see, for instance, Percival and Richards (1982) or Henrard (1993)). In the case of Hamiltonian (5.75) with $\lambda = \epsilon t$, this theory starts from the observation that the action I of the frozen pendulum is a constant of the motion for $\epsilon = 0$ and that for ϵ small a new 'constant of the motion' J close to I might be computed by perturbation theory in ϵ. This constant is just an approximate one since the motion is non-integrable; indeed the perturbation expansion is asymptotic.

The calculation works properly provided that, over the time interval of interest, the energy h of the orbit for the frozen pendulum keeps its absolute value bounded from zero (i.e. from the separatrix energy) by a quantity of order ϵ (Elskens and Escande 1991). On an orbit with action I and $\theta = \theta_0$ when $\lambda = \lambda_0$, at lowest order in ϵ the frozen angle is found to evolve according to

$$\theta(I, \lambda) = \theta_0 + \epsilon^{-1} \int_{\lambda_0}^{\lambda} \Omega(I, \lambda') \, d\lambda' \tag{5.77}$$

where $\lambda = \epsilon t$ and

$$\Omega(I, \lambda) \equiv \frac{\partial H_{\mathrm{ps}}(I, \lambda)}{\partial I}. \tag{5.78}$$

This is shown using the approximate expression $\dot{\theta} = \Omega(I, \lambda)$.

In fact, even if the series defining J converges slowly, the adiabatic idea may still be used. This idea is that the frozen system dynamics may be used to approximate the true orbit during one period of the motion of the frozen system, if the time-dependent system changes weakly during this time (Tennyson *et al* 1986, Cary *et al* 1986). Then the condition for adiabaticity is $\Omega(I, \lambda) \gg \epsilon$. As Ω

diverges only logarithmically as a function of the energy[21] $h = H_3$, this condition excludes only a domain in $|h|$ of order $e^{-c/\epsilon}$ where c is a positive constant. This enables one to derive the so-called neo-adiabatic theory which works for $|h| \lesssim \epsilon^{2/3}$ (Elskens and Escande 1993), i.e. over a domain which overlaps with the adiabatic domain.

Neo-adiabatic theory rests on the fact that Hamilton equations treat energy and time as another pair of conjugate variables. Solving $H(p, q, t) = h$ in the form $p = P(h, t, q)$, with the new timelike variable q, yields the equations of motion in the form $dh/dq = -\partial_t P$, $dt/dq = \partial_h P$. This enables one to turn rapidly changing functions into slowly varying ones. Of course, one may also use p as a timelike variable and $q = Q(h, t, p)$ as its conjugate Hamiltonian.

For the modulated pendulum, neo-adiabatic theory identifies the intersections of the orbit with the p- and q-axes (let us call these points vertices) and computes the change of frozen energy and the time delay when the orbit goes from one vertex to the next one. This enables us to show that formula (5.77) still holds in the neo-adiabatic energy domain.

As X corresponds to $h = 0$, neo-adiabatic theory cannot describe the part of the (un)stable manifold closest to X. Fortunately linear theory suffices to compute the orbit explicitly there with a WKB approximation (Bender and Orszag 1978). This solution holds in a region of typical extent ϵ about X which overlaps the domain where neo-adiabatic theory holds. Matching the solutions in the two domains shows that an arc of the unstable manifold starting at X, with length of order 1 in the Poincaré section at time $t = \lambda_0/\epsilon$, $\lambda_0 = 0$ (respectively π), is defined by

$$\theta(I) = \theta_v + \epsilon^{-1} \int_{\lambda_0}^{\lambda(I)} \Omega(I, \lambda') \, d\lambda' \tag{5.79}$$

where (I, θ) are the action-angle variables for untrapped (respectively trapped) orbits, θ_v is the value of θ (mod 2π) when crossing the p- (respectively q-) axis for $\lambda_0 = 0$ (respectively π), and $\lambda(I)$ is defined by

$$H_{ps}(p, q, \lambda(I)) = 0 \tag{5.80}$$

with $\lambda_0 < \lambda(I) < \lambda_0 + \pi$. This formula holds for both branches of the unstable manifold. The stable manifold is computed by the same formula, with a minus sign in front of the integral. These formulae may be recovered easily by assuming the validity of classical adiabatic theory over the whole arcs of the manifolds.

Formula (5.80) and similar ones for the stable manifold show that the arcs of these manifolds extend over the whole domain in action swept by the pulsating separatrix and that they rotate (libration or rotation) $O(\epsilon^{-1})$ times in phase space. As $\theta(I)$ is monotonically increasing, the arcs of the stable and of the unstable

---

[21] For the pendulum (1.12), $\Omega(H) = \pi(A/m)^{1/2}/K(m)$, where $K(m) \simeq \frac{1}{2} \ln[16/(1-m)]$ for $m \lesssim 1$ is an elliptic integral of the first kind and $m = 2(H/A + 1)^{-1}$. For more general systems, logarithmic divergence also occurs, as it merely follows from the linearization of the motion close to an X-point (exercise 5.9 and Cary *et al* (1986)).

manifold make $O(\epsilon^{-1})$ transverse homoclinic intersections. They form together a trellis whose meshes are small parallelograms of area $O(\epsilon)$. As their sides cannot intersect any invariant torus, this rules out the existence of KAM tori of the passing type inside S and forces possible islands inside S to be inside the meshes, which limits their maximum size. Each node of the trellis is a transverse homoclinic point with a topologically chaotic Smale horseshoe. This chaos may influence all nearby orbits and cause the seemingly chaotic motion over S.

All these results constitute, in fact, three theorems which apply to any Hamiltonian system with a slowly pulsating separatrix whose enclosed area does not vanish (Elskens and Escande 1991). They back up the numerical evidence reported in the previous subsection; in particular, that about the smallness of the islands in the chaotic sea. However, surprisingly the total area of the islands does not go to zero as $\epsilon \to 0$ for (5.71); however, it does in more asymmetric systems (Neishtadt *et al* 1997).

The condition for a non-vanishing area enclosed by the pulsating separatrix in the theorems of Elskens and Escande (1991) is rather technical in origin and could be removed. However, for Hamiltonian (5.71) with $A = 0$ to enable the pulsating area to vanish twice per time period, the X-point of the frozen pendulum exchanges its nature with the O-point also twice per period when passing through the free-particle case, which complicates the construction of the homoclinic tangle and the following separatrix crossings. Numerical simulations show Poincaré sections similar to the case $A > \mu > 0$, and the adiabatic invariant can be constructed similarly (Menyuk 1985). A system with a self-consistently pulsating separatrix of this type occurs in a model for transport in a two-dimensional fluid flow and in one-dimensional plasma dynamics with electron 'holes' and 'clumps' (del-Castillo-Negrete and Firpo 2002).

On letting ϵ grow, the homoclinic trellis progressively disentangles, and at some moment the heteroclinic intersection between the primary resonances of Hamiltonian (5.71) disappears. This is related to the closing of the holes in a cantorus which becomes a KAM torus. In section 5.2, large-scale chaos appeared to be due to the dramatic break-up of the most robust KAM tori. In this section, order appears to be due to the dramatic disentangling of manifolds. The large- and small-amplitude limits of the waves provide complementary limits where the geometrical aspects of chaos and order are simple; and shed light on chaos and order in intermediate regimes.

Finally, when defining Hamiltonian (5.71), we assumed μ to be of the order of unity. If we consider the overlap of a large resonance with resonances whose trapping widths remain of the order of the phase velocity mismatch, then $\mu = O(\epsilon^2)$, and (5.75), (5.76) show that the amplitude of the perturbation to the large resonance vanishes, and that the chaotic domain vanishes on the scale of this resonance. This is in agreement with the observations in section 5.2 on the resonance overlap with large resonances.

Chapter 6 considers the limit of Hamiltonian chaos where many resonances are present with a strong local, but not global, overlap. This leads to a diffusive behaviour for the chaotic orbits.

5.6 Historical background and further comments

The commented picture gallery by Abraham and Shaw (1992) offers a unique approach to simple and complex orbits, and to homoclinic tangles in chaotic dynamics.

Tools other than those we have already discussed for characterizing Hamiltonian chaos are, for example, frequency analysis (Laskar 1993, 1999), singularities in complex time for $(q(t), p(t))$ (Conte 1999), classification of periodic orbits and dynamics around them (Gutzwiller 1991), Ruelle–Pollicott resonances (Ruelle 1989), symbolic dynamics (Alekseev and Yakobson 1981, Guckenheimer and Holmes 1983) etc. This is a rapidly evolving research topic, contributing to and benefiting from advances in computing methods and in pure and applied mathematics, as well as to the wealth of applications in celestial mechanics (satellite motion, space probes), fluid mechanics (mixing, dye or pollutant dispersion), plasma physics etc. While the ideas of transport are in tune with the views of classical mechanics and hydrodynamics (Ottino 1989), the concepts of scaling and universality invoked in this chapter are distinctive of statistical physics.

It is worth noting that, in spite of the importance of the conjugacy of momenta p to coordinates q, which is the foundation of Hamiltonian dynamics, a major physical concept is the resonance, which manifests itself by the condition on velocities that $|\dot{q} - \Omega(p)|$ be small. This apparent inconsistency is better understood if one recalls that the resonance condition requires a comparison of velocities on nearby orbits, i.e. the smallness is estimated in terms of the twist (the 'gradients' of velocities in phase space, i.e. the Hessian matrix $\partial \Omega_i / \partial I_j = \partial^2 H / (\partial I_j \partial I_i)$ in section 5.3.2).

Let us also note that non-autonomous continuous-time dynamics with 1.5 degrees of freedom differs significantly from dynamics with more degrees of freedom. Indeed, invariant tori for integrable dynamics with N degrees of freedom have dimension N, and so do the corresponding KAM tori, in a $2N$-dimensional phase space, whereas when there are 1.5 degrees of freedom, tori are two dimensional in a three-dimensional space. Thus, in higher dimension, tori do not separate the phase space into disconnected domains[22]. Moreover, in higher dimension, resonance conditions involve generically a single combination $k \cdot \Omega = 0$ (with $k \in \mathbb{Z}^N$, $k \neq 0$) but double or higher resonances are possible (e.g. $\Omega_1 = \Omega_2 = \Omega_3$) resulting in stronger chaos. However, KAM theory also applies to N-degrees-of-freedom Hamiltonian dynamics and a renormalization scheme has also been developed to describe their break-up (Chandre and MacKay (2000) and references therein).

---

[22] However the dynamics may be sensitive to KAM tori corresponding to the local approximation of the dynamics by a two-resonance Hamiltonian (Escande *et al* 1994).

Historically, chaos was identified by the non-existence of 'nice' invariant manifolds in the phase space for typical Hamiltonian systems by Poincaré, around 1900. However, this non-existence of smooth, uniformly defined constants of the motion allows the preservation of a non-smooth family of invariant subsets in the phase space as proved by Kolmogorov, Arnol'd and Moser (1956–63). Even in the chaotic domain, motion can preserve perturbed actions for an exponentially long time, as shown by Nekhoroshev in 1972–3. Three different viewpoints (1, deal with all phase space at all times; 2, deal separately with each torus, characterized by its own velocity, for all times; and 3, deal with the conservation of action up to some tolerance, over a finite time) yield three different characterizations of chaos. A fourth viewpoint, enabling further progress at present, focuses on estimating the statistics for how far orbits wander in a given time, i.e. on transport problems. These problems are illustrated in the next chapters.

The actual calculation of transport from the geometry of homoclinic tangles for the quantitative theory of transport requires the construction of secondary lobes and higher-order homoclinic and heteroclinic intersections (Meiss 1992, Wiggins 1992, Rom-Kedar 1994). Even in the adiabatic limit, successive separatrix crossings are not independent, which significantly affects transport (Bruhwiler and Cary 1989, Cary and Skodje 1989).

Analysing chaotic orbits for the large-overlap limit for discrete time is easier than for continuous time. Aubry and Abramovici (1990) introduced a Lagrangian formalism to prove the existence of orbits of arbitrary types for the standard map in this limit. However, proving whether chaotic orbits form a positive measure set in phase space, for a set of parameter values with positive measure, is still a challenge and a topic of active research.

Chapter 6

Diffusion: the case of the non-self-consistent dynamics

In contrast to chapter 5, which introduced the basic concepts of Hamiltonian chaos, this chapter deals with the diffusive motion present in Hamiltonian dynamics in general. The part of this diffusion related to chaos is essential to describing important nonlinear aspects of the self-consistent dynamics. The physical problem of interest here is the motion of one particle in a prescribed broad spectrum of waves with random phases. In fact this dynamics is a good approximation to that occurring in many Hamiltonian systems. Therefore, the concepts and techniques described in this chapter apply to issues as different as the chaos of magnetic field lines, the heating of particles by cyclotronic waves and chaos of rays in geometrical optics.

In section 6.2, diffusion is first shown to occur, on the basis of numerical calculations, for a simple case of this system defined in section 6. Two diffusion regimes are seen to exist. The first, with a diffusion coefficient termed quasilinear, is present for small times for any amplitude A of the waves. The second one exists for larger times when A is large enough, it is chaotic and may have a different diffusion coefficient. However, this coefficient takes on the quasilinear value when A is large enough.

Then the origin of chaotic diffusion is heuristically explained in section 6.3 by using a rigorous principle of locality: only waves with a phase velocity sufficiently close to the particle velocity are relevant for describing transport features. This gives a simple scaling property to the finite-time dynamics (section 6.4). Sections 6.5 and 6.6 are devoted to the physical interpretation of these results, calling on further numerical evidence.

In the second part of this chapter, which can be read independently of the previous sections, it is shown that the properties of the dynamics of the simple case also hold for a broader class of dynamics, including that of a particle in a set of prescribed Langmuir waves. In particular, chaotic quasilinear diffusion is derived analytically for this more general case in section 6.7 and the following sections. The initial quasilinear diffusion is first explained in section 6.8. It is initially non-chaotic and is shown to last over a time $A^{-2/3} \ln A$ for a large amplitude A of the wave. On this basis, quasilinear diffusion is proved to exist over asymptotic times in section 6.9. Then in section 6.10 we compute the drag on a particle related to chaotic diffusion, and the classical Landau formula for the non-chaotic case is shown to extend to the chaotic quasilinear case. These results will be extended in chapter 7 to the self-consistent dynamics.

6.1 Model Hamiltonian

The problem of diffusion in differentiable Hamiltonian systems is a complex one. Therefore, we start with a simple model dynamical system which makes the introduction of the basic concepts easier. This paradigm is defined by the Hamiltonian

$$H(p, q, t) = \frac{p^2}{2} + A \sum_{m=-M}^{M} \cos(q - mt + \varphi_m) \tag{6.1}$$

where M is a large integer and the φ_m are random variables (independent with uniform distribution on $[0, 2\pi]$). It describes the one-dimensional motion of a particle in a set of longitudinal waves. These waves have equally spaced phase velocities and have the same wavenumber and amplitude. This is expected to give some homogeneity to the dynamics along the p-axis for $|p|$ small enough with respect to M. Section 6.7 shows that the dynamics of a particle in a set of Langmuir waves may be locally governed by a Hamiltonian of type (6.1). The equations of motion for this Hamiltonian are

$$\dot{q} = p \tag{6.2}$$

$$\dot{p} = A \sum_{m=-M}^{M} \sin(q - mt + \varphi_m). \tag{6.3}$$

The right-hand side of the equation for $\dot{p}$ is the instantaneous force acting on the particle. For large values of M and A, and for arbitrary choices of the phases φ_m, the force oscillates rapidly in time. This makes the motion of the particle complicated and the presence of large phase velocities (from $-M$ to M typically) makes the numerical simulation of this system demanding. In the limit $M \to \infty$ and when all the φ_m are 0, these equations have a simple limit, which corresponds to the standard map with parameter $K = 4\pi^2 A$ defined by (5.59) and (5.60).

6.2 Diffusion as a numerical fact

When A is small enough, the motion is expected to be regular.

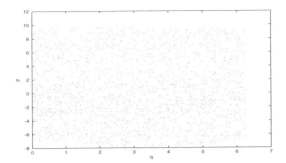

Figure 6.1. Poincaré section of an orbit of (6.1) for $A = 0.2$, $M = 8$, and phases generated by a pseudo-random generator with uniform distribution on $[0, 2\pi]$.

Exercise 6.1. Show that the two-resonance approximation overlap parameter is

$$s_{\text{overlap}} = 4A^{1/2} \tag{6.4}$$

and estimate that the threshold for large-scale transport in this model should be near $A_c \approx 0.03$ by using the resonance overlap criterion (5.17). Note that the presence of many waves makes this estimate quite rough, as one computes it neglecting the influence of all but two waves.

When A is large enough, the motion is chaotic in the range of velocities $-M \leq p \leq M$ and the numerically computed Poincaré map (see figure 6.1) displays a chaotic sea without sizable islands. In such a dynamical regime, for appropriate statistics, the orbits are numerically seen to have a diffusive behaviour over finite time durations. This means that, if a set of orbits has its velocity p distributed according to the distribution function $f(p, t)$ at time t, the evolution equation for f is well approximated by

$$\frac{\partial f}{\partial t} = D \frac{\partial^2 f}{\partial p^2} \tag{6.5}$$

where D is the diffusion coefficient.

We consider a statistics over the orbits where all averages are performed with respect to the φ_n. It thus corresponds to a set of orbits with a given value of $p(0)$ and $q(0)$. We now show numerically that diffusive behaviour exists for such a statistics. If $f(p, 0) = \delta(p - p_0)$, then (6.5) implies that $f(p, t)$ is a Gaussian

$$f(p, t) = \frac{1}{\sqrt{2\pi}\sigma} \exp\left(-\frac{(p - p_0)^2}{2\sigma^2}\right) \tag{6.6}$$

where

$$\sigma = \sqrt{\langle \Delta p^2(t) \rangle} \tag{6.7}$$

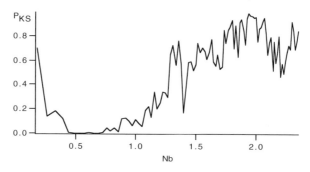

Figure 6.2. Probability P_{KS} for the distribution to be Gaussian versus N_b for $A = 0.8$. (After Bénisti (1995).)

with

$$\Delta p(t) = p(t) - p_0. \tag{6.8}$$

We use the thorough numerical study of Bénisti (1995) and Bénisti and Escande (1997, 1998b) which considers a set of 9000 different orbits, all starting at $p = q = 0$, but with a different random choice of φ_m for each of them. The Gaussianity of the numerical $f(p, t)$ is checked using the classical Kolmogorov–Smirnov test (see, e.g., Feller 1970). This test compares two experimental distribution functions and yields the probability P_{KS} for them to describe the same random variable. Here the comparison is made with a Gaussian. This Kolmogorov–Smirnov test allows one to directly visualize when a diffusion equation is appropriate to model the statistical properties of the dynamics defined by (6.1). Since diffusion is expected, it is also expected that the set of orbits considered in the numerical simulations spreads out along the p-axis as time goes on. For reasons which will become clear later, we focus here on how the diffusive nature of the dynamics of (6.1) evolves as a function of this spreading along the p-axis. To characterize this spreading we define

$$N_b(t) = \frac{\langle p_{max}(t) - p_{min}(t) \rangle}{2\Delta p_R} \tag{6.9}$$

where $p_{max}(t)$ and $p_{min}(t)$ are, respectively, the maximum and minimum value of $p(t')$ for $0 \le t' \le t$, and

$$\Delta p_R = 1 + \text{Int}(5A^{2/3}) \tag{6.10}$$

where $\text{Int}(x)$ denotes the integer part of x. The reason for these definitions will become clear in section 6.4. The probability P_{KS} for the distribution at time t to be Gaussian is plotted versus N_b for a series of values of A.

The result of the test for $A = 0.8$ is shown in figure 6.2. This plot reveals that there is an initial diffusion for the dynamics, and that the dynamics then ceases to be diffusive up to about $N_b = 1.3$, where a new diffusion sets in. To

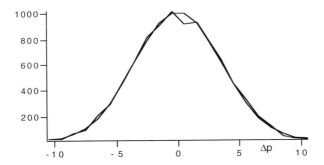

Figure 6.3. Distribution function at the beginning of the simulation for $A = 0.8$ when $P_{KS} = 0.7$ (continuous line) compared to the Gaussian with the same variance (dotted line). (After Bénisti (1995).)

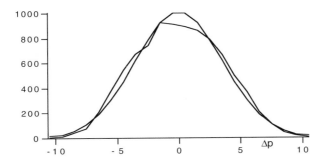

Figure 6.4. Distribution function for $A = 0.8$ when $N_b = 0.7$ with $P_{KS} = 0.0016$ (continuous line) compared to the Gaussian with the same variance (dotted line). (After Bénisti (1995).)

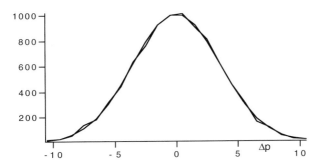

Figure 6.5. Distribution function at the end of the simulation for $A = 0.8$ when $P_{KS} = 0.85$ (continuous line) compared to the Gaussian with the same variance (dotted line). (After Bénisti (1995).)

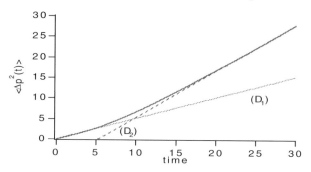

Figure 6.6. $\langle \Delta p^2(t) \rangle$ versus time for $A = 4/\pi^2 \simeq 0.4$. Line D_1 has a slope $2D_{\text{QL}}$, and line D_2 fits the asymptotics of $\langle \Delta p^2(t) \rangle$. (After Bénisti (1995).)

make the information provided by the Kolmogorov–Smirnov test more concrete, figures 6.3–6.5 display the numerical $f(p, t)$ at various values of P_{KS} compared to the Gaussian with the same variance.

Figure 6.6 displays $\langle \Delta p^2(t) \rangle$ versus t for $A \simeq 0.4$. A diffusion coefficient D is estimated by locally fitting $\langle \Delta p^2(t) \rangle$ versus t with the formula

$$\langle \Delta p^2(t) \rangle = 2D(t - t_0) \tag{6.11}$$

by an appropriate choice of D and t_0. Two diffusion regimes show up through two different values of D. The first value of D cannot be distinguished from the so-called quasilinear value

$$D_{\text{QL}} = \frac{\pi A^2}{2} \tag{6.12}$$

the motivation for the name being given in section 6.8.1. The second one is larger than D_{QL}. The same diffusion coefficients are found with $p(0) \neq 0$ provided that $|p(0)| \ll M$. This motivates the choice of Hamiltonian (6.1) as a model. The duration of the initial quasilinear diffusion stage can be estimated as being $t_{\text{QL}} \simeq 4$ in figure 6.6.

Figure 6.7 displays $\langle \Delta p^2(t) \rangle / A^2$ versus time for $A \simeq 1.5$ and 5. The initial quasilinear diffusion is seen to hold up to $t_{\text{QL}} \simeq 1.5$. For higher values of A, measuring t_{QL} becomes more difficult. Indeed, when A increases, the decrease in P_{KS} between the initial and the final diffusion regimes becomes more fuzzy; and for $A > A_0 \simeq 11.5$, diffusion is seen to hold for all times with the unique diffusion coefficient D_{QL}.

Figure 6.8 displays P_{KS} versus N_{b} for $A = 5$ and $A = 11.5$, respectively. A unique value of D, the quasilinear one, is found for all times, as shown in figure 6.9 which corresponds to figure 6.6 but for $A = 11.5$.

Figure 6.10 displays D/D_{QL} versus $K = 4\pi^2 A$. This ratio reaches a maximum value about 2.3 for $A \simeq 0.45$, and goes to 1 for large A (Cary

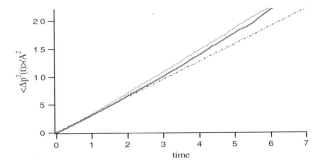

Figure 6.7. $\langle \Delta p^2(t)\rangle / A^2$ versus time for $A = 15/\pi^2 \simeq 1.5$ (full line) and $A = 50/\pi^2 \simeq 5$ (dotted line). The broken line has the quasilinear slope. All three lines match each other until $t \simeq 1.5$. (After Bénisti and Escande (1997).)

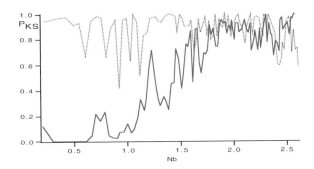

Figure 6.8. Probability for the distribution to be Gaussian versus N_b for $A = 5$ (full line) and for $A = 11.5$ (dotted line). (After Bénisti (1995).)

et al 1990, Helander and Kjellberg 1994)[1]. It starts from zero at the large-scale stochasticity threshold (which will be shown to be quasi-uniform in p in section 6.3.3) and grows when A increases up to the previous maximum, before slowly relaxing to 1. If the average is performed over initial positions for one typical choice of the random phases, the dynamics does not look diffusive (Bénisti and Escande 1998b).

Finally, consider the standard map (5.59), (5.60), which amounts to taking $M \rightarrow \infty$ and all $\varphi_m = 0$ for all m (a highly non-typical case). Figure 6.11

---

[1] Some numerical simulations indicate that the maximum would be just a transient (Ishihara *et al* 1993, Ragot 1998); the issue is not yet settled, but might be related to the non-symplectic character of the integration schemes in these simulations. Cary *et al* (1990) gave an analytical support to the permanent character of the maximum by also discussing the approximation of the true dynamics by a discrete mapping.

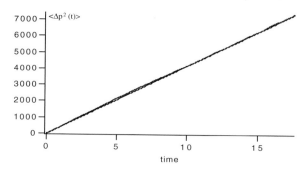

Figure 6.9. $\langle \Delta p^2(t) \rangle$ versus time for $A = 11.5$ (full line). The dotted line has a slope $2D_{QL}$. (After Bénisti (1995).)

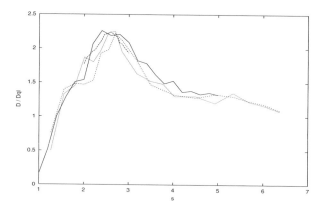

Figure 6.10. Ratio of the diffusion coefficient to the quasilinear value versus s. For (6.1): full line, 1000 samples of phases and two initial-particle data; long dashes, 1000 samples of phases and 200 initial-particle data; short dashes, data from Cary *et al* (1990). For Hamiltonian (6.48): dots, data from Cary *et al* (1990).

displays P_{KS} versus $\tau \equiv tA^{2/3}$ for this case. Gaussianity sets in at $\tau \simeq 100$ which is similar to that for the random phases even though the average is performed only over initial positions: the 'standard' map has non-standard transport properties (Bénisti and Escande 1998b)! Furthermore, the equivalent of figure 6.10 for this map displays an oscillatory relaxation toward 1 for large A (Rechester and White 1980).

6.3   Concept of resonance box

When studying the motion of a particle in a prescribed wave (section 1.2), we have already found through perturbation theory that, however large the amplitude of the

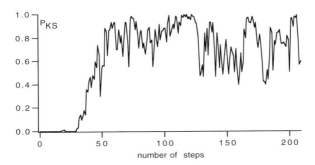

Figure 6.11. Probability for the distribution to be Gaussian versus time for the standard map with $K = 32$. (After Bénisti (1995).)

wave is, a particle far from being resonant with the wave is weakly perturbed by it. The denominators found in the perturbative approach of section 5.3.1 indicate that this idea extends to the case where the particle is acted upon by a series of waves. We now make this idea more concrete.

6.3.1 Heuristic analytical approach

We first use an intuitive but approximate approach. As the particle velocity may be subject to chaotic transport, it may drift away from its initial value. Therefore, being non-resonant with a given wave is an instantaneous concept and the following calculations are meant to hold over a finite time. We focus on the orbit which is at $p = 0$, $q = q_0$ at $t = 0$. We want to define a scale μ such that the resonances with $|n| \geq \mu$ of Hamiltonian (6.1) have a negligible influence on the particle transport as long as p stays close to zero.

As a first step we estimate the scale μ such that the orbit feels a negligible perturbation due to the resonances of Hamiltonian (6.1) where resonances $|n| < \mu$ are suppressed. For two prescribed positive integers μ and M with

$$M \gg \mu \gg 1 \tag{6.13}$$

let

$$\mathcal{D} = [-M, -\mu] \cup [\mu, M] \qquad \mathcal{D}' = [-\mu + 1, \mu - 1] \tag{6.14}$$

and consider the dynamics defined by the 'non-resonant' Hamiltonian

$$H_{\mathrm{NR}}(p, q, t) = \frac{p^2}{2} + A \sum_{m \in \mathcal{D}} \cos(q - mt + \varphi_m) \tag{6.15}$$

where the φ_m are random phases. We compute, to first order in A, the perturbation of the orbit of interest. This yields

$$p^{(1)}(t) = A \sum_{m \in \mathcal{D}} \frac{1}{m} \cos(q_0 - mt + \varphi_m) \tag{6.16}$$

plus a possible constant we set to zero to have $\langle p^{(1)}(t) \rangle = 0$. As for motion in a single wave, we require (for perturbation theory to apply) that the velocity modulation be smaller than the mismatch μ between the velocities of the particle and the slowest waves (non-resonance). Since many random phases are present in (6.16), the velocity modulation is estimated through the variance of $p^{(1)}$:

$$\langle (p^{(1)}(t))^2 \rangle = A^2 \sum_{m=\mu}^{M} \frac{1}{m^2} \simeq A^2 \int_{m=\mu-1/2}^{M+1/2} \frac{dm}{m^2} \simeq \frac{A^2}{\mu} \qquad (6.17)$$

where condition (6.13) is used in the last estimates. Then the non-resonance condition $\langle (p^{(1)}(t))^2 \rangle \ll \mu^2$ gives

$$\mu \gg A^{2/3}. \qquad (6.18)$$

The presence of many waves makes this condition more stringent for large A than $\mu \gg A^{1/2}$, which was the one we found for the case of a unique wave in section 1.2. For A large, the collective action of many waves gives them a resonant domain broader than that of a single wave.

As a second step we compare the dynamics of Hamiltonian (6.1) with that of the same Hamiltonian where only the resonances with $|n| < \mu$ are retained. We now want to compare two orbits such that the first one follows the dynamics defined by Hamiltonian (6.1) and the second follows that of the 'strongly resonant' Hamiltonian

$$H_{SR}(p, q, t) = \frac{p^2}{2} + A \sum_{m \in \mathcal{D}'} \cos(q - mt + \varphi_m) \qquad (6.19)$$

where condition (6.13) holds; this Hamiltonian has a narrower set of resonances than Hamiltonian (6.1). We consider orbits such that $q(0) = q_0$, $p(0) = p_0$, where $|p_0| < \eta\mu$ with $\eta \ll 1$, and follow them over a time t_η such that $p(t)$ stays in the same interval, i.e.

$$|p(t)| < \eta\mu \qquad (6.20)$$

for all $0 \leq t \leq t_\eta$, for all choices of the phases φ_m. We compute the orbit $(q(t), p(t))$ of dynamics (6.1) as a perturbation of the orbit $(q^{(SR)}(t), p^{(SR)}(t))$ of dynamics (6.19); the perturbation appears in the Hamiltonian through the same potential as in Hamiltonian (6.15) and

$$\dot{p} - \dot{p}^{(SR)} = A \sum_{m \in \mathcal{D}} \sin(q(t) - mt + \varphi_m)$$

$$+ A \sum_{m \in \mathcal{D}'} (\sin(q(t) - mt + \varphi_m) - \sin(q^{(SR)}(t) - mt + \varphi_m)).$$

$$(6.21)$$

The terms in the first sum oscillate rapidly (as $|\dot{q}| = |p| \ll |m|$ for $m \in \mathcal{D}$) and the terms in the second sum depend on the departure of q from $q^{(SR)}$. We make the following ansatz for $p^{(1)}(t)$:

$$p^{(1)}(t) = A \sum_{m \in \mathcal{D}} \frac{1}{m - p^{(SR)}(t)} \cos(q^{(SR)}(t) - mt + \varphi_m) \qquad (6.22)$$

which would correspond to first-order perturbation theory for Hamiltonian (6.15) for a constant $p^{(SR)}(t)$. Deriving this expression with respect to time we get

$$\dot{p}^{(1)} = A \sum_{m \in \mathcal{D}} \sin(q^{(SR)}(t) - mt + \varphi_m)$$
$$+ A \sum_{m \in \mathcal{D}} \frac{\dot{p}^{(SR)}(t)}{(m - p^{(SR)}(t))^2} \cos(q^{(SR)}(t) - mt + \varphi_m). \qquad (6.23)$$

Our analysis relies on the observation that $q^{(SR)}(t)$ depends only on the initial condition (q_0, p_0) and on the phases $\{\varphi_m\}$ with $m \in \mathcal{D}'$; and is independent of the phases $\{\varphi_m\}$ with $m \in \mathcal{D}$. Thus expectations of functions of $q^{(SR)}(t)$ with respect to the phases $\{\varphi_m\}$ with $m \in \mathcal{D}$ are elementary.

The first term in the right-hand side of (6.23) is the estimate for $\dot{p} - \dot{p}^{(SR)}$ in our perturbation expansion (neglecting the difference between $q(t)$ and $q^{(SR)}(t)$ to this order of calculations). It has a vanishing expectation and its variance is $(M - \mu + 1)A^2$. The second term also has a vanishing expectation and its variance is

$$A^2 |\dot{p}^{(SR)}(t)|^2 \sum_{m \in \mathcal{D}} \frac{1}{2(m - p^{(SR)}(t))^4} \approx \frac{A^2 |\dot{p}^{(SR)}(t)|^2}{3\mu^3} \approx \frac{2}{3} A^4 \mu^{-2} \qquad (6.24)$$

using (6.20) and estimating

$$|\dot{p}^{(SR)}(t)|^2 = \left| A \sum_{m \in \mathcal{D}'} \sin(q^{(SR)}(t) - mt + \varphi_m) \right|^2$$

by $(2\mu - 1)A^2$, as if the position $q^{(SR)}(t)$ were independent of the phases $\{\varphi_m\}$ with[2] $m \in \mathcal{D}'$. Combining condition (6.18) and the first half of condition (6.13) shows that the first term in (6.23) is much larger than the square root of the variance (6.24). It is also much larger than the covariance of the two terms in the right-hand side of (6.23), which is bounded by the geometric mean of the two variances. This justifies ansatz (6.22), and we may write:

$$\dot{p} - \dot{p}^{(SR)} - \dot{p}^{(1)} = A \sum_{m=-M}^{M} (\sin(q(t) - mt + \varphi_m) - \sin(q^{(SR)}(t) - mt + \varphi_m))$$

---

[2] The absolute bound $|\dot{p}^{(SR)}(t)|^2 \leq (2\mu - 1)^2 A^2$ holds but approaching it is highly untypical. According to section F.3, our approximation is reasonable at least for short times.

$$+ \frac{A^2}{2} \sum_{m \in \mathcal{D}} \sum_{n \in \mathcal{D}'} \frac{1}{(m - p^{(SR)}(t))^2}$$

$$\times \sin(2q^{(SR)}(t) - (m + n)t + \varphi_m + \varphi_n)$$

$$+ \frac{A^2}{2} \sum_{m \in \mathcal{D}} \sum_{n \in \mathcal{D}'} \frac{1}{(m - p^{(SR)}(t))^2}$$

$$\times \sin((m - n)t + \varphi_m - \varphi_n). \tag{6.25}$$

The frequency of the (m, n) term in this equation is $m \pm n$. For a given m, it lies in the same range of frequencies as that of Hamiltonian (6.19) for, at most, $2\mu - 1$ values of n. We should not include such terms in the sequel to the perturbation calculation: in order to avoid small denominators, we should perform second-order perturbation theory for the same range of frequencies as we did at first order. Therefore we find that at second order the low-frequency dynamics of Hamiltonian (6.1) is ruled by a supplementary force with respect to that of Hamiltonian (6.19). The variance of this supplementary force is readily estimated to be A^4/μ^2. It is negligible with respect to that of the force defining $\dot{p}^{(SR)}$, μA^2, if condition (6.18) holds. Then it is also negligible with respect to that of the force defining $\dot{p}^{(1)}$, MA^2.

As a result, the calculation to this order suggests that the dynamics of Hamiltonian (6.1) is ruled by almost the same force as that of Hamiltonian (6.19) as long as $|p(t)| < \eta\mu$, i.e. for $0 < t < t_\eta$. This dynamics includes the chaotic transport existing when there is resonance overlap. Therefore, condition (6.18) ensures altogether that the particle is far from resonance with the waves of Hamiltonian (6.15) and that their contribution to beating resonances may be neglected in the dynamics of Hamiltonian (6.1)—at least to second-order perturbation.

For A large, one may choose η such that $\eta\mu > 1$ and our analysis may be naturally extended to any time t_0 at which the velocity p verifies

$$|M - |p(t_0)|| \gg \mu \tag{6.26}$$

which means that the instantaneous velocity is far enough from the extreme phase velocities of the resonances. To this end we introduce a truncated Hamiltonian of the type

$$H_{\text{loc}}(p, q, t | p(t_0)) = \frac{p^2}{2} + A \sum_{|m - p(t_0)| \leq \mu} \cos(q - mt + \varphi_m). \tag{6.27}$$

Then, for times t such that $|p(t) - p(t_0)| < \eta\mu$ with $\eta \ll 1$ for all φ_m, H_{loc} defines almost the same dynamics as Hamiltonian (6.1), provided that condition (6.18) is satisfied. As a result, we find that only the waves close enough to the instantaneous velocity of the particle are active for producing chaotic transport. This property is called *locality* (Cary *et al* 1990) in the following, and we say that Hamiltonian (6.27) with condition (6.18) defines a *reduced dynamics* close to time t_0 of the full dynamics defined by Hamiltonian (6.1).

A further extension of this result can be obtained with the requirement that $|p(t) - p(t_0)| < \eta\mu$ for most φ_m only. The greater the number of excluded φ_m, the more stringent condition (6.18) becomes, but the larger the time interval close to t_0 over which H_{loc} rules the full dynamics. Indeed exceptional cases, e.g. where all φ_m are close to zero, give the dynamics the jumplike behaviour occurring for the standard map (5.59), (5.60), which makes the time interval t_η small for A large.

6.3.2 Numerical check of the concept

Numerical simulations enable one to check the concept of locality (Bénisti and Escande 1998a). Orbits defined by Hamiltonians (6.1) and (6.27) for various choices of μ are computed by the same symplectic algorithm (see appendix D) for 9000 outcomes of the set of random φ_m. Parameter M is chosen such that condition (6.26) is verified through all the calculations. The distribution functions $f(p, t)$ obtained for the two Hamiltonians are compared by the Kolmogorov–Smirnov test. They are found to be close to each other and close to a Gaussian for $A \geq 0.2$ (i.e. $s_{\text{overlap}} \gtrsim 1.8$) and $\alpha \geq 5$ where

$$\alpha = \mu A^{-2/3}. \tag{6.28}$$

At this stage a test, which turns out to be more sensitive than the Kolmogorov–Smirnov test, is the comparison of the variances of the two distribution functions, or that of the corresponding diffusion coefficients D and D'. The numerical error when evaluating D and D' is about 5%. When increasing μ, $D - D'$ becomes also 5% for some value μ_{m} of μ.

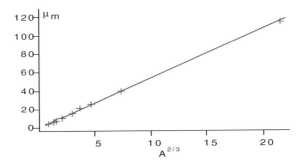

Figure 6.12. μ_{m} versus $A^{2/3}$. The straight line through the origin is a good interpolation of the numerical points. (After Bénisti (1995).)

Figure 6.12 displays μ_{m} versus $A^{2/3}$. The straight line going through the origin with slope

$$\alpha_0 = 5.4 \tag{6.29}$$

is a good interpolation of the numerical points. This yields a numerical clue for the locality property and quantifies how strong the inequality of relation (6.18) which introduced the $A^{-2/3}$ scaling needs to be. This scaling is seen numerically to hold for

$$A \geq 0.8 \qquad (6.30)$$

though diffusion already holds for large times for $A \geq 0.2$.

6.3.3   Rigorous approach*

The simple analytical argument for locality is far from being rigorous since it does not tell us what happens at higher order in perturbation theory. A rigorous result fills in this gap (Bénisti and Escande 1998a):

> *Theorem of phase velocity width compression*:
> For any value of $\Delta v \geq 2^{3/8}e^{-1} \simeq 0.47 \ldots$ and for any $0 < \epsilon \leq 1$, let $A = \epsilon \Delta v^{3/2}$. Then, for any P^* such that $|P^*| \leq M - \Delta v - 1$, there exists a canonical change of variables $(q, p) \mapsto (Q, P)$, which
>
> (i)   is close to the identity except for p close to P^* for a set of phases whose normalized measure can be made arbitrarily close to 0 by decreasing ϵ;
> (ii) transforms Hamiltonian (6.1) into the Hamiltonian
>
> $$H_R(Q, P, t) = p^2/2 + \epsilon K(Q, P, t) + R(Q, P, t, \epsilon) \qquad (6.31)$$
>
> where K contains only (a) resonances with phase velocities v_ϕ such that $|v_\phi - P^*| \leq \Delta v$ and (b) terms which do not depend on Q and which oscillate with an angular frequency ω such that $|\omega| < \Delta v$.
>
> Moreover, the remainder $R(Q, P, \epsilon)$ is such that
>
> $$\sqrt{\langle R^2 \rangle} < 5\epsilon^{(2^{-9/8}\epsilon^{-1/4})}. \qquad (6.32)$$

This theorem implies that, for a finite time, in variables (Q, P) the dynamics defined by (6.1) is similar to the reduced dynamics provided that ϵ is small enough. However, eliminating resonances by perturbation theory yields higher-order resonances. Therefore Hamiltonian H_R contains more than a mere subset of resonances of Hamiltonian (6.1), and cannot be identified with Hamiltonian (6.27).

This theorem may also be applied to Hamiltonian (6.27), defining a Hamiltonian $H_{R\text{loc}}$ with the corresponding K_{loc} and R_{loc}. If (6.26) holds, then for ϵ small enough, A large enough, and

$$C \equiv \mu/\Delta v \qquad (6.33)$$

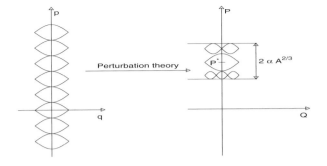

Figure 6.13. Schematic representation of the resonances of Hamiltonians H_R and H_{Rloc}. (After Bénisti (1995).)

large enough, it can be shown (Bénisti and Escande 1998a) that Hamiltonians H_R and H_{Rloc} have sets of resonances which can be made arbitrarily close statistically. This is sketched in figure 6.13 where all resonances have the same phase and appear as trapping domains.

The statistical closeness of H_R and H_{Rloc} is characterized by comparing their n-time correlation functions or the 'power spectra' of their resonances. More precisely, for this second technique we write H_R in the form $H_R(q, p, t) = p^2/2 + \sum_{v_\phi} h(q - v_\phi t, p, v_\phi) \Delta v_\phi$, compute the mean intensity of the component $a^2(v_\phi) = \langle h^2(q - v_\phi t, p, v_\phi) \rangle$ and consider the cumulative function

$$\sigma_R(v) = \int_0^v a^2(v') \, dv'. \tag{6.34}$$

Defining similarly σ_{Rloc}, one estimates the statistical discrepancy between H_R and H_{Rloc} using

$$\Delta(H_R, H_{Rloc}, v) = \left| \frac{d}{dv}(\sigma_R - \sigma_{Rloc}) \right| = |a_R^2 - a_{Rloc}^2|.$$

Then one shows that for all ϵ' there is a (large) value of C such that for all A and for all $M > \mu$, $\Delta(H_R, H_{Rloc}, v) < \epsilon'$. One can even consider the relative intensity of the spectra associated with H_R and H_{Rloc} and define $\Gamma_R(v_\phi) \equiv \Delta(H_R, H_{Rloc}, v)/a_{Rloc}^2$, which also tends to zero in the same limit. As may be expected, the maximum relative discrepancy Γ_R occurs for the extreme phase velocities present in $K(Q, P, t)$.

The conditions on (ϵ, A, C) giving the statistical closeness of K and K_{loc} imply the statistical closeness of the two canonical changes of variables. This justifies that, over a finite time T_{loc}, the full dynamics can be made arbitrarily close statistically to the reduced one provided that conditions $A \gg 1$, (6.18) and (6.26) hold. The finite time T_{loc} is at least the time such that

$$|P(t) - P^*| < 1 - 1/\Delta v \tag{6.35}$$

for all φ_m and all $0 \leq t \leq T_{\text{loc}}$. As $P(t) - p(t)$ can be made arbitrarily small (in a statistical sense) for ϵ small enough, the condition on $p(t)$ is almost the same in this limit.

For larger times, the statistical closeness of the dynamics defined by H_R and H_{Rloc} can also be approximately established, by splitting the time interval of interest into small enough intervals where the dynamics is close to some P^* in the sense of (6.35). We shall further discuss this splitting in section 6.6. The condition $C \gg 1$ giving the statistical closeness of K and K_{loc} means that the number of primary resonances acting on the particle is large enough. We show later that violating this condition strongly modifies the reduced dynamics.

The $\gg$ signs in (6.18) and in $A \gg 1$ contribute to ensuring sufficient conditions for the statistical closeness of the full and reduced dynamics. Actually, they are seen to be far too stringent by the numerical test of locality in section 6.3.2, which suggests replacing them with conditions $\alpha \geq \alpha_0 \approx 5.4$ and (6.30), where α is defined by (6.28).

This rather small value of α_0 can be substantiated analytically as follows. The maximum of the relative discrepancy $\Gamma_R(v_\phi)$ between resonance spectra is computed to second order in ϵ, as well as $\langle (P(t) - p(t))^2 \rangle$. This yields $C(\epsilon)$ (see section 5 of Bénisti and Escande (1998a)). In principle, ϵ should be defined by assessing the maximum size of the two R which perturbs the diffusion coefficient by less than 5% and by an estimate of $R(\epsilon)$. As this is not yet feasible, ϵ is evaluated by requiring the change of variables $(q, p) \mapsto (Q, P)$ to be one-to-one for most phase realizations through condition $\langle (\partial^2 \Phi / \partial q \partial P)^2 \rangle \leq 1$, where $\Phi(q, P)$ is the generating function. Then $\mu(\Gamma)$ can be computed and turns out to have a shape similar to the same function computed numerically. $\Gamma_R = 5\%$ yields the estimate $\alpha = 4.2$ where α is defined by (6.28); this is close to the $\alpha = 5.4$ of (6.29). This reasoning implicitly identified Γ_R to the relative error on diffusion coefficients, which is reasonable in the quasilinear case but still needs to be verified in the more general ones.

6.4   Scaling properties of finite-time dynamics

The *resonance box* is the name given to the set of neighbouring resonances of Hamiltonian (6.27) centred at velocity p_0 within an interval of phase velocities $p_0 - \Delta v_{\text{box}} \leq p \leq p_0 + \Delta v_{\text{box}}$, with

$$\Delta v_{\text{box}} = 5.4 A^{2/3} \simeq 4.6 D_{\text{QL}}^{1/3} \tag{6.36}$$

called the resonance half-width. The factor 5.4 comes from (6.29). Δp_R defined by (6.10) is an approximation to Δv_{box} which is correct for intermediate values of A; and it is useful to give a simple physical interpretation to the change to resonance box. As a result $N_b(t)$ as defined by (6.9) is the average number of resonance boxes visited by the set of orbits of interest up to time t. As assumed previously we see that the threshold for large-scale stochasticity is defined by resonances present within a resonance box. It is, therefore, statistically uniform in the domain $[-M + \Delta v_{box}, M - \Delta v_{box}]$.

Let β be a number larger than α_0 and let $\lambda_\beta = 1 + \text{Int}(\beta A^{2/3})$ where $\text{Int}(x)$ stands for the integer part of x. As long as $|p(t) - p(0)| < \lambda_\beta - \Delta v_{box}$, the property of locality implies the statistical closeness of the dynamics defined by Hamiltonians (6.1) and

$$H_\beta(p, q, t) = \frac{p^2}{2} + A \sum_{m=-\lambda_\beta}^{\lambda_\beta} \cos(q - mt + \varphi_m). \tag{6.37}$$

Using new variables

$$p_1 = A^{-2/3} p \tag{6.38}$$

$$\tau = A^{2/3} t \tag{6.39}$$

the dynamics defined by (6.37) is ruled by the Hamiltonian

$$H'_\beta(p_1, q, \tau) = \frac{p_1^2}{2} + C(\tau) \cos q + S(\tau) \sin q \tag{6.40}$$

where

$$C(\tau) + iS(\tau) = A^{-1/3} \sum_{m=-\lambda_\beta}^{\lambda_\beta} \exp(im\tau A^{-2/3} - i\varphi_m). \tag{6.41}$$

C and S are stochastic processes which are shown in section F.1 to have statistical properties which do not depend on A for A large. The origin of this fact can be understood by computing the two-time correlation function

$$\langle C(\tau_1)C(\tau_2) \rangle = \tfrac{1}{2} A^{-2/3} \sum_{m=-\lambda_\beta}^{\lambda_\beta} \cos[m(\tau_1 - \tau_2)A^{-2/3}]. \tag{6.42}$$

This is a Riemann sum which may be turned into an integral provided that

$$\tau \leq t_u A^{2/3} \tag{6.43}$$

where $t_u \ll 1$ is a constant independent of A. In this limit, (6.42) becomes

$$\langle C(\tau_1)C(\tau_2) \rangle = \frac{1}{2} \int_{-\beta}^{\beta} \cos[(\tau_1 - \tau_2)x] \, dx = \frac{\sin[\beta(\tau_1 - \tau_2)]}{\tau_1 - \tau_2} \tag{6.44}$$

where the final expression is assumed to take on the limiting value β if $\tau_1 = \tau_2$. This expression is independent of A. The n-time correlation function is dealt with in the same way and is also found to be independent of A. Then one may formally write

$$C(\tau) = \int_{-\beta}^{\beta} \cos[\tau x - \varphi(x)] \, dx \tag{6.45}$$

where $\varphi(x)$ is a white noise with a uniform probability distribution over the circle. The same approach enables us to show that $S(\tau)$ verifies a similar equation with a sine replacing the cosine.

Let C_0, S_0 and H_β'' be the limits of C, S and H_β' for $A \to \infty$. When condition (6.43) is fulfilled, H_β' has the same statistical properties as H_β''. In particular, if the force ($\simeq dp_1/d\tau$) of Hamiltonian H_β'' decorrelates at time τ_c, so does that of Hamiltonian H_β' provided that A is large enough for τ_c to verify condition (6.43). This implies that, for A large enough, the force defined by Hamiltonian H_β decorrelates after a time shorter than t_u and scaling like $A^{-2/3}$. Then two diffusion-like (diffusion is not properly guaranteed: we focus on the second moment) coefficients D may be defined by (6.11) with $t_0 = 0$ for the dynamics defined by Hamiltonians H_β' and H_β. The first one (D_β') is independent of A and the second one (D_β) scales like A^2 as a result of (6.38), (6.39). This shows the existence of scaling properties, for A large, of the finite-time dynamics of Hamiltonian (6.1). Indeed we prove in section 6.8.2 that, for A large, the dynamics is diffusive over a time $\tau_{QL} \sim A^{-2/3} \ln A$, with a quasilinear diffusion coefficient scaling like A^2.

6.5   Origin of the force decorrelation

6.5.1   Locality and resonance boxes for random phases

We must distinguish between two decorrelation regimes. The force decorrelates in the initial quasilinear regime as a mere consequence of the width $2M$ of the frequency spectrum. However, figure 6.6 shows the existence of a second decorrelation regime for the force, which translates into the existence of the 'diffusion coefficient' computed through (6.11). This decorrelation is less obvious, as it may occur at time-scales beyond the initial quasilinear one and its coefficient D may differ from the initial one. Observationally, it is monitored by the relaxation of $\langle \Delta p^2(t) - C \rangle / (2Dt)$ toward 1 (with C and D estimated by a best fit). Figure 6.14 displays this quantity versus $N_b(t)$, defined by (6.9), for $A \simeq 1.5$, and shows that the second decorrelation occurs after visiting a little less than the width of one resonance box. This holds over the whole range $0.2 < A < 10$. For higher values of A, there appears to be a unique diffusion regime and the second decorrelation, though enabling diffusion, does not appear in the graph of $\langle \Delta p^2(t) \rangle$.

The concept of locality gives a simple rationale for the force to decorrelate after the visit of one resonance box (i.e. after p has shifted by the width $2\Delta v_{box}$ of one resonance box). Indeed, since the reduced dynamics defined by Hamiltonian (6.27) with condition (6.18) gives the diffusion coefficient of the full dynamics, the corresponding 'reduced force' (defined by this Hamiltonian) has the statistical properties of the 'full force' (of the full dynamics) which are relevant to transport. When the motion of a particle shifts its velocity by an amount equal to the width

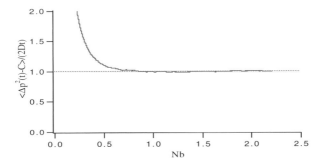

Figure 6.14. $\langle \Delta p^2(t) - C \rangle / (2Dt)$ versus $N_b(t)$ for $A = 15/\pi^2 \simeq 1.5$. (After Bénisti (1995).)

of a resonance box, then the reduced force acting on this particle contains none of the phases which were present in the reduced force before the shift. These phases are present in $q(t)$ though. However, as the impact of a single phase on the motion is weak over the time necessary to visit a resonance box, and as any two phases are statistically independent, these two forces are decorrelated.

This rationale is not sufficient to show that no recorrelation occurs when the particle velocity comes back to its first value. To this end, we must show that the particle spends enough time out of the resonance box centred at $p(t_0)$ for its position to be scattered by an amount larger than 2π when $p(t)$ comes back to $p(t_0)$. This may be checked by the following approximate calculation. If the motion is ruled by a diffusion in p with coefficient D, it will be shown by (6.78) that the shift $\Delta q(t)$ of q with respect to the ballistic orbit verifies $\langle \Delta q(t)^2 \rangle = \frac{2}{3} Dt^3$. Therefore the variances in p and q occurring over a finite time Δt are related through

$$2D\Delta t = \langle \Delta p^2(\Delta t) \rangle = (12D^2 \langle \Delta q^2(\Delta t) \rangle)^{1/3}. \tag{6.46}$$

With the definition of Δv_{box}, for A large it is natural to define $N_b(t)$, the number of resonance boxes visited during time t, through (6.9) where Δp_R is substituted with Δv_{box}. A random-walk argument given in section F.2 enables one to estimate $N_b(t)$ as

$$N_b(t) = \sqrt{\frac{2\langle \Delta p^2(t) \rangle}{\pi \Delta v_{\text{box}}^2}} \tag{6.47}$$

where Δv_{box} is defined by (6.36). This estimate and (6.46) yield $N_b \simeq 0.5$ for $\langle \Delta q^2(\Delta t) \rangle = 4\pi^2$ if one approximates D by its quasilinear value. As a result, after the time necessary for the orbit to visit one resonance box, the position is scattered enough for the force to be decorrelated if the particle velocity is back to its original value.

This result suggests that the growth of the variance of q might be the basic ingredient for the force decorrelation. Yet this is not true. Numerical simulations with an 'over-reduced dynamics', where $\alpha = 3$, show that this variance grows faster than with the reduced one but the force decorrelation occurs later (Bénisti and Escande 1997). This happens because the narrower spectrum gives more coherence to the dynamics. Quasi-trapping is observed, which increases the influence of one phase on the orbit when it is inside one narrow resonance box. Therefore, the previous decorrelation of the force due to the change of resonance box no longer occurs. Decreasing α further suppresses the force decorrelation even though chaotic unbounded motion exists. Thus the change of resonance box is a leading factor to provide the force decorrelation. The large width of the frequency spectrum in Hamiltonian (6.1) and the growth of the variance of q are secondary but necessary elements[3]. This will be confirmed by the fact that the characteristic time τ_{box} to visit a resonance box is about ten times the characteristic time τ_{spread} for the standard deviation of the position to become 2π, according to (6.119).

We now know that the force of the full dynamics decorrelates after a finite time corresponding to the visit of one resonance box. The existence of the rescaled dynamics defined by Hamiltonian (6.40) implies that this time scales like $A^{-2/3}$. When A increases, this decorrelation time decreases faster than the time $\tau_{QL} \sim A^{-2/3} \ln A$ of duration of the initial quasilinear regime which will be defined by (6.92) in section 6.8.2. Therefore, a cross-over occurs between these two regimes for A large enough and the chaotic motion is then characterized by a quasilinear diffusion coefficient. This explains the convergence of the numerical diffusion coefficient toward the quasilinear value when A grows, and the cross-over between the initial and the final diffusion regimes evidenced by the Kolmogorov–Smirnov test in section 6.2. The presence of the logarithm in τ_{QL} makes this convergence slow though. Another important consequence of the force decorrelation after visiting one resonance box is that, when the velocity distribution spreads by more than one resonance box width, there is no way for it to reduce its width later on.

6.5.2 Correlated phases

If the φ_m are correlated over a finite range, when A increases the diffusion coefficient may not converge toward its quasilinear value. Indeed section 6.8.2 will evaluate τ_{QL} by using the weak effect of one phase on the initial dynamics and the independence of the effects of two different phases. Therefore, when the correlation between neighbouring phases increases, τ_{QL} is bound to decrease since the independence of the phases is a key argument in finding the logarithmic correction to the simple $A^{-2/3}$ scaling. Then the cross-over with the chaotic correlation time scaling like $A^{-2/3}$ is no longer possible. This scaling still holds as the previous argument leading to the resonance width can be modified to account

---

[3] This argument for the force decorrelation uses the concept of chaos in a weak sense: the velocity is considered to wander through the whole interval $[-M, M]$; in particular, the exponential divergence of nearby orbits is not used.

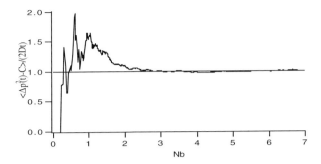

Figure 6.15. $\langle \Delta p^2(t) - C \rangle / (2Dt)$ versus $N_b(t)$ for $A = 5$. (After Bénisti (1995).)

for the finite range in m of the correlation between the phases. We can thus expect a chaotic diffusion to occur for A large, and its coefficient to scale like A^2. Indeed Pesme *et al* (1993) find the diffusion coefficient to be half of the quasilinear one for the correlation of phases they consider.

6.5.3 Random initial positions

In the case of random initial positions, Bénisti (1995) and Bénisti and Escande (1997, 1998b) have shown the existence of a chaotic 'diffusion coefficient'. Figure 6.15 corresponds to figure 6.14 in this case. The force decorrelation is seen to occur after the visit of two resonance boxes instead of one in the case of random phases. Among the necessary ingredients for this decorrelation, the case with random initial positions still has the broad spectral width of the force and the incoherent growth of $\langle \Delta q^2(\Delta t) \rangle$. As the estimates of section 6.3 also hold in the case of random initial positions, the concepts of locality and of resonance box are also relevant in this case. However, a mere average over initial positions is not sufficient to decorrelate the force when the velocity is shifted by $2\Delta v_{\text{box}}$ as in the case of random phases. As the decorrelation occurs later, we may conclude that the change of resonance box is the leading factor for the decorrelation of the force in the case of random phases. Furthermore, the chaotic decorrelation time for the standard map is seen numerically to decrease more slowly than $A^{-2/3}$ (Bénisti and Escande 1998b). As this map may be considered as a limiting case of (6.1), this is one more fact showing the importance of the statistics over phases in obtaining a fast decrease in correlations in the chaotic regime.

6.6 Origin of the chaotic diffusion

The concept of resonance box also provides a rationale for the origin of diffusion. Let us split the time interval $[0, t]$ into subintervals $[t_i, t_{i+1}]$, sufficiently short for $\Delta p_i = p(t_{i+1}) - p(t_i)$ to be smaller than Δv_{box} for most phases. Let B_i be the resonance box constructed about $p(t_i)$. A straightforward extension of the previous reasoning about the decorrelation due to a change of box yields that Δp_i is independent of any Δp_j such that B_i and B_j do not intersect. Therefore $\Delta p(t)$ may be viewed as the sum of variables which are correlated only over a finite range. When the number of resonance boxes visited by the orbit increases, the central limit theorem applies and the distribution function of $\Delta p(t)$ tends to a Gaussian. Returning into an already visited resonance box also typically brings an uncorrelated increment to p, as the particle has spent enough time out of this box for the distribution of q to be scattered enough. A similar reasoning allows one to show the Gaussianity of q during the same time interval.

When t grows, so does the number of visits an orbit makes to a given resonance box. Then it is important to make sure that the correlation of the force within a resonance box decays fast enough with time in order to prove Gaussianity for long times. Calculating this function brings in terms like $\langle\cos[q_0(t_1) \pm q_0(t_2) - m_1 t_1 \mp m_2 t_2 + \varphi_{m_1} \pm \varphi_{m_2}]\rangle$ whose absolute value may be bounded by $\langle\cos[q_0(t_1) \pm q_0(t_2)]\rangle$ since $\varphi_{m_1} \pm \varphi_{m_2}$ adds randomness to the argument of the cosine. Assuming that $q_0(t_1) \pm q_0(t_2)$ is a Gaussian variable, one can show (equation (F.26)) that $\langle\cos[q_0(t_1) \pm q_0(t_2)]\rangle$ decreases faster than exponentially. Therefore the convergence of $q_0(t_1) \pm q_0(t_2)$ toward Gaussianity for moderate values of $t_1 - t_2$ due to the changes of resonance box produces a fast decrease in correlations which, in turn, yields a fast convergence toward Gaussianity of $q_0(t_1) \pm q_0(t_2)$ for longer time intervals.

Let us now estimate the time after which the distribution of $p(t)$ is practically Gaussian. Equation (6.47) shows that the number of resonance boxes visited by the initial set of chaotic orbits grows like the square root of time. Thus, since the force decorrelates for $N_b = 1$, then for $N_b = 2$ four correlation times have elapsed, and Δp has undergone four independent increments, a number which, practically speaking, already yields a nearly Gaussian shape to the distribution function of the sum. Therefore, we may expect $N_b = 2$ to be the typical number of resonance boxes to visit for Gaussianity to set in. This agrees with the value $N_b \sim 1.5\text{--}2$ which may be deduced from figures 6.2 and 6.8 for Gaussianity to appear again for intermediate values of A in numerical calculations of the orbits. For large values of A, the cross-over between the initial quasilinear regime and the chaotic one hides the contribution of the change in resonance box to Gaussianity.

The lack of Gaussianity in the case of random initial positions (Bénisti and Escande 1998b) shows that the decorrelation due to the change of resonance box is crucial in the case of random phases. In plasma experiments, quantities are often averaged over time and the plasma is continuously regenerated. This makes likely the existence of an average over phases and therefore the occurrence of diffusion. As shown in chapter 7, this prejudice is weakened by the existence of self-consistency between waves and particles when the evolution time of the waves is comparable to the time of spreading for the particles.

6.7   Locality and diffusion for more general Hamiltonians

The locality property may be extended to more general Hamiltonians than (6.1), which include the class describing the motion of a particle in a prescribed set of Langmuir waves or van Kampen-like eigenmodes. They are of the type

$$H(p, q, t) = \frac{p^2}{2} + \sum_{m=1}^{M} A_m \cos(k_m q - \omega_m t + \varphi_m) \qquad (6.48)$$

where the φ_m are random variables and the (A_m, k_m, ω_m) are prescribed triplets of parameters. A rigorous result like the theorem of phase velocity width compression is not yet available for this case but a heuristic argument yields an upper bound to the width of the resonance box (Bénisti and Escande 1998a). If all A_m scale like A, this width scales again like $A^{2/3}$. However, the heuristic argument needs the introduction of two small parameters: one reminiscent of parameter ϵ in the phase velocity width compression theorem; and one which involves the amplitudes of resonances out of the resonance box. The latter shows that one may fail to recover the scaling in $A^{2/3}$, depending on the structure of the Hamiltonian.

For the application to Langmuir waves, it is interesting to specify Hamiltonian (6.48) to the case where A_m, k_m and ω_m may be considered as slowly varying smooth functions of the real parameter m. Then the phase velocity of resonance m of Hamiltonian (6.48) is separated from those of resonances $m \pm 1$ by

$$\Delta v_m = \frac{1}{k_m} \frac{\partial \omega_m}{\partial m} - \frac{\omega_m}{k_m^2} \frac{\partial k_m}{\partial m} = \frac{v_{gm} - v_{\phi m}}{k_m} \frac{\partial k}{\partial m}. \qquad (6.49)$$

Consider the dynamics close to time t_0 of a particle whose velocity $p(t_0)$ is closest to the phase velocity of resonance l of Hamiltonian (6.48). Assume that, for m close to l, A_m, k_m and $\Delta w_{ml} \equiv (\omega_m/k_m) - (\omega_l/k_l)$ vary on a scale $\lambda \gg 1$. Consider the Hamiltonian obtained from (6.48) by restricting the sum to the interval $[l - \mu, l + \mu]$, where μ is an integer satisfying $0 < \mu \ll \lambda$. Then we approximate A_m and k_m by constants A_* and k_*, and Δw_{ml} by its lowest-order Taylor expansion $(m - l)\Delta v_l$. Let $A = A_*/\Delta v_l^2$. Rescaling position, time and momentum, so that $k_l = 1$ and $\Delta v_l = 1$, yields the Hamiltonian

$$H'(p', q', t') = \frac{p'^2}{2} + A \sum_{-\mu}^{\mu} \cos(q' - m't' + \varphi_{m'+l}). \qquad (6.50)$$

Exercise 6.2. Estimate the error made in approximating the equations of motion generated by (6.48) by the dynamics of (6.50), using expansions $k_m = k_*(1 + \kappa(m - l)/\lambda + \cdots)$, $A_m = A_*(1 + \alpha(m - l)/\lambda + \cdots)$, etc for $\lambda \gg 1$.

As this Hamiltonian is similar to (6.1), locality also holds for its dynamics. Let Δv_{box} be the resonance half-width, as defined by (6.36). If $\Delta v_{\text{box}} \ll \lambda$ we may take $\mu = \Delta v_{\text{box}}$, Hamiltonian (6.50) defines the reduced dynamics of (6.48) for t

close to t_0, and locality also holds for Hamiltonian (6.48). Since the coefficients of this Hamiltonian are slowly varying, the diffusion coefficient may be defined locally in p by including only the contribution of the instantaneous resonance box.

We now define a quasilinear diffusion coefficient for Hamiltonian (6.48). Let

$$v_m = \frac{\omega_m}{k_m} \tag{6.51}$$

$$\Delta v_m = v_{m+1} - v_m \tag{6.52}$$

and let p_0 be a velocity in the range of the v_m. Let

$$\Omega_m = k_m p_0 - \omega_m \tag{6.53}$$

$$\Delta \Omega_m = \Omega_{m+1} - \Omega_m. \tag{6.54}$$

We assume that $\Delta \Omega_m$ and Δv_m have a sign independent of m, which is natural for Langmuir waves and for the dynamics of Hamiltonian (6.1). In contrast with v_m, Δv_m, k_m and ω_m, note that Ω_m and $\Delta \Omega_m$ also depend on p_0. Let

$$D_m = \frac{\pi A_m^2 k_m}{2|\Delta v_m|}. \tag{6.55}$$

D_m and Δv_m may fluctuate with m but we assume that for some $L \geq 0$ (which should not be confused with a possible spatial length of the system)

$$\bar{D}_m \equiv \sum_{j=-L}^{L} \frac{D_{m+j}|\Delta v_{m+j}|}{|v_{m+L+1} - v_{m-L}|} \tag{6.56}$$

and

$$\overline{\Delta v_m} \equiv \frac{v_{m+L+1} - v_{m-L}}{2L+1} \tag{6.57}$$

are slowly varying functions of p_0; $L = 0$ corresponds to non-fluctuating D_m and Δv_m. $\bar{D}_m$ may be considered as the value at v_m of a smooth function $D_{QL}(p)$ interpolating the $\bar{D}_m$:

$$D_{QL}(v_m) \equiv \bar{D}_m. \tag{6.58}$$

$D_{QL}(v)$ is called the quasilinear coefficient related to velocity p. This definition coincides with (6.12) for Hamiltonian (6.1). Later the related quantities

$$\mathcal{D}(v_m, p_0) \equiv \frac{\pi A_m^2 k_m^2}{2|\Delta \Omega_m|} \tag{6.59}$$

and

$$\bar{\mathcal{D}}(v_m, p_0) \equiv \sum_{j=-L}^{L} \frac{D_{m+j}|\Delta \Omega_{m+j}|}{|\Omega_{m+L+1} - \Omega_{m-L}|} \tag{6.60}$$

will be useful.

Exercise 6.3. Show that, for $v_m \simeq p_0$, D_m is close to $\mathcal{D}(v_m, p_0)$. How much do (6.55) and (6.59) differ if $v_m \neq p_0$? Similarly, estimate and discuss the difference between the velocity-average (6.56) and the relative-frequency-average $\bar{\mathcal{D}}(v_m, p_0)$ for

(i) m such that $|v_m - p_0|$ is minimum and
(ii) arbitrary m and p_0.

Hamiltonian (6.48) has been studied in the literature with more general parameters than those of Hamiltonian (6.1); and, for large A_m, the diffusion coefficient was numerically found to take on the quasilinear value (Rechester *et al* 1979, Cary *et al* 1990, Ragot 1998). In the remainder of this chapter we prove this to be true[4]. The average over $M \gg 1$ random phases is central to our proof. Such an average is natural when noting that the transport in (6.48) is much less diffusion-like if one averages only over initial conditions (p_0, q_0) (Bénisti and Escande 1998b). The large-A limit (dynamically speaking, the limit of strong resonance overlap) corresponds to the limit of a continuous spectrum often encountered in physics.

An easy way to characterize our scalings is to introduce the characteristic frequency, called the resonance-broadening frequency by Dupree (1966),

$$\gamma_{Dm} \equiv \tau_{Dm}^{-1} \equiv k_m^{2/3} D_m^{1/3}. \tag{6.61}$$

We shall see by (6.79) that it is of the order of the reciprocal of the time-scale over which particles initially at the same position become spread over a whole period $2\pi/k_n$ of wave n.

It is convenient to compare this frequency with the frequency of wave $m + 1$ in the frame comoving with wave m. The latter frequency defines the discretization time

$$\tau_{\text{discr}m} \equiv |k_m \overline{\Delta v_m}|^{-1}. \tag{6.62}$$

These frequencies and characteristic times enable one to define a dimensionless ratio

$$\mathcal{B}_m \equiv \frac{k_m |\overline{\Delta v_m}|}{\gamma_{Dm}} = \frac{\tau_{Dm}}{\tau_{\text{discr}m}}. \tag{6.63}$$

As these quantities are smooth functions of v_m, we associate them with interpolating functions $\gamma_D(p)$, $\tau_D(p)$, $\tau_{\text{discr}}(p)$ and $\mathcal{B}(p)$. Let $\mathcal{B}$ be the typical size of $\mathcal{B}(p)$. $\mathcal{B}^{-1}$ gives the typical number of resonances in a resonance box.

The small-$\mathcal{B}$ limit can be viewed as the limit of dense wave spectrum $\Delta v \to 0$ or as the large-amplitude limit for the waves. We measure these amplitudes by the parameter

$$E = [2D_{\text{QL}}/(\pi \tau_{\text{discr}})]^{1/2} \tag{6.64}$$

[4] Pesme and Brisset (private communication 1981, Pesme and DuBois 1982, Pesme 1994) assert to have proved a quasilinear diffusion equation for a spatially averaged velocity distribution function using a diagrammatic technique.

which corresponds to the typical electric field $k_m A_m$ of a wave. We define the overlap parameter between waves m and $m + 1$ by

$$s_{ov}(v_m) \equiv 2(A_{m+1}^{1/2} + A_m^{1/2})/\Delta v_m. \tag{6.65}$$

Exercise 6.4. For a smooth wave spectrum, show that the ratio (6.63) is related to $s_{ov}(v_m)$ by

$$\mathcal{B}_m = 8\pi^{-1/3} s_{ov}(v_m)^{-4/3}. \tag{6.66}$$

(i) Evaluate the typical value s of $s_{ov}(v_m)$ for $\mathcal{B} = 7$ and evaluate $\mathcal{B}$ for $s = 2.5$. To what values of A for Hamiltonian (6.1) do these values correspond?

(ii) What is the value of E for Hamiltonian (6.1)? Estimate the typical values of s and $\mathcal{B}$ in terms of E, k_n, τ_{discr} and M.

(iii) Show that if one requires that D_m, $s_{ov}(v_m)$ and γ_{Dm} be all independent from m, Hamiltonian (6.48) reduces to (6.1).

In the following, we are interested in the dense-spectrum, i.e. the strong-resonance-overlap, limit. To ensure a genuine scaling, we consider families of dynamics (6.48) where $A_n = a_n E$ and the reference amplitudes a_n are constant while $E \to \infty$, or $\mathcal{B}(v_n) = b_n \mathcal{B}$ and the coefficients b_n are constant and O(1) while $\mathcal{B} \to 0$. Similarly, we characterize an interpolating function like $\gamma_D(p)$ by a representative value γ_D while we shall keep the shape of the function invariant in the limit $\mathcal{B} \to 0$.

In agreement with most of the literature on quasilinear transport, the analysis is performed here in terms of quadratic means and not of probability distribution functions but we indicate at the end of section 6.9 how our technique could be used to prove the Gaussianity of such functions.

6.8 Initial quasilinear diffusion

We define the correlation time of the wave spectrum as seen by a particle with velocity p_0:

$$\tau_c = (\Omega_{max} - \Omega_{min})^{-1} \tag{6.67}$$

where Ω_{max} and Ω_{min} are the maximum and the minimum value of Ω_m as defined by (6.53). Note that $M\tau_c = \tau_{discrm}$ for (6.1). Let $k_{max} = \max_m k_m$. Apart from the dimensionless ratio $\mathcal{B}$, our model is characterized by a Kubo number $\mathcal{K}_c$, which we define by (6.80).

Whatever $A \neq 0$ is, the initial dynamics displays the same features. First, on a time short with respect to τ_c, all sines in (6.3) may be considered as constant and p evolves linearly with time (this is hidden in the previous Kolmogorov–Smirnov plots because of insufficient resolution at small N_b). This 'constant acceleration regime' lasts a time $\sim \tau_c$ for any choice of phases $\{\varphi_m\}$.

Exercise 6.5. What is the probability distribution of the initial acceleration $\dot{p}(0)$ if the phases φ_m are uniformly distributed on $[0, 2\pi]$ and independent? For $0 \leq t \lesssim \tau_c$, assume that the acceleration is constant over $[0, t]$ and let $\Delta p(t) = p(t) - p(0)$. For $n \geq 0$ and $n' \geq 0$, estimate $\langle p(t)^n q(t)^{n'} \rangle$ and $\langle e^{iup(t)} \rangle$. For $0 \leq t' \leq t$, estimate $\langle \Delta p(t') \Delta p(t) \rangle$. What is the probability distribution of $p(-t)$ (in the 'past')? Is the scaling $\tau_c \sim M^{-1}$ important?

When t becomes large enough with respect to τ_c, a diffusive regime sets in. We now show that it is due to the broad spectrum of frequencies in (6.3) and to the weak dependence of the orbit on every single phase. As seen numerically, the corresponding diffusion coefficient has the so-called quasilinear value. For later time, the nonlinear character of the dynamics becomes dominant and the subsequent motion depends on A. Diffusion is recovered only for A large enough as a consequence of the chaotic motion.

We focus on the orbit being at (p_0, q_0) at $t = 0$, and we first consider t to be small enough for the orbit to stay close to the unperturbed orbit

$$q^{(0)}(t) = q_0 + p_0 t. \tag{6.68}$$

Let

$$\Delta q(t) = q(t) - q^{(0)}(t). \tag{6.69}$$

We evaluate

$$\Delta p(t) = p(t) - p_0 \tag{6.70}$$

and compute its statistical properties when averaging over all φ_m.

The following subsections estimate $\langle \Delta p(t)^2 \rangle$ by perturbation expansion with respect to the wave amplitudes. A similar calculation was made in section 4.1.4, under different assumptions. There the wave amplitudes were small, the resulting deviations from ballistic motion were small and the average was performed over the particles' initial positions. Here, the wave amplitudes may be large (this is a minor difference, as we compute only over short times) and the average is performed over the waves' phases. Elaborating this argument will enable us to extend the estimates to much longer times and more complicated cases later on.

6.8.1 Non-chaotic quasilinear diffusion

First-order perturbation expansion in the wave amplitudes yields

$$\Delta p(t) = \sum_{m=1}^{M} \frac{A_m k_m}{\Omega_m} [\cos(k_m q_0 + \varphi_m) - \cos(\Omega_m t + k_m q_0 + \varphi_m)] \tag{6.71}$$

with Ω_m defined by (6.53). If $\Omega_m = 0$ for some m, one replaces the corresponding term in (6.71) by its limit for $\Omega_m \to 0$. At this order

$$\langle \Delta p(t)^2 \rangle = \sum_{m=1}^{M} \left(\frac{A_m k_m}{\Omega_m} \right)^2 [1 - \cos(\Omega_m t)]. \tag{6.72}$$

Let $\Delta\Omega_{LM} = \max_m |\Omega_{m+L+1} - \Omega_{m-L}|$ and

$$\tau_{\text{discr}}(p_0) = \Delta\Omega_{LM}^{-1} \simeq (2L+1)^{-1}\tau_{\text{discr}m} \qquad (6.73)$$

be the discretization time of the wave spectrum as seen by the particle with velocity p_0. For $t \ll \tau_c$, Δp grows linearly with time, and $\langle \Delta p^2 \rangle$ grows quadratically, as all modes act with a constant force on the orbit.

We now assume

$$t \ll \tau_{\text{discr}} \qquad (6.74)$$

$$t \gg \tau_c. \qquad (6.75)$$

Because of the first inequality the sum in (6.72) may be turned into an integral and because of the second one the bounds for integration may be considered as infinite, which, using (B.15), yields

$$\langle (\Delta p(t))^2 \rangle = \frac{2}{\pi} \int_{-\infty}^{\infty} D\left(p_0 - \frac{\Omega}{k}, p_0\right) \frac{1 - \cos(\Omega t)}{\Omega^2} \, d\Omega$$

$$\simeq \frac{2t}{\pi} \int_{-\infty}^{\infty} D\left(p_0 - \frac{\alpha}{kt}, p_0\right) \frac{1 - \cos\alpha}{\alpha^2} \, d\alpha$$

$$\simeq 2t\bar{D}(p_0, p_0) = 2D_{\text{QL}}(p_0)t \qquad (6.76)$$

where $D(p, p_0)$ is given by (6.59) and $\bar{D}(p, p_0)$ by (6.60). As a result $\langle (\Delta p(t))^2 \rangle$ has a diffusion-like behaviour and the diffusion coefficient takes on the quasilinear value. The adjective *quasilinear* used to characterize this diffusion coefficient is related to the fact that $\Delta p(t)$ is computed by a linear calculation in this estimate. The original reason is given in section 7.7.

Exercise 6.6. Estimate the errors in (6.76) due to the approximations. Show that, with the same accuracy, $\langle \Delta p(t) \rangle = O(ct)$ for some $c > 0$, so that $\langle \Delta p(t) \rangle^2 \ll \langle (\Delta p(t))^2 \rangle$ in this initial regime. Estimate higher moments. See also section 6.10.

The two-time correlation function of Δp is similarly estimated as

$$\langle \Delta p(t_1)\Delta p(t_2) \rangle \simeq 2D_{\text{QL}} \min(t_1, t_2). \qquad (6.77)$$

A similar calculation for q yields $\langle \Delta q(t) \rangle = 0$ and, using (6.77),

$$\langle (\Delta q(t))^2 \rangle = \int_0^t \int_0^t \langle \Delta p(t_1)\Delta p(t_2) \rangle \, dt_1 \, dt_2 = \tfrac{2}{3}D_{\text{QL}}t^3. \qquad (6.78)$$

In (6.76) we notice that the quasilinear estimate stems from an integral over α where t plays a negligible role. For small values of α, $(1 - \cos\alpha)/\alpha^2$ is close to $1/2$; for large values of α, $(1 - \cos\alpha)/\alpha^2$ decreases like $1/\alpha^2$ and gives a small contribution to the integral. Therefore most of the contribution to the integral comes from the range $|\alpha| \lesssim 1$. As a result the range of m contributing to the diffusion corresponds to modes acting with a nearly constant force over time t. When t increases this range shrinks like $1/t$. Such a description makes sense for $\tau_c \ll t \ll \tau_{discr}$. In this time range the linear growth of $\langle \Delta p(t)^2 \rangle$ in (6.76) can be understood as due to the sum of $O(1/t)$ squared contributions of independent constant forces to the change in the particle velocity: $\langle \Delta p(t)^2 \rangle \sim (1/t)t^2$. Thus the non-chaotic quasilinear diffusion is the result of the cooperative action of a decreasing number of waves when time increases. It is exactly the opposite for chaos, as the resonance box concept suggests, and as the analytical computation of the corresponding diffusion coefficient will confirm in section 6.9.

The range of t is further restricted by the condition for the orbit to remain close to the unperturbed one. This is traditionally translated into condition $\langle k_{max}^2 \Delta q^2(t) \rangle \ll 4\pi^2$, namely $t \ll \tau_{spread}$ where τ_{spread} is the spreading time

$$\tau_{spread} = (6\pi^2 k_{max}^{-2} D_{QL}^{-1})^{1/3} \simeq 4(k_{max}^2 D_{QL})^{-1/3} = 4\left(\frac{k_n}{k_{max}}\right)^{2/3} \tau_{Dn}. \quad (6.79)$$

Though k_n and τ_{Dn} depend on n, this dependence is not central to our arguments, as we focus on a dense spectrum limit in which k and γ_D are smooth functions of the wave phase velocity. In other words, for all n the characteristic time (6.79) is of the same order of magnitude, which we denote simply by τ_{spread}. This condition on t and condition (6.75) require

$$\mathcal{K}_c \equiv \tau_c/\tau_{spread} \ll 1 \quad (6.80)$$

where we define the Kubo number $\mathcal{K}_c$, comparing the short correlation time of the waves with the chaotic time-scale.

It is shown in section F.3 that $\Delta p(t)$ is a Brownian motion with a Gaussian distribution function for $t \ll \min(\tau_{discr}, \tau_{spread})$. Therefore $\Delta p(t)$ diffuses over this time domain. If $\tau_{discr} \leq t \ll \tau_{spread}$ the motion is no longer diffusive but quasi-periodic. Finally, we note that if all k_m are distinct, a mere average over q_0 yields (6.72) and thus the quasilinear estimate even if the φ_m are fixed.

6.8.2   Initial quasilinear diffusion up to chaos

In this section, we evaluate $\Delta p(t)$ as did Bénisti and Escande (1997) by formally integrating the equation of motion for p. This yields $\langle \Delta p(t) \rangle = 0$ (to dominant order, see section 6.10) over the range $0 \leq t \ll \tau_{QL}$ defined by (6.92), and $\langle \Delta p^2(t) \rangle = \Delta_0 + \Delta_+ + \Delta_-$, with

$$\Delta_j = -\epsilon_j \int_0^t \int_0^t \sum_{m_1=1}^M \sum_{m_2=1}^M \frac{A_{m_1} k_{m_1} A_{m_2} k_{m_2}}{2} \langle \cos[\Phi_{m_1}(t_1) + \epsilon_j \Phi_{m_2}(t_2)] \rangle \, dt_1 \, dt_2$$

$$(6.81)$$

where

$$\Phi_m(t) = k_m \Delta q(t) + \Omega_m t + k_m q_0 + \varphi_m \quad (6.82)$$

with $\epsilon_\pm = \pm 1$ and $\epsilon_0 = -1$, and under the condition $m_1 \neq m_2$ for $j = -$, and condition $m_1 = m_2$ for $j = 0$. Let $t_- = t_1 - t_2$ and $t_+ = (t_1 + t_2)/2$.

For $t_- \ll \tau_{\text{spread}}$, $\langle \exp[ik_m(\Delta q(t_+ + t_-/2) - \Delta q(t_+ - t_-/2))]\rangle$ may be considered equal to one. Therefore the support in t_- of the integrand in Δ_0 is of the order of τ_c. Since we assumed $\tau_c \ll \tau_{\text{spread}}$ by (6.80), the integration domain in t_- may be restricted to $|t_-| \leq \nu \tau_c$ where ν is a few units. In the limit where

$$\nu \tau_c \ll t \ll \tau_{\text{discr}} \tag{6.83}$$

we obtain

$$\Delta_0 \simeq \sum_{m=1}^{M} \int_0^t 2 D_{\text{QL}}(p_0) \pi^{-1} \int_0^{\nu \tau_c} \langle \cos[\Omega_m t_-]\rangle \Delta \Omega_m \, dt_- \, dt_+$$

$$= 2 D_{\text{QL}}(p_0) \sum_{m=1}^{M} (\pi \Omega_m)^{-1} \langle \sin[\Omega_m \nu \tau_c]\rangle \Delta \Omega_m t = 2 D_{\text{QL}}(p_0) t \tag{6.84}$$

where the discrete sum over m was approximated by an integral.

For $t \ll \tau_{\text{spread}}$ we may approximate $q(t)$ again by its unperturbed value $q^{(0)}(t)$. As this orbit does not depend on the phases, the averaged cosines in (6.81) are zero for $j = \pm$; and so are the $\Delta_\pm$. $\langle \Delta q^2(t)\rangle$ too may be computed by integrating the equation of motion. This involves calculating $\langle \Delta p(t_1) \Delta p(t_2)\rangle$, in the same way as $\langle \Delta p^2(t)\rangle$, and one recovers the estimates (6.77) and (6.78). This provides a way for introducing the condition $t \ll \tau_{\text{spread}}$ without resorting to the traditional perturbative approach and shows that the usual quasilinear diffusion coefficient may be recovered independently by our second approach.

Exercise 6.7. Estimate $\langle \Delta p(t_1) \Delta p(t_2)\rangle$, $\langle \Delta q(t_1) \Delta p(t_2)\rangle$ and $\langle \Delta q(t_1) \Delta q(t_2)\rangle$ for $0 \leq t_1 \leq t_2 \ll \tau_{\text{discr}}$.

In fact our second approach is much more powerful (Escande and Elskens 2001). As pointed out by Bénisti and Escande (1997), $\Delta_\pm$ vanishes provided that the dependence of Δq on any $N_\varphi = 2$ phases with all other phases fixed is weak, a condition far less stringent than the previous condition $N_\varphi = M$ which led to $t \ll \tau_{\text{spread}}$.

We now sketch how to estimate the time τ_{QL} over which Δq depends weakly on any $N_\varphi = 2$ phases with all other phases fixed. The full derivation is given in sections F.4 and F.5. To avoid too many heavy formulae, this is first done for $N_\varphi = 1$ and then extended to $N_\varphi = 2$ afterwards. To estimate the spreading of the orbits as a function of the $N_\varphi = 1$ phase, we study how the orbit which is at (q_0, p_0) at $t = 0$ is modified when the phase φ_n changes from zero to a finite value. Let $(q_{/\!\!/}(t), p_{/\!\!/}(t))$ be the orbit for $\varphi_n = 0$, let $\delta q_n(t) = q(t) - q_{/\!\!/}(t)$ and $\delta p_n(t) = \delta \dot{q}_n(t) = p(t) - p_{/\!\!/}(t)$. We assume t to be small enough so that $k_{\max}|\delta q_n(t)| \ll \pi$. As $\delta q_n(t)$ is small, we linearize the equation of motion in the form

$$\delta \ddot{q}_n(t) = \delta \dot{p}_n(t) \simeq F(t) \delta q_n(t) + A_n k_n (\sin \Psi_n(t) - \sin \Psi_{n0}(t)) \tag{6.85}$$

where we introduce the acceleration gradient

$$F(t) = \sum_{m=1}^{M} k_m^2 A_m \cos \Psi_m(t) \qquad (6.86)$$

with

$$\Psi_m = k_m q_{\tilde{n}}(t) - \omega_m t + \varphi_m \qquad (6.87)$$

and

$$\Psi_{n0} = k_n q_{\tilde{n}}(t) - \omega_n t. \qquad (6.88)$$

Then (6.85) and initial conditions $(\delta q_n(0), \delta p_n(0)) = (0, 0)$ imply

$$\delta q_n(t) = \int_0^t (t - t'') F(t'') \delta q_n(t'') \, dt'' + \delta q_{n0}(t), \qquad (6.89)$$

where

$$\delta q_{n0}(t) = A_n k_n \int_0^t \int_0^{t'} (\sin \Psi_n(t'') - \sin \Psi_{n0}(t'')) \, dt'' \, dt'. \qquad (6.90)$$

In the short-time limit, the dominant term in this expression for δq_n is δq_{n0} but over longer times the first term may self-amplify and overtake the second one. We show in section F.4 that $\langle \delta q_n(t) \rangle \simeq 0$ for $t \ll \tau_{\text{discr}}$ and

$$k_n^2 \langle \delta q_n(t)^2 \rangle \le 0.14 \mathcal{B}_n (e^{4^{1/3} \gamma_{Dn} t} - 1 + 2g(4^{1/3} \gamma_{Dn} t)) \qquad (6.91)$$

where $\mathcal{B}_n$ is given by (6.63) and where $g(x) = e^{-x/2} \cos(x\sqrt{3}/2) - 1$. The resulting estimate for the standard deviation of $\delta q_n(t)$ starts from zero at $t = 0$ and diverges exponentially for $t \to \infty$. Its exponentiation time-scale $\tau_{\text{Lyap}} = 2^{1/3} \tau_{Dn} \sim \tau_{\text{spread}}$ is the reciprocal of the Lyapunov characteristic instability rate. The coefficient in front of the exponential goes to zero as $E \to \infty$, so that the time needed by our upper estimate for $k_n^2 \langle \delta q_n(t)^2 \rangle$ to reach unity is of the order of

$$\tau_{\text{QL}} = \gamma_{Dn}^{-1} \ln(\mathcal{B}^{-1}) \qquad (6.92)$$

with $\mathcal{B}$ a typical value for the $\mathcal{B}_n$. Though this time τ_{QL} goes to zero as $E \to \infty$, it is $O(|\ln \mathcal{B}|)$ times larger than the characteristic time τ_{spread} over which the initial quasilinear approximation was justified in the previous section[5]. The estimate (6.92) is a lower bound for the physically relevant time, because (6.91) is an upper estimate for $\langle \delta q_n(t)^2 \rangle$. However, this upper estimate should be realistic in the limit $E \to \infty$ of interest. In order to satisfy the second of the inequalities (6.83) up to $t = \tau_{\text{QL}}$ one needs $\tau_{\text{QL}} \le \tau_{\text{discr}}$ which is naturally verified for $\mathcal{B} \ll 1$.

---

[5] The time τ_{QL} plays the same role here as the time t_{QL} did in the discussion of numerical simulations in section 6.2. However, their distinct definitions (analytical versus numerical, E large versus intermediate) imply that we use distinct notation for them.

Exercise 6.8. Prove directly that the τ_{Lyap} just introduced is indeed the Lyapunov time by simplifying the second step of the calculation in section F.4 in the following way: substitute $q_{\eta}(t)$ with the orbit being at (q_0, p_0) at $t = 0$ without specifying a special value for the phases; substitute $\delta q_n(t)$ with a small perturbation of this orbit which does not vanish at $t = 0$; let the averages be over all phases.

The scaling (6.92) of τ_{QL} for large times may be recovered without the previous calculations on the basis of the following qualitative argument. Over the time τ_{spread}, $\langle \delta q_n(t)^2 \rangle$ may be computed by perturbation theory and is $\mathcal{B}$ times $\langle \Delta q(t)^2 \rangle$ since $\mathcal{B}^{-1}$ gives the typical number of resonances in a resonance box. Therefore it grows like t^3. As soon as $\langle \delta q_n(t)^2 \rangle$ has a finite size, the chaotic motion forces an exponential divergence of the true orbit on the basis of this finite size. So the finite size of $\langle \delta q_n(t_r)^2 \rangle$ for a given time t_r yields an exponential growth like $\exp 2(t - t_r)/\tau_{\text{Lyap}}$ where τ_{Lyap}^{-1} is the (yet to be defined) Lyapunov exponent of the chaotic dynamics. This yields an overall scaling $\langle \delta q(t)^2 \rangle \sim t_r^3 \exp[2(t - t_r)/\tau_{\text{Lyap}}]$ which is maximum for $t_r = 3\tau_{\text{Lyap}}/2$. We conclude from this that $\langle \delta q_n(t)^2 \rangle \sim \mathcal{B} \exp(2t/\tau_{\text{Lyap}})$ for large times. Section 6.4 showed that all characteristic times of chaos must scale like τ_{spread}; so does τ_{Lyap}. Then setting $\langle \delta q_n(t)^2 \rangle \sim 1$ yields an estimate of τ_{QL} confirming the scaling (6.92).

We define a time τ_* such that $\tau_{\text{spread}} \ll \tau_* \ll \tau_{\text{QL}}$. The result of this discussion is that '$q(t)$ depends poorly on any given phase over a time τ_*' (section F.5). For $M \gg 1$, the argument is easily strengthened into '$q(t)$ depends poorly on any two given phases over a time τ_*'. To this end $(q_{\eta_1, \eta_2}(t), p_{\eta_1, \eta_2}(t))$ and $(\delta q(t), \delta p(t))$ are defined starting from $\varphi_{m_1} = \varphi_{m_2} = 0$, and a third term similar to the second one is added to the right-hand side of (6.85). The requested result is obtained by analysing the new evolution equation as was done for (6.85). Time τ_* is introduced for the sake of mathematical rigour. For physicist estimates it may be substituted with τ_{QL}.

Figure 6.16(*a*) displays the numerical calculation of $\langle \delta q^2(t) \rangle$ versus time for $A = 2, 5$ and 10 for Hamiltonian (6.1). This quantity becomes of order 1 for $t \simeq 1$; this time is similar to the time $t_{\text{QL}} \simeq 1.5$ deduced from figure 6.7 yielding the time limit for the initial quasilinear diffusion. The tendency of τ_{QL} to decrease when A grows is visible in figure 6.16(*b*).

As a result, for $t \ll \tau_{\text{QL}}$, the non-quasilinear terms $\Delta_{\pm}$ are negligible since q has a small dependence on any given pair of phases in this time range. However, as they might scale unfavourably with M, these terms are estimated in section F.5 by expliciting in the argument of the cosine of (6.81) the main dependence on φ_{m_1} and φ_{m_2} through estimates $\delta\Phi_{m_1}$ and $\delta\Phi_{m_2}$ of the type $k_m \delta q_{n0}$ for both phases, and by expanding to second order in these $\delta\Phi$. Such estimates hold for $t \ll \beta\tau_*$ with $0 < \beta < 1$ for E large enough and yield with dimensionless constants c_j of the order of unity:

$$\Delta_+ \sim c_2 \gamma_D^6 k^{-2} \tau_c^2 t^2 \qquad (6.93)$$

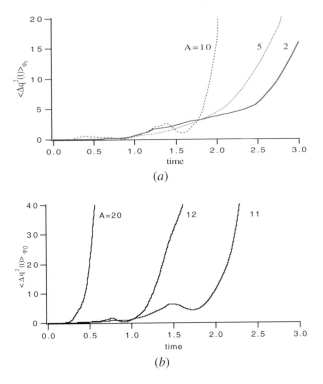

Figure 6.16. $\langle \delta q^2(t) \rangle$ versus time for (*a*) $A = 2, 5$ and 10 and (*b*) $A = 11, 12$ and 20 for Hamiltonian (6.1). (After Bénisti (1995).)

$$\Delta_- \sim c_2 \gamma_D^6 k^{-2} \tau_c^2 t^2 + c_5 k^{-1} |\Delta v| \gamma_D^6 t^5 \tag{6.94}$$

which are negligible with respect to Δ_0 in the time interval of interest. Note that the ratios $\Delta_\pm / \Delta_0$ remain small as $M \to \infty$, although there are $2M^2 - M$ 'off-diagonal' terms and only M 'diagonal' terms.

The proof that $\Delta p(t)$ diffuses for $0 \le t < \tau_{QL}$ follows that in section 6.8.1[6]. This ends the proof of existence of the initial quasilinear diffusion regime for the case of random phases. This regime is initially non-chaotic but for $B \ll 1$ it lasts over a time larger than τ_{spread} and chaotic effects may be present after this time; indeed, the term giving the exponential divergence of orbits was accounted for in the evaluation of the effect of one phase on the dynamics with the characteristic time (recall (5.7))

$$\tau_{Lyap} = 2^{1/3} \tau_{Dn} \ll \tau_{QL} \tag{6.95}$$

i.e. the reciprocal of the exponential rate in (6.91) for $\langle \delta q_n^2 \rangle^{1/2}$.

This reasoning relies strongly upon the fact that there are many ($M \gg 1$) random phases. To the contrary Bénisti (1995) and Bénisti and Escande (1997,

---

[6] Equation (F.14) is recovered and the calculation of $G(t)$ is similar to that in section F.3.

1998b) found, for one typical choice of random phases but random q_0, a lack of convergence of $\langle \Delta p^2(t) \rangle_{q_0}$ versus t toward a straight line even though working with ten times more orbits than in the case with random phases.

6.9   Diffusion coefficient in the chaotic regime

Let $p_{\min} = \min(v_m)$ and $p_{\max} = \max(v_m)$. We are about to show that in the velocity domain $[p_{\min}, p_{\max}]$ the dynamics is chaotic enough for a typical orbit to be unconfined within this domain for E large but we assume that the time of interest is smaller than the time for the orbit to reach the boundaries of the chaotic domain. Therefore we set the condition

$$\min\{(p_0 - p_{\min})^2, (p_0 - p_{\max})^2\}$$
$$\gg 2D_{QL}\tau_{QL} \sim (D_{QL}/k)^{2/3} |\ln \mathcal{B}| \sim E^{4/3} k^{-2/3} \tau_{\mathrm{discr}}^{2/3} |\ln \mathcal{B}| \quad (6.96)$$

to compute now the diffusion coefficient due to chaotic motion when M and s (or $1/\mathcal{B}$) are large. We define

$$\delta q(\tau | p, q, t) = q(t + \tau) - q - p\tau \quad (6.97)$$

where $q(t')$ is the position at time t' of an orbit being at (p, q) at time t, so that $\delta q(\tau | p, q, t)$ tells the departure of this orbit from the free motion during the time interval τ. For $|\tau| \leq \tau_*$, $\delta q(\tau | p, q, t)$ is in the initial quasilinear regime. It is a Gaussian variable whose standard deviation is large at $|\tau| = \tau_*$ since $\mathcal{B} \ll 1$, which implies the expectation of its cosine to vanish. Its variance is given by (6.78) with τ substituted for t.

Integrating formally the equation of motion for p yields

$$\langle \Delta p^2(t) \rangle = -\sum_{m,n=1}^{M} \sum_{\epsilon=\pm 1} \epsilon \frac{A_m k_m A_n k_n}{2} \int_0^t \int_0^t \langle \cos[(k_m + \epsilon k_n) q(t'')$$
$$+ k_m p(t'')(t' - t'') - \omega_m t' - \epsilon \omega_n t''$$
$$+ k_m \delta q(t' - t'' | p(t''), q(t''), t'') + \varphi_m + \epsilon \varphi_n] \rangle \, dt' \, dt''. \quad (6.98)$$

Note that the expectation of $\langle \cos[k_m \delta q(t' - t'' | p(t''), q(t''), t'')] \rangle$ is independent from $q(t'')$, as all phases φ_m are drawn independently from a uniform distribution on the circle.

We introduce the probability distribution $P(\delta p, t | p_0)$ of $\delta p = p(t) - p_0$ for an orbit started at p_0 at $t_0 = 0$, and we write the integrand of (6.98) for diagonal $(m = n, \epsilon = -1)$ terms:

$$\langle \cos[k_m \delta q(t' - t'' | p(t''), q(t''), t'') + k_m p(t'')(t' - t'') - \omega_m(t' - t'')] \rangle$$
$$= \Re \int \langle e^{i[k_m \delta q(t' - t'' | p_0 + \delta p, q, t'') + (\Omega_m + k_m \delta p)(t' - t'')]} \rangle_* P(\delta p, t'' | p_0) \, d\delta p$$

$$
= \Re \int \left(\langle e^{ik_m \delta q(t'-t''|p_0,q,t'')} \rangle_* e^{i(\Omega_m + k_m \delta p)(t'-t'')} \right.
$$

$$
\left. + \delta p \frac{\partial}{\partial p_0} \langle e^{ik_m \delta q(t'-t''|p_0,q,t'')} \rangle_* e^{i(\Omega_m + k_m \delta p)(t'-t'')} + \cdots \right)
$$

$$
\times P(\delta p, t''|p_0) \, d\delta p \tag{6.99}
$$

where the star subscript means the average conditioned by the constraint $p(t'') = p_0 + \delta p$ and the Taylor expansion in δp was explicitly written to first order only. Such an expansion makes sense if $\langle \exp i[k_m \delta q(t'-t''|p_0+\delta p, q, t'')] \rangle_*$ is weakly dependent on δp over most of the support of P in this variable. As δq is computed with the knowledge of p and q at time t'', which sets only two conditions on a set of many phases[7], this average may be computed using the initial quasilinear estimate at time $|t'-t''| \leq \tau_*$. Hence the function $\langle e^{ik_m \delta q(t'-t''|p,q,t'')} \rangle_*$ is correctly evaluated by the previous quasilinear estimate over its whole support in $t' - t''$ as $\tau_{QL} \gg \tau_{spread}$. As a result, the Taylor expansion is justified if the width of $P(\delta p, t|p_0)$ in δp is smaller than the scale on which $D_{QL}(p_0)$ varies.

We compute the contribution to $\langle \Delta p^2(t) \rangle / (2t)$ of the successive terms in the Taylor expansions. For the first term,

$$
B \equiv \lim_{t \to \infty} \sum_{m=1}^{M} \frac{(A_m k_m)^2}{4t} \Re \int_0^t \int_0^t \int \langle e^{ik_m \delta q(t'-t''|p_0,q,t'')} \rangle_*
$$

$$
\times e^{i(\Omega_m + k_m \delta p)(t'-t'')} P(\delta p, t''|p_0) \, d\delta p \, dt' \, dt''
$$

$$
= \lim_{t \to \infty} \sum_{m=1}^{M} \frac{(A_m k_m)^2}{8t} \Re \int_0^t \int_{-t''}^{t-t''} \langle e^{ik_m \delta q(\tau|p_0,q,t'')} \rangle_*
$$

$$
\times \tilde{P}(k_m \tau, t''|p_0) \exp[i\Omega_m \tau] \, d\tau \, dt'' \tag{6.100}
$$

where we used the Fourier transform of P

$$
\tilde{P}(\alpha, t''|p_0) = \int_{-\infty}^{\infty} P(\delta p, t''|p_0) \exp(i\alpha \delta p) \, d\delta p. \tag{6.101}
$$

The width of P defined in section A.3 is a characteristic scale $w(P, t'')$ for δp, which is of the order of $\langle \delta p^2 \rangle^{1/2}$ for dimensional reasons. This width is a measure of the size of 'most' of the support of P, and is proportional to the standard deviation $\langle \delta p^2 \rangle^{1/2}$ if P is Gaussian (exercise A.1). Up to $t = \tau_*$, the proof of the initial quasilinear diffusion[8] implies that the width of P grows like $(2D_{QL}t)^{1/2}$. Assume that this quasilinear growth holds until $t = t_* \geq \tau_*$. The part of the integration domain over τ where $\tilde{P}$ takes appreciable values in (6.100) is $w_\tau(t'') = k^{-1}w(\tilde{P}, t'') \simeq \pi/(kw(P, t''))$. This formula and

---

[7] Actually, the q-condition is irrelevant as the phases are uniformly distributed.
[8] It even shows that P is close to Gaussian for E large (see the Kolmogorov–Smirnov test on figure 6.8 or focus on the lower moments in sections 6.8.1, 6.8.2 and 6.10).

equations (6.76) and (6.79) imply that, for $t \le t_*$, requiring $w_\tau \ll \tau_{\mathrm{spread}}$ means $t \gg \tau_{\mathrm{spread}}/12$. Therefore, for $t \ge t_*$ over most of the time-integration interval, $\tilde{P}$ is narrow enough in $k\tau$ for the spread of δq to be negligible in the part of the integration domain over τ where $\tilde{P}$ takes appreciable values in (6.100). Then $\langle \exp[ik_m \delta q(\tau | p_0, q, t'')]\rangle_* \simeq 1$ and

$$B = \lim_{t \to \infty} \sum_{m=1}^{M} \frac{\pi A_m^2 k_m}{2t} \int_0^t P(v_m - p_0, t'' | p_0) \, dt''$$

$$= \int_0^t \sum_{m=1}^{M} \frac{D_m \Delta v_m}{2t} P(v_m - p_0, t'' | p_0) \, dt'' \qquad (6.102)$$

where the inverse Fourier transform was provided by the integral over τ.

Now, we assume τ_* to be so large that P is almost constant over the range $[v_{m-L}, v_{m+L+1}]$ for all m over most of the integration interval for $t \ge \tau_*$. Then we use (6.56) and (6.58) and substitute the sum over v_m by an integral

$$2Bt = \int_0^t \int D_{\mathrm{QL}}(v) P(v - p_0, t'' | p_0) \, dv \, dt'' = 2D_{\mathrm{QL}}(p_0)t \qquad (6.103)$$

provided $t \ll \Delta v_D^2 / 2D_{\mathrm{QL}}(p_0)$ in order to avoid feeling the gradient of $D_{\mathrm{QL}}(p)$ (where we denote by $\Delta v_D = D_{\mathrm{QL}}/|\partial_p D_{\mathrm{QL}}|$ the characteristic scale for this gradient). This reasoning holds as long as t_* is small enough for the support of P not to hit the $p_{\min}$ or $p_{\max}$ values.

We now show that this is the only restriction on t_*. The acceleration produced by the M waves has its amplitude bounded by $a_{\max} = \sum_{m=1}^{M} k_m A_m$. Then dividing the width $w(P, t_*)$ by two needs at least a time $\Delta\tau = w(P, t_*)/(2a_{\max})$. Therefore the reasoning introducing (6.102) applies for $t \le t_* + \Delta\tau$ and the quasilinear growth of $w(P, t)$ can be proved until $t'_* = t_* + \Delta\tau$ using (6.103). Then t'_* may be used as a new t_*. Since $\Delta\tau$ is an increasing function of t_*, this reasoning can be iterated until t_* is long enough for the support of P to hit the $p_{\min}$ or $p_{\max}$ values.

The contribution from $B' \equiv \partial_{p_0} \langle e^{ik_m \delta q(t'-t''|p_0,q,t'')}\rangle_*$ is estimated with the same two successive Fourier transforms as before. The first one produces terms like $\partial_\alpha \tilde{P}(\alpha, t'' | p_0)$ with $\alpha = k_m \tau$. As δq is computed with the knowledge of p and q at time t'' which sets only two conditions on a set of many phases, its spread may be computed by using the initial quasilinear estimate for time $\tau < \tau_*$. As for the computation of B, this condition is satisfied in the useful range of the integration domain provided that t is large enough. Therefore $B' \simeq \partial_{p_0} \exp[-k_m^2 D_{\mathrm{QL}}(p_0)\tau^3/3] \simeq -\tau^3 \gamma_D^2 (\partial_v \gamma_D) e^{-\tau^3 \gamma_D^3/3}$. Then the second Fourier transform is applied to a function $\alpha^3 \partial_\alpha \tilde{P}(\alpha | p_0, t'')$, and yields an expression where the sum over m involves the function $R(u) \equiv d^3[u P(u, t'' | p_0)]/du^3$ computed at $u = v_m - p_0$. More generally, the nth-order term in the Taylor expansion of (6.99) involves functions $R_j(u) \equiv d^{3j}[u^j P(u, t'' | p_0)]/du^{3j}$ with

$1 \leq j \leq n$. This introduces a scaling factor $\sim u^{-2j} \sim [D_{\text{QL}}(p_0)t'']^{-j}$ in the integral over t'' which makes small the contributions of the nth ($n \geq 1$)-order term in the Taylor expansion with respect to that with $n = 0$ for t large.

Exercise 6.9. Elaborate this argument more quantitatively.

The general term in (6.98) can be estimated by a similar calculation. A sequence of two Fourier transforms is again recovered. After the first one, averages like $\langle \exp i[k_m \delta q(\tau | p_0, q, t'') + \varphi_m + \epsilon \varphi_n] \rangle_*$ are found. They vanish as the constraint $q(t'') = q$, $p(t'') = p_0 + \delta p$ leaves the average over any two phases almost free and δq is negligible for τ small. Therefore B is the dominant contribution to $\langle \Delta p^2(t) \rangle$ which thus grows in a quasilinear way. This ends our proof of the quasilinear estimate for asymptotic times.

Note that the conditional probability P permits us to use our knowledge of initial quasilinear diffusion for proving it over asymptotic times only because we first proved that $\tau_{\text{QL}} \gg \tau_{\text{spread}}$. As a result, our two previous conditions on t to have a large spread of P in δp reduce to

$$t \gg \tau_D \tag{6.104}$$

or, according to (6.79), $t \gg \tau_{\text{spread}}$. In contrast with the initial non-chaotic quasilinear regime where it scaled like $1/t$, the number of modes acting on the particle scales like $t^{1/2}$. Indeed, the chaotic diffusion makes the particle visit a growing velocity domain which increases the number of modes acting resonantly on it.

As many Hamiltonian systems may be locally reduced to case (6.48) (Escande 1985), this further extends its range of applicability and shows that the universality class of quasilinear diffusion is broad.

Higher-order moments of Δp could be estimated using a similar technique. Indeed, the initial quasilinear diffusion regime holds over a time of order $\tau_* \gg \tau_{\text{spread}}$. This ensures that the joint conditional probability density for $p(t) - p_0$ and $p(t') - p_0$ with $t' > t$ given p_0 factorizes into $P(p(t) - p_0, t | p_0) P(p(t') - p(t), t' | p(t))$ provided $|t' - t| \gg \tau_{\text{spread}}$. This generalizes to the conditional probability of the set of the $p(t^i) - p_0$ with $1 \leq i \leq n$ provided that B^{-1} is large enough. Therefore the use of conditional probabilities enables one to retain, after Fourier transforms, the same terms for a moment of order $2n$ as in the case where $q(t)$ is weakly dependent on any phase provided that B^{-1} is large enough. This again yields a Gaussian estimate for the chaotic case in the limit where B is small.

Exercise 6.10. Compute the moment of order 4 with this technique.

6.10 Drag coefficient

If $D_{\text{QL}}(p)$ is not a constant, chaotic motion also induces a drag on particles. Let t_* be a time much larger than the correlation time τ_c of the field (6.67). We now compute for E large the average of $\Delta p \equiv p(t_* + \Delta t) - p(t_*)$ for all orbits such that $p(t_*)$ takes a given value p_* and $q(t_*)$ takes a given value q_*. Denote by $\langle \bullet \rangle_*$ the average over all phases restricted by this constraint. We compute $\langle \Delta p \rangle_*$ by formally integrating (6.3) but we make more explicit the dependence of q on wave m. This can be done using the Picard fixed-point equation (in function space) obtained by formally iterating the equations of motion (6.2), (6.3)

$$q(t) = q^{(0)}(t) + \Delta q(t) \tag{6.105}$$

where

$$q^{(0)}(t) = q_* + (t - t_*)p_* \tag{6.106}$$

$$\Delta q(t) = \sum_{m=1}^{M} \Delta q_m(t) \tag{6.107}$$

$$\Delta q_m(t) = A_m k_m \int_{t_*}^{t} \int_{t_*}^{t'} \sin[k_m q(t'') - \omega_m t'' + \varphi_m]\, dt''\, dt' \tag{6.108}$$

with conditions $p(0) = p_0,\, q(0) = q_0,\, p(t_*) = p_*$ and $q(t_*) = q_*$.

Assuming that Δt is small enough for

$$|k_m \Delta q_m(t_* + \Delta t)| < \eta \tag{6.109}$$

where η is a small quantity, we approximate $\langle \Delta p \rangle_*$ to first order in η

$$\langle \Delta p \rangle_* \simeq \sum_{m=1}^{M} A_m k_m \int_{t_*}^{t_* + \Delta t} \left\langle \sin \Phi_m(t) \right.$$

$$\left. + (\cos \Phi_m(t)) A_m k_m^2 \int_{t_*}^{t} \int_{t_*}^{t'} \sin \Psi_m(t'')\, dt''\, dt' \right\rangle_* dt$$

$$= \sum_{m=1}^{M} A_m k_m \int_{t_*}^{t_* + \Delta t} \langle \sin \Phi_m(t) \rangle_*\, dt$$

$$+ \sum_{\epsilon = \pm 1} \sum_{m=1}^{M} \frac{A_m^2 k_m^3}{2} \int_{t_*}^{t_* + \Delta t} \left\langle \left(\int_{t_*}^{t} \int_{t_*}^{t'} \sin[\Psi_m(t'') + \epsilon \Phi_m(t)]\, dt''\, dt' \right) \right\rangle_* dt \tag{6.110}$$

where (6.108) was used and

$$\Psi_m(t) = k_m q(t) - \omega_m t + \varphi_m \tag{6.111}$$

$$\Phi_m(t) = \Psi_m(t) - k_m \Delta q_m(t). \tag{6.112}$$

If $t_* \ll \tau_{\text{spread}}$, then $p \simeq p_*,\, q \simeq q_*$, the phase $\Phi_m(t)$ is uniformly distributed as is φ_m and $\langle \sin \Phi_m(t) \rangle_*$ vanishes. If $t_* \gg \tau_{\text{spread}}$ and $\tau_{\text{QL}} \gg \Delta t \gg \tau_{\text{spread}}$

(which is possible in the limit $\mathcal{B} \to 0$), the variance of $\Delta q(t)$ is large over most of the integration interval. So is the variance of $\Delta q(t) - \Delta q_m(t)$ as many waves contribute to the spread of $\Delta q(t)$. Therefore the spread of $\Phi_m(t)$ is also large and $\langle \exp[-i\Phi_m(t)] \rangle_*$ vanishes. Thus the first term in (6.110) may be neglected.

We now show that the average of the second term for $\epsilon = 1$ vanishes in both time limits of interest. If $t_* \ll \tau_{\text{spread}}$, the randomness of the argument in the sine comes from a term $2\varphi_m$ only. Therefore the average of the sine vanishes. If $t_* \gg \tau_{\text{spread}}$, the contribution of $\Phi_m(t)$ to the phase $\Psi_m(t'') + \Phi_m(t)$ involves large random values of $\Delta q(t) - \Delta q_m(t)$ over most of the integration interval in t, which cannot be cancelled by possibly large random values of $\Delta q(t'')$: if t'' is close to t these random values add, and otherwise $\Delta q(t'')$ has a smaller size (estimated, e.g., by its standard deviation) than $\Delta q(t)$. Thus $\langle \sin[\Psi_m(t'') + \Phi_m(t)] \rangle_*$ vanishes.

We identify Ψ_m and Φ_m, which brings an error $\sim \eta$ in $\Psi_m(t'') - \Phi_m(t)$. Then by inverting the order of the integrals over t' and t'' and introducing $\tau = t'' - t$, (6.110) becomes

$$
\langle \Delta p \rangle_* = -\sum_{m=1}^{M} \frac{A_m^2 k_m^3}{2} \int_{t_*}^{t_*+\Delta t} \int_{t_*-t}^{0} \tau \langle \sin[\Omega_m \tau + k_m(\Delta q(t+\tau)
$$
$$
- \Delta q(t))] \rangle_* \, d\tau \, dt \tag{6.113}
$$

where

$$
\Omega_m = k_m p_* - \omega_m. \tag{6.114}
$$

Because of the large number of phases, we identify $\langle \bullet \rangle_*$ and $\langle \bullet \rangle$ in the right-hand side of (6.113), which reads, using (6.59),

$$
\langle \Delta p \rangle_* = -\int_{t_*}^{t_*+\Delta t} \int_{t_*-t}^{0} \sum_{m=1}^{M} \frac{\mathcal{D}(v_m, p_*)}{\pi} k_m \tau
$$
$$
\times \langle \sin[\Omega_m \tau + k_m(\Delta q(t+\tau) - \Delta q(t))] \rangle |\Delta\Omega_m| \, d\tau \, dt. \tag{6.115}
$$

The wide spectrum of frequencies implies that the sum over m is small, unless $|\tau| \lesssim \tau_c$. As $\tau_c \ll \tau_{\text{spread}}$, this implies that, in the dominant contribution to the integral, $k_m(\Delta q(t+\tau) - \Delta q(t))$ must be small and

$$
\frac{\langle \Delta p \rangle_*}{\Delta t} = -\frac{1}{\Delta t} \int_{t_*}^{t_*+\Delta t} \int_{t_*-t}^{0} \sum_{m=1}^{M} \frac{D_m}{\pi} k_m \tau \sin(\Omega_m \tau) |\Delta\Omega_m| \, d\tau \, dt
$$
$$
= \frac{dD(p)}{dp} \tag{6.116}
$$

with

$$
D(p) = \sum_{m=1}^{M} \frac{D_m}{\pi} \int_{t_*-t}^{0} \cos(\Omega_m \tau) \, d\tau \, |\Delta\Omega_m| = \sum_{m=1}^{M} \frac{D_m}{\pi \Omega_m} \sin(\Omega_m t) |\Delta\Omega_m|
$$
$$
\simeq \int \frac{D_{\text{QL}}(p)}{\pi \Omega} \sin(\Omega t) \, d\Omega = D_{\text{QL}}(p) \tag{6.117}
$$

where $\tau_c \ll t \ll \tau_{\mathrm{discr}}$ was used, with τ_{discr} defined by (6.73). It must be noted that neither the initial conditions (q_0, p_0) at time $t = 0$ nor the current time t_* and position q_* appear in (6.116). Indeed, only the current particle velocity p_* appears in this equation. The other data were introduced in our argument to enable us to estimate the average forces in a regime where the particles are spread in position and momentum. Thus, if the diffusion coefficient is not a constant, an average drag is exerted on particles and the force $\partial D(p)/\partial p$ is called the drag coefficient (see exercise E.8).

Equations (6.116) and (6.117), though derived in a different way, correspond to the classical result of Landau (1937) for non-chaotic diffusive Hamiltonian motion. It thus appears that they also hold in the chaotic case, thanks to the loss of correlations over long time-scales.

Exercise 6.11. Recall that the microscopic dynamics is time reversible, for the mapping $(q', p', t', \varphi'_m, \omega'_m, k'_m, A'_m) = (q, -p, -t, \varphi_m, -\omega_m, k_m, A_m)$. Show that the previous calculations hold only if $t > 0$, whereas for $t < 0$ one actually finds $\langle \dot{p}(t) \rangle_* = -\partial D_{\mathrm{QL}}(p)/\partial p$. Show it first for $|t| \ll \tau_{\mathrm{spread}}$ and then for large $|t|$. Discuss this result in terms of trajectories in (q, p, t) space.

We now make explicit the condition $|k_m \Delta q_m(t_* + \Delta t)| < \eta$, i.e. (6.109). The maximal value of $|\Delta q_m(t_* + \Delta t)|$ occurs if the particle stays exactly resonant with wave m over the time interval $[t_*, t_* + \Delta t]$. This yields

$$|k_m \Delta q_m(t_* + \Delta t)| \leq A_m k_m^2 \Delta t^2/2 \sim N_0^2 s^2 (\tau_{\mathrm{spread}}/\tau_{\mathrm{discrm}})^2 \sim N_0^2 s^{-2/3} \quad (6.118)$$

where τ_{discrm} is defined by (6.62), where $N_0 = \Delta t/\tau_{\mathrm{spread}}$ is a large number and where s is the typical overlap parameter. As a result, condition (6.109) is rigorously satisfied by taking s large enough.

A non-rigorous estimate with a smaller exponent for N_0 can be obtained as follows. For E large, the diffusion is quasilinear for all times. Thus the force decorrelation discussed in section 6.5.1 occurs after a time interval τ_{spread}. Therefore, in order to estimate $\Delta q_m(t_* + \Delta t)$, we split Δt into subintervals of size $\delta t \simeq \tau_{\mathrm{spread}}$, so that $\Delta q_m(t_* + \Delta t) = \sum_n \Delta q_m(t_* + (n-1)\delta t, t_* + n\delta t)$. If wave m is in the resonance box of the particle, during a time τ_{spread} the corresponding maximum coherent contribution $\Delta q_m(t_* + (n-1)\delta t, t_* + n\delta t)$ is $A_m k_m \tau_{\mathrm{spread}}^2/2$. When wave m is not in the resonance box of the particle, $\Delta q_m(t_* + \Delta t)$ just gets a perturbative fluctuating contribution from wave m which may be neglected according to the locality principle. For Hamiltonian (6.1) we may define a typical time spent in one resonance box by $\tau_{\mathrm{box}} \equiv (2\Delta v_{\mathrm{box}})^2/(2D_{\mathrm{QL}}) \simeq 44 D_{\mathrm{QL}}^{-1/3}$, which scales like $E^{-2/3} \sim B$ in agreement with section 6.4. For the more general Hamiltonian (6.48), the rescaling of section 6.7 yields

$$\tau_{\mathrm{box}} \simeq 40(k_m^2 D_{\mathrm{QL}})^{-1/3} = 11\tau_{\mathrm{spread}}. \quad (6.119)$$

Assume that wave m is in the initial resonance box of the particle. According to section F.2, after a time $n\tau_{\mathrm{spread}}$ with $n \gg 1$, the particle has typically visited $[8n\tau_{spread}/(\pi\,\tau_{box})]^{1/2} \simeq \sqrt{n/4}$ resonance boxes and has the probability $[2\tau_{box}/(\pi n\tau_{spread})]^{1/2} \simeq \sqrt{7/n}$ to be in the initial box. Therefore, during the next interval the maximal coherent change of $\Delta q_m^2(t_* + \Delta t)$ is $A_m^2 k_m^2 \tau_{spread}^4/\sqrt{2n}$. For waves out of the initial resonance box of particle l, the estimate is lower. Summing these (independent random) contributions for $n = 1, 2, \ldots, t/\tau_{spread}$ and taking the square root yields the estimate

$$|k_m \Delta q_m(t_* + \Delta t)| \simeq [60 k_m \Delta v_m (\tau_{spread}\Delta t)^{1/2}]^{1/2} \simeq 25 N_0^{1/4} s^{-2/3}$$

(6.120)

where Δv_m is given by (6.52).

Exercise 6.12. Show that (6.116) can be recovered by removing the integrals $\int_{t_*}^{t_*+\Delta t} dt$ of this section, setting then $t_* = 0$ and substituting Δt with t in all expressions. Then in (6.120) Δt must be substituted with t, which is much more stringent.

Exercise 6.13. Estimate the moment of order 3.

6.11   Fokker–Planck equation

In both previous sections we had to consider times $t \gg \tau_{spread}$ in order to find a convergence of the moments for $\tau_{QL} \gg \tau_{spread}$. This decorrelation is crucial in establishing the validity of the diffusion approximation over the large time-scales. It ensures that the particle motion is Markovian on the scales much larger than τ_{spread}. In section 6.6 we estimated that two resonance boxes had to be visited for the chaotic diffusion to set in. This is again reasonable since $2\tau_{box}/\tau_{spread} \simeq 22$ which is a fairly large number. This, in turn, implies that, over times larger than τ_{box}, the probability distribution function $f(p)$ for p satisfies the Fokker–Planck equation (section E.3)

$$\frac{\partial f}{\partial t} = \frac{\partial}{\partial p}\left(D_{QL}(p)\frac{\partial f}{\partial p}\right).$$

(6.121)

In the case of a uniform diffusion coefficient ($\partial_p D_{QL} = 0$), the fundamental solution to the Fokker–Planck equation is the Gauss distribution with variance $2D_{QL}t$, and the higher moments $\langle \Delta p(t)^k \rangle$ factorize accordingly.

The Fokker–Planck equation (6.121) can be easily recovered after computing the quasilinear diffusion coefficient, if we assume that the chaotic motion is a diffusive one. Then since $f(p)$ is a density in p space, a flux $D_{QL}(p_0)\partial_p f(p_0)$ of particles is passing at $p = p_0$ and the conservation of the number of particles implies we recover equation (6.121). The Fokker–Planck equation implies that there is a drag on particles given by (6.116).

Finally we note that condition $\tau_{QL} \gg \tau_{spread}$ needs a high resonance overlap to be satisfied because of the logarithm in (6.92). This is consistent with the slow convergence of the diffusion coefficient toward its quasilinear value in figure 6.10.

6.12   Historical background

The idea of locality was already present in the resonance broadening concept introduced by Dupree (1966). His work was an improvement upon the too simple idea of the original quasilinear theory in plasma physics that the particles are subjected only to the field of the wave with which they are resonant. Dupree assumed that instead of a perturbed ballistic motion, the particle motion is diffusive since $t = 0$. He found that the waves to be included in order to compute the diffusion coefficient are in a range $[D/(3k)]^{1/3} \sim A^{2/3}|\Delta v|^{1/3}$ about the particle velocity. In this chapter, locality is found without assuming diffusion and exists even before the asymptotic diffusion regime sets in for intermediate values of A but the scaling of D may differ from $A^{2/3}$ for some wave spectra.

A locality principle was used by Chirikov when computing the stochasticity threshold in some part of phase space as corresponding to the overlap of the two closest resonances, even if the Hamiltonian of interest has many of them (Chirikov, 1979).

Kubo numbers are associated with the competition between stochastic and deterministic time-scales in Markov processes (van Kampen 1992).

Chapter 7

Self-consistent dynamics in the diffusive regime

This chapter deals with the particle's diffusive motion and the growth of Langmuir waves in self-consistent dynamics. This is a fundamental phenomenon observed experimentally in plasmas and travelling wave tubes. The self-consistent evolution of waves and particles when the waves have a small amplitude was presented in chapter 4. Here we study this evolution in a more general setting including the case where the particles' motion is essentially chaotic.

Four time-scales are important for our analysis: the correlation time τ_c of the wave spectrum, the typical time τ_w for the growth or decay of the wave amplitudes, the nonlinear time-scale τ_{spread} for the spreading of nearby orbits to become important and the discretization time $\tau_{\mathrm{discr}} = (k\Delta v)^{-1} \sim M\tau_c$ characterizing the tightness of the wave spectrum. To compare the second and third time-scales, we introduce the dimensionless number

$$\mu_{\mathrm{ws}} \equiv \frac{\tau_w}{\tau_{\mathrm{spread}}}. \tag{7.1}$$

As we are interested in a turbulent regime with strong overlap between neighbouring resonances, the ratio $\mathcal{B} \sim \tau_{\mathrm{discr}}/\tau_{\mathrm{spread}}$ defined by (6.63) is also small in this chapter.

Our present approach enables us to describe the self-consistent dynamics in both limits $\tau_w \gg \tau_{\mathrm{spread}} \gg \tau_c$ (so that $\mu_{\mathrm{ws}} \to \infty$ along with $\mathcal{B} \to 0$) and $\tau_{\mathrm{spread}} \gg \tau_w \gg \tau_c$ (then $\mu_{\mathrm{ws}} \to 0$ along with $\mathcal{B} \to 0$). The latter limit is that of the previous small-amplitude analysis (chapter 4). Here we consider a large number of particles and waves with a large spread Δu in particle and phase velocities.

Rigorously speaking, the presence of both waves and particles in the problem should impose the definition of two correlation times: one relevant to particle motion (the reciprocal of the spread of $k_j p_l - \omega_j$ when j is varied for a typical value of p_l) and one relevant to wave motion (the reciprocal of the spread of

$k_j p_l - \omega_j$ when l is varied for a typical value of j). In fact the self-consistency of the problem matches the relevant ranges of p_l and ω_j / k_j, and both correlation times have the same order of magnitude, which we denote by

$$\tau_c = (k \Delta u)^{-1} \tag{7.2}$$

where k is a typical wavenumber.

Several techniques used in this chapter are strongly inspired from those used for the non-self-consistent diffusive motion. Our techniques are elementary, in that they only refer to the motion of actual particles, to the evolution of waves given these motions and to explicit estimates for the dominant contributions in the limit $N \to \infty$. We assume the length L of the system to be large, so that M and the number of waves resonating with the particles are large.

Throughout this chapter we estimate averages $\langle \bullet \rangle$ for initial particle positions x_{l0} and initial wave phases θ_{j0} drawn independently from uniform distributions on the circle (with length L for positions, length 2π for the phases). Then we use the fact that, for $\tau_w \ll \tau_{spread}$, the particle motion is approximately ballistic; for $\tau_w \gg \tau_{spread}$, the waves' average amplitudes evolve slowly. However, as the wave phases and intensities fluctuate, the particle motions differ from those of the non-self-consistent case. In particular, self-consistency between waves and particles may prevent waves from having independent random phases for all times, even if phases are independent and random initially. Moreover, the chaotic particle motion generates not only the average wave growth but also a fluctuation of each wave amplitude with a typical time-scale corresponding to the particle chaotic correlation time τ_{spread}.

This chapter is organized as follows. We first compute a diffusion coefficient for the self-consistent dynamics. After a few natural approximations the derivation follows the same lines as section 6.9 and we find this coefficient to take on the quasilinear value (section 7.1). On this basis we give a heuristic derivation of the joint quasilinear equations for the particles and the waves in both time limits of interest (section 7.2). Then we turn to a more rigorous derivation of these equations. In a first step we compute the evolution of waves in the weak turbulence regime (section 7.3) using, in particular, a technique like that of section 6.10. In a second step a similar calculation for the drag on particles leads to the evolution of the particle velocity distribution function (section 7.4). We confirm the surprising result of the heuristic derivation that the statistical behaviour of the system is the same for both time-scale orderings of interest and that it is described by the quasilinear equations, whose general properties are briefly examined in section 7.5. As an example, we apply our analytical results to the saturation of the weak-warm-beam instability in section 7.6. The chapter ends with a brief account of the longstanding controversy about quasilinear theory and related experimental results (section 7.7).

7.1 Quasilinear diffusion coefficient

Evolution equations generated by the self-consistent Hamiltonian (2.109) are given by system (2.111)–(2.114). It is convenient to define a coupling constant

$$\beta'_j = \varepsilon \beta_j / k_j \tag{7.3}$$

and the complex wave envelope

$$z_j = Z_j \exp(\mathrm{i}\omega_{j0} t). \tag{7.4}$$

The evolution equation for z_j is

$$\dot{z}_j = \mathrm{i}\beta'_j \sum_{l=1}^{N} \mathrm{e}^{\mathrm{i}(\omega_{j0} t - k_j x_l)}. \tag{7.5}$$

As for the non-self-consistent case, we compute how the variance of the velocity of particles with given initial velocity p_0 grows with time. An initial, not intrinsically chaotic, quasilinear diffusion can be recovered over a time of the order of $\tau_{\mathrm{QL}} = (\pi \tau_{\mathrm{discr}} k^2 E^2 / 2)^{-1/3} \ln(\mathcal{B}^{-1})$ (equation (6.92)), where E is the typical electric field of the waves and $\mathcal{B} = (kE\tau_{\mathrm{discr}}^2)^{-2/3}$ is the corresponding current value of the ratio (6.63) of the discretization time τ_{discr} defined by (6.62) (here its typical value is $k_j / [\omega_j(k_{j+1} - k_j)]$) to the spreading time. The argument is quite similar to the non-self-consistent case, because the averages may be split into averages involving the z_j and averages involving the x_l thanks to the weak correlation between any wave and any particle. Knowledge of the initial quasilinear diffusion regime is sufficient to deal with the case $\tau_{\mathrm{w}} \ll \tau_{\mathrm{spread}}$, because in that case the particle motions are nearly ballistic on the time-scale over which the wave changes and the perturbation to these ballistic motions was found in chapter 4 to obey the quasilinear equations. Thus we now focus on the case $\tau_{\mathrm{w}} \gg \tau_{\mathrm{spread}}$.

Formal integration of the evolution equation of p_l in terms of the z_j yields its variance as

$$\langle \Delta p_l^2(t) \rangle = -\frac{1}{4} \sum_{j=1}^{M} \sum_{j'=1}^{M} \int_0^t \int_0^t \beta'_j k_j \beta'_{j'} k_{j'} (\langle z_j(t') z_{j'}(t'') \mathrm{e}^{\mathrm{i}[\Psi_{lj}(t') + \Psi_{lj'}(t'')]} \rangle$$

$$- \langle z_j(t') z_{j'}^*(t'') \mathrm{e}^{\mathrm{i}[\Psi_{lj}(t') - \Psi_{lj'}(t'')]} \rangle) \, \mathrm{d}t'' \, \mathrm{d}t' + \mathrm{CC} \tag{7.6}$$

where 'CC' means 'the complex conjugate of all preceding terms' and

$$\Psi_{lj}(t) = k_j x_l(t) - \omega_{j0} t. \tag{7.7}$$

As for the non-self-consistent case, the evolutions of $x_l(t')$ and $x_l(t'')$ are correlated only for $|t'' - t'| < \Delta t \sim \tau_{\mathrm{spread}} \ll \tau_{\mathrm{QL}}$. This bounds the relevant integration domain.

We first consider the second term in (7.6) with $j = j'$ for $t \gg \Delta t$ and call it

$$B = \frac{1}{4} \sum_{j=1}^{M} (\beta'_j k_j)^2 \int_0^t \int_0^t \langle z_j(t') z_j^*(t'') e^{i[\Psi_{lj}(t') - \Psi_{lj}(t'')]} \rangle \, dt'' \, dt' + \text{CC}. \quad (7.8)$$

We compute the integrand by expressing $z_j(t'')$ in terms of $z_j(t')$ by formal integration of the evolution equation (7.5) for z_j:

$$F \equiv \langle z_j(t') z_j^*(t'') e^{i[\Psi_{lj}(t') - \Psi_{lj}(t'')]} \rangle$$
$$= \langle |z_j(t')|^2 e^{i[\Psi_{lj}(t') - \Psi_{lj}(t'')]} \rangle$$
$$- i\beta'_j \sum_{l'=1}^{N} \int_{t'}^{t''} \langle z_j(t') e^{i[\Psi_{lj}(t') - \Psi_{lj}(t'') + \Psi_{l'j}(t_1)]} \rangle \, dt_1. \quad (7.9)$$

We first discuss the first term in (7.9). For $|t'' - t'| > \Delta t$, $\Psi_{lj}(t'')$ is uncorrelated with $\Psi_{lj}(t')$. Since many waves are acting on a particle, this is also true for the conditional distribution of Ψ given a fixed value of $|z_j(t')|^2$. Therefore, in the first term of (7.9), $\langle e^{i\Psi_{lj}(t'')} \rangle$ may be factorized out and it vanishes over most of the integration interval of t''. For $|t'' - t'| \leq \Delta t$, the effect of $z_j(t')$ upon $\Psi_{lj}(t') - \Psi_{lj}(t'')$ may be computed perturbatively. It is small in the limit of a continuous spectrum where the wave energy is shared among an infinite number of waves. As a result we may factorize the average $\langle |z_j(t')|^2 e^{i[\Psi_{lj}(t') - \Psi_{lj}(t'')]} \rangle$ into $\langle |z_j(t')|^2 \rangle \langle e^{i[\Psi_{lj}(t') - \Psi_{lj}(t'')]} \rangle$. Anticipating that the second term of (7.9) is negligible, we are then left with

$$B \simeq \frac{1}{2} \sum_{j=1}^{M} (\beta'_j k_j)^2 \int_0^t \int_0^t \langle |z_j(t')|^2 \rangle \langle \cos[\Psi_{lj}(t') - \Psi_{lj}(t'')] \rangle \, dt'' \, dt'. \quad (7.10)$$

As $\tau_w \gg \tau_{\text{spread}}$, we consider a time t such that $\tau_w \gg t \gg \tau_{\text{spread}}$. Then $\langle |z_j(t')|^2 \rangle$ may be considered as a constant and this equation has exactly the same structure as equation (6.98) obtained for the non-self-consistent case. The contribution of the first term in (7.9) to $B/(2t)$ then yields the quasilinear diffusion coefficient.

Now we consider the second term in (7.9). If $l' = l$, since $t \gg \tau_{\text{spread}}$, in most of the integration interval for t' and t'' the three phases in the exponential cannot cancel and, at fixed $z_j(t')$, the argument of the exponential has large fluctuations. Therefore, the expectation is small and this contribution is negligible.

Finally we consider the last term of (7.9) for $l' \neq l$. Then

$$\langle z_j(t') e^{i[\Psi_{lj}(t') - \Psi_{lj}(t'') + \Psi_{l'j}(t_1)]} \rangle = \langle z_j(t') e^{i[\Psi_{lj}(t') - \Psi_{lj}(t'')]} \langle e^{i\Psi_{l'j}(t_1)} \rangle_* \rangle \quad (7.11)$$

where the star denotes conditional averaging with fixed $z_j(t')$ and fixed increment $\Psi_{lj}(t') - \Psi_{lj}(t'')$, which sets three conditions over $N + M$ random phases. The value of $\Psi_{l'j}(t_1)$ follows from an integration over a long time interval so that it is uniformly distributed (modulo 2π) even under the two conditions over $N + M$ random phases and thus $\langle e^{i\Psi_{l'j}(t_1)} \rangle_* = 0$. Hence (7.11) vanishes and in (7.9) only the first term must be taken into account.

The other terms in (7.6) are negligible, as shown by a reasoning similar to that in the non-self-consistent case for non-diagonal terms. As a result the particles diffuse again according to (6.55), (6.56), (6.58) and (6.76) provided that, in the definition of $D_{QL}(p)$, $D_m(p_0)$ is substituted with

$$D_m = \frac{\pi(\varepsilon\beta_m)^2 \langle |z_m(t')|^2 \rangle}{2k_m \Delta v_m} = \frac{\pi\varepsilon^2\beta_m^2 \langle I_m(t') \rangle}{k_m \Delta v_m} \tag{7.12}$$

where ω_{m0} is substituted for ω_m in the definition of Δv_m. In the limit of a continuous wave spectrum, this diffusion coefficient takes the form (4.34)

$$D_{QL}(p) = \frac{\pi\varepsilon^2\beta^2}{k} \langle I(p) \rangle \tag{7.13}$$

where $\beta(p)$, $k(p)$ and $\omega(p)$ interpolate respectively, β_j, k_j and ω_{j0} (with j chosen such that $|p - \omega_{j0}/k_j|$ is minimum), and where $I(p)$ is defined by $\sum_j I_{j0} \bullet \to \int \bullet I(p) \mathrm{d}p$.

7.2   Simple derivation of the quasilinear equations

We now give a simple derivation of the quasilinear equations assuming that, for $\tau_w \gg \tau_{spread}$, τ_{spread} is the decorrelation time of the dynamics, i.e. that the decorrelation time for a single particle is also the decorrelation time for the waves. A more rigorous, though more technical, derivation is given in the next two sections. In the limit $\tau_w \ll \tau_{spread}$ of small wave amplitudes, (4.39) provided the Fokker–Planck equation

$$\frac{\partial f}{\partial t} = \frac{\partial}{\partial p}\left(D_{QL}(p)\frac{\partial f}{\partial p}\right) - \frac{\partial}{\partial p}(F_c(p)f) \tag{7.14}$$

for the evolution of the particle distribution function. Extending to the self-consistent case the arguments about the quasi-Gaussianity of the velocity distribution function in the chaotic regime, we deduce that f is also diffusing when $\tau_w \gg \tau_{spread}$. Since f is a density in p space, a flux $D_{QL}(p_0)\partial_p f(p_0)$ of particles is passing at $p = p_0$ and the conservation of the number of particles implies we recover (7.14) with only its first term in the right-hand side (as in section 6.11). Naturally we miss the second one which corresponds to a drag on particles related to spontaneous emission: indeed this effect is not related to particle diffusion. However, it corresponds to the sum of the individual (Cherenkov-like) emission by particles and we may also expect it to occur in the strongly nonlinear limit. This justifies (7.14) also in this limit.

In section 4.1.1 we computed the quasilinear diffusion coefficient at velocity p by taking into account waves with a phase velocity within several γ_{jL}/k_j of p. Similarly, we computed the Landau growth rate of a wave with phase velocity p by considering particles with velocities within several γ_{jL}/k_j of p. This was because, statistically, the wave–particle momentum was locally conserved in p. In the present case of strong resonance overlap, the momentum exchange between particles and waves may also be viewed as a local process, with the 'interaction range' in velocity being now defined by extending the concept of resonance box.

In section 4.1.1 the local conservation of momentum turned out to relate a Landau growth to a quasilinear drag $\partial_p D_{QL}(p)$. In the chaotic case we find again a quasilinear diffusion coefficient. Then the Fokker–Planck equation (7.14) implies the existence of a quasilinear drag as well; and the local conservation of momentum therefore implies a Landau growth for the waves. This collective effect must be complemented again with spontaneous emission.

Thus the wave dynamics should be ruled statistically in both limits $\tau_w \gg \tau_{spread}$ and $\tau_w \ll \tau_{spread}$ by

$$\langle \dot{I}_j(t) \rangle = 2\gamma_{jL}\langle I_j \rangle + S_j \tag{7.15}$$

where

$$\gamma_{jL} = \frac{\pi}{2} \frac{N\varepsilon^2 \beta_j^2}{k_j^2} \frac{\partial f}{\partial p}(v_j, t) \tag{7.16}$$

is the Landau exponentiation rate and

$$S_j = \frac{\pi N\varepsilon^2 \beta_j^2}{k_j^3} f(v_j, t) \tag{7.17}$$

is the source term due to spontaneous emission, in agreement with (4.16).

The reason for obtaining the same quasilinear equations in the two time limits of interest can be understood as follows. The calculation of the diffusion coefficient shows that the correlations between waves and particles are too weak to modify it with respect to the non-self-consistent case which yields a quasilinear estimate both when chaos is negligible and when it is strong, provided resonance overlap is strong too. Once the diffusion coefficient is known, the quasilinear friction coefficient on particles is a natural consequence. So is the Landau formula for waves, thanks to the local conservation of wave–particle momentum.

In both regimes of interest quasilinear diffusion is an effect due to the compound action of many waves. However, the validity of the Landau formula is due to different physics in the two cases: non-chaotic quasilinear diffusion for $\tau_w \ll \tau_{spread}$ and chaotic quasilinear diffusion of particles over a few resonance box widths ($\sim(k_j \tau_{spread})^{-1}$) for $\tau_w \gg \tau_{spread}$.

It is important to note that spontaneous emission is like a Cherenkov emission where particles act individually, but that the stimulated (Landau) emission is a collective phenomenon which only involves the average drag on particles, be it due to a regular or to a chaotic motion. Note that chaos does not decrease the stimulated emission as it could since there is some spatial coherence between particles with the same velocity in the regular case.

A priori one could have wondered whether the fluctuating amplitude of the waves would not change their diffusive action on particles. The absence of such a change can be understood as follows. The fluctuation of the wave amplitudes is due to a mechanism similar to trapping where the equivalent of the trapping time is the spreading time which bounds the duration of the coherent motion of a bunch of particles. Therefore a wave has an effective frequency spectrum with a width $\sim \tau_{\text{spread}}^{-1}$, which amounts to an effective phase velocity uncertainty or spreading $\sim (k_j \tau_{\text{spread}})^{-1}$. But, over the time τ_{QL} involved in the asymptotic estimate of the diffusion coefficient, a particle is acted upon by waves in a phase velocity domain $\sim (D_{\text{QL}} \tau_{\text{QL}})^{1/2} \gg (k_j \tau_{\text{spread}})^{-1}$, so that the finite width of a single wave frequency spectrum does not bar a wave from contributing with its full quadratic mean amplitude to the diffusion coefficient.

At this point the reader may either go to section 7.5 or work out the next two sections which provide a more rigorous derivation of equations (7.14) and (7.15) in the $\tau_w \gg \tau_{\text{spread}}$ regime.

7.3 Evolution of waves*

We consider a time t_r and focus on the time interval $[t_r, t_r + \Delta t]$, with $t_r \gg \Delta t \gg \tau_c$. For $\mu_{\text{ws}} \ll 1$, let $\tau_c \ll \Delta t \ll t_r \ll \tau_{\text{spread}}$; for $\mu_{\text{ws}} \gg 1$, let $\tau_w \gg t_r \gg \tau_{\text{QL}} \gg \Delta t \gg \tau_{\text{spread}} \gg \tau_c$. The formal solution to the equations of motion yields the system of integral equations

$$x_l(t) = x_l^0(t) + \delta x_l(t) \tag{7.18}$$

$$z_j(t) = z_j(t_r) + i\beta_j' \int_{t_r}^{t} \sum_{l=1}^{N} e^{i[\omega_{j0} t' - k_j x_l(t')]} \, dt' \tag{7.19}$$

where

$$x_l^0(t) = x_l(t_r) + p_l(t_r)(t - t_r) \tag{7.20}$$

$$\delta x_l(t) = \sum_{j=1}^{M} \delta x_{lj}(t) \tag{7.21}$$

$$\delta x_{lj}(t) = \int_{t_r}^{t} \int_{t_r}^{t'} \Re(i\beta_j' k_j z_j(t'') e^{i[k_j x_l(t'') - \omega_{j0} t'']}) \, dt'' \, dt'. \tag{7.22}$$

We now want to compute $\langle \Delta |z_j^2(t_r, \Delta t)| \rangle \equiv \langle |z_j^2(t_r + \Delta t)| - |z_j^2(t_r)| \rangle$. Equation (7.5) yields

$$\langle \Delta |z_j^2(t_r, \Delta t)| \rangle = i\beta_j' \sum_{l=1}^{N} \int_{t_r}^{t_r + \Delta t} \langle z_j^*(t) e^{i(\omega_{j0} t - k_j x_l(t))} \rangle \, dt + \text{CC}. \tag{7.23}$$

To compute the average, we must make the dependence of x_l on wave j more explicit. This can be done using the Picard functional fixed-point equations (7.18) and (7.19). To this end we define

$$\Phi_{lj}(t) = \Psi_{lj}(t) - k_j \delta x_{lj}(t). \tag{7.24}$$

We assume that Δt is small enough for $|k_j \delta x_{lj}(\Delta t)| < \eta$ for all (l, j), where η is a small quantity (this condition will be made more precise by (7.42) at the end of this section). On this basis, we approximate $\langle \Delta | z_j^2(t_r, \Delta t)| \rangle$ by its first-order expansion in η:

$$
\begin{aligned}
\langle \Delta | z_j^2(t_r, \Delta t)| \rangle = i\beta_j' \sum_{l=1}^{N} \int_{t_r}^{t_r+\Delta t} & \left\langle z_j^*(t) e^{-i\Phi_{lj}(t)} \right. \\
& \left. \times \left(1 - i \int_{t_r}^{t} \int_{t_r}^{t'} \Re[i\beta_j' k_j^2 z_j(t'') e^{i\Psi_{lj}(t'')}] \, dt'' \, dt' \right) \right\rangle dt + \mathrm{CC} \\
= i\beta_j' \sum_{l=1}^{N} \int_{t_r}^{t_r+\Delta t} & \left[\langle z_j^*(t_r) e^{-i\Phi_{lj}(t)} \rangle \right. \\
& - i\beta_j' \int_{t_r}^{t} \sum_{n=1}^{N} \langle e^{i[\Psi_{nj}(t') - \Phi_{lj}(t)]} \rangle \, dt' \bigg] dt \\
+ \frac{1}{2}\beta_j' k_j^2 \sum_{l=1}^{N} \int_{t_r}^{t_r+\Delta t} & \int_{t_r}^{t} \int_{t_r}^{t'} (i\beta_j' \langle z_j^*(t) z_j(t'') e^{i[\Psi_{lj}(t'') - \Phi_{lj}(t)]} \rangle \\
& - i\beta_j' \langle z_j^*(t) z_j^*(t'') e^{-i[\Psi_{lj}(t'') + \Phi_{lj}(t)]} \rangle) \, dt'' \, dt' \, dt + \mathrm{CC} \tag{7.25}
\end{aligned}
$$

using (7.22) and (7.19) successively. We now evaluate each term in the last expression.

First we write $\langle z_j^*(t_r) \exp[-i\Phi_{lj}(t)] \rangle = \langle z_j^*(t_r) \langle \exp[-i\Phi_{lj}(t)] \rangle_* \rangle$ where $\langle \rangle_*$ denotes averaging with fixed $z_j(t_r)$. If $\tau_w \ll \tau_{\mathrm{spread}}$, for $t \leq t_r + \Delta t$, $x_l(t)$ is almost uniformly distributed as x_{l0}, $\delta x_{lj}(t)$ is negligible and $\langle \exp[-i\Phi_{lj}(t)] \rangle_*$ vanishes. If $\tau_w \gg \tau_{\mathrm{spread}}$, we note that, for a fixed $z_j^*(t_r)$, the distribution of $x_l(t)$ for $t \geq t_r$ is broad since there are many waves in a resonance box and since t_r is much larger than the time $\sim \tau_{\mathrm{spread}}$ to visit a resonance box. Then $\langle \exp[-i\Phi_{lj}(t)] \rangle_*$ vanishes[1]. Therefore, the first term may be neglected in both time limits of interest.

In either limit we may also neglect the $n \neq l$ contribution to the second term in (7.25). Indeed, if $\tau_w \ll \tau_{\mathrm{spread}}$, the motion of both particles is almost ballistic, and the randomness of their initial positions makes the average vanish in the second term. If $\tau_w \gg \tau_{\mathrm{spread}}$, particle l feels the action of many waves and

---

[1] Note that this reasoning would not hold if $z_j^*(t_r)$ were substituted with $z_j^*(t)$ since $\langle z_j^*(t) \exp[-i\Phi_{lj}(t)] \rangle$ at various t are strongly correlated, which prevents their expectations from being computed independently.

the motion of particle n provides information only about a single combination of them. Then

$$\langle e^{i[\Psi_{nj}(t')-\Phi_{lj}(t)]}\rangle = \langle e^{i\Psi_{nj}(t')}\langle e^{-i\Phi_{lj}(t)}\rangle_*\rangle \simeq 0 \qquad (7.26)$$

where the inner expectation, being conditioned by a fixed $\Psi_{nj}(t')$, vanishes for the t of interest, as this condition has negligible impact.

Then we estimate the average in the third term of (7.25) as

$$\langle z_j^*(t)z_j(t'')e^{i[\Psi_{lj}(t'')-\Phi_{lj}(t)]}\rangle$$

$$= \langle z_j^*(t_r)z_j(t_r)e^{i[\Psi_{lj}(t'')-\Phi_{lj}(t)]}\rangle$$

$$+ i\beta_j'\left\langle z_j^*(t_r)\int_{t_r}^{t''}\sum_{n=1}^{N}\langle e^{i[\Psi_{lj}(t'')-\Phi_{lj}(t)-\Psi_{nj}(t_1)]}\rangle_* \, dt_1\right\rangle$$

$$- i\beta_j'\left\langle z_j(t_r)\int_{t_r}^{t}\sum_{n=1}^{N}\langle e^{i[\Psi_{lj}(t'')-\Phi_{lj}(t)+\Psi_{nj}(t_2)]}\rangle_* \, dt_2\right\rangle$$

$$+ \beta_j'^2\int_{t_r}^{t''}\int_{t_r}^{t}\sum_{n=1}^{N}\sum_{m=1}^{N}\langle e^{i[\Psi_{lj}(t'')-\Phi_{lj}(t)+\Psi_{nj}(t_2)-\Psi_{mj}(t_1)]}\rangle \, dt_2 \, dt_1. \quad (7.27)$$

In the first term in the right-hand side of (7.27) we identify Ψ_{lj} and Φ_{lj}, which brings an error $\sim \eta$ into $\Theta \equiv \Psi_{lj}(t'') - \Phi_{lj}(t)$. The study of the non-self-consistent case shows that over the time interval $|t - t''|$ such that the fluctuations of Θ remain moderate, the contribution of wave j to Θ comes mainly from $k_j[\delta x_{lj}(t'') - \delta x_{lj}(t)]$ which is bounded by η. Therefore, we may split the average in the first term into one for the wave envelope and one for the exponential.

In the second term of (7.27) we may compute the average in two steps by first fixing the value of $z_j^*(t_r)$ which sets only two constraints on the many initial phases and positions of the system. Then the average vanishes for the same reason as for the second term of (7.9). So does the third term of (7.27) for similar reasons. The fourth term is more involved to compute, but will also turn out to be negligible. We neglect it from now on and will prove this to be right after gaining further practice in estimating averages, after (7.38).

The fourth term in (7.25) can be estimated with the same approximations as those used for the third one. All terms which are negligible for the third one are also negligible for the fourth one. With these approximations, the fourth term in (7.25) reduces to the analogue of the first one in (7.27), i.e. $\langle z_j^*(t_r)z_j^*(t_r)\rangle\langle e^{-i[\Psi_{lj}(t'')+\Phi_{lj}(t)]}\rangle$. Then the phase of the exponential involves a dominant fluctuating contribution $2k_jx_l(t_r)$, which gives a vanishing value to the average. Therefore the fourth term in (7.25) may be neglected altogether.

Thus (7.25) reduces to

$$\langle\Delta|z_j^2(t_r, \Delta t)|\rangle = 2\beta_j'^2\int_{t_r}^{t_r+\Delta t}\int_{t_r}^{t}\sum_{l=1}^{N}\langle\cos[\Psi_{lj}(t') - \Psi_{lj}(t)]\rangle \, dt' \, dt$$

$$- (\beta'_j k_j)^2 \sum_{l=1}^{N} \int_{t_r}^{t_r+\Delta t} \int_{t_r}^{t} \int_{t_r}^{t'} \langle |z_j(t_r)|^2 \rangle$$

$$\times \langle \sin[\Psi_{lj}(t'') - \Psi_{lj}(t)] \rangle \, dt'' \, dt' \, dt \tag{7.28}$$

where we identified again $\Phi_{lj}(t)$ and $\Psi_{lj}(t)$, bringing a relative error η into the estimate.

On reversing the order of integrals over t' and t'', and introducing the probability density $P_l(p, x, t)$ for particle l to be at (p, x) at time t, (7.28) becomes

$$\langle \Delta |z_j^2(t_r, \Delta t)| \rangle = 2\beta_j'^2 \sum_{l=1}^{N} \int_{t_r}^{t_r+\Delta t} \int \int P_l(p, x, t)$$

$$\times \int_{t_r-t}^{0} \langle \cos(\Omega_j \tau + \delta \Psi_{lj}(\tau | p, x, t)) \rangle_* \, d\tau \, dp \, dx \, dt$$

$$+ (\beta'_j k_j)^2 \sum_{l=1}^{N} \int_{t_r}^{t_r+\Delta t} \int \int P_l(p, x, t) \int_{t_r-t}^{0} \langle |z_j(t)|^2 \rangle$$

$$\times \tau \langle \sin[\Omega_j \tau + \delta \Psi_{lj}(\tau | p, x, t)] \rangle_* \, d\tau \, dp \, dx \, dt \tag{7.29}$$

where

$$\Omega_j = k_j p - \omega_{j0}, \tag{7.30}$$

$$\delta \Psi_{lj}(\tau | p, x, t) = \sum_{j'=1}^{M} k_j \int_{t}^{t+\tau} \int_{t}^{t_1} \Re(i\beta'_{j'} k_{j'} z_{j'}(t_2) e^{i[k_{j'} x_l(t_2) - \omega_{j'0} t_2]}) \, dt_2 \, dt_1 \tag{7.31}$$

with the constraints $x_l(t) = x$ and $p_l(t) = p$ and where $\langle \bullet \rangle_*$ is the average conditioned with the same constraints. As in the non-self-consistent case, $\delta \Psi_{lj}(\tau | p, x, t)$ spreads with time and takes on appreciable values for a time of the order of τ_{spread} defined as in the non-self-consistent case by (6.79), with D_{QL} defined by (7.13).

In (7.29) the sums over l and p provide a spectrum of Ω_j with a large width $k_j \Delta u$. For the same reasons as for the estimates of Ξ in section F.5, the value of these sums is small unless $|\tau| \leq \tau_c$. As for the non-self-consistent case we assume

$$\tau_c \ll \tau_{\text{spread}} \tag{7.32}$$

and we consider times $t \gg \tau_c$ (this is equivalent to condition (4.13)). Therefore, $\delta \Psi_{lj}(\tau | p, x, t)$ is negligible in the dominant contribution of the integral over τ. Integrating by parts the second integral over p in (7.29) yields

$$\langle \Delta |z_j^2(t_r, \Delta t)| \rangle = 2\beta_j'^2 \sum_{l=1}^{N} \int_{t_r}^{t_r+\Delta t} \int \int P_l(p, x, t)$$

$$\times \int_{t_r-t}^{0} \langle \cos(\Omega_j \tau) \rangle \, d\tau \, dp \, dx \, dt$$

$$+ \beta_j'^2 k_j \sum_{l=1}^{N} \int_{t_r}^{t_r+\Delta t} \iint \frac{\partial P_l(p,x,t)}{\partial p} \langle |z_j(t_r)|^2 \rangle$$

$$\times \int_{t_r-t}^{0} \langle \cos(\Omega_j \tau) \rangle \, d\tau \, dp \, dx \, dt \qquad (7.33)$$

where we have identified $\langle \bullet \rangle_*$ and $\langle \bullet \rangle$ because of the large number of particles. We define the velocity distribution function of the N particles as

$$f(p,t) = \frac{1}{N} \sum_{l=1}^{N} \int P_l(p,x,t) \, dx \qquad (7.34)$$

and assume it evolves over a characteristic time much larger than Δt. As $\Delta t \gg \tau_c$, we use (B.16). Then the integrand in the integral over t may be considered as a constant corresponding to the derivative of $|z_j^2(t)|$ and (7.33) yields

$$\langle \dot{I}_j(t) \rangle \equiv \frac{\langle \Delta |z_j^2(t_r, \Delta t)| \rangle}{2\Delta t} = 2\alpha_j \langle I_j(t) \rangle \frac{\partial f}{\partial p}(v_j,t) + \frac{2\alpha_j}{k_j} f(v_j,t) \qquad (7.35)$$

where we revert to $I_j = |z_j|^2/2$, where t_r is substituted with t, where

$$\alpha_j = \frac{\pi}{2} N \beta_j'^2 = \frac{\pi}{2} \frac{N\varepsilon^2 \beta_j^2}{k_j^2} \qquad (7.36)$$

as in (4.17) and where

$$v_j = \frac{\omega_{j0}}{k_j} \qquad (7.37)$$

$$\Delta v_j = v_{j+1} - v_j. \qquad (7.38)$$

Equation (7.35) may be rewritten as (7.15). Our final equations (7.35) and (7.15) are valid in both limits $\tau_w \gg \tau_{\text{spread}}$ and $\tau_w \ll \tau_{\text{spread}}$. They are identical to (4.14) which was established in the second limit only.

Finally, we must verify that the fourth term in (7.27) is negligible. If $n \neq m$, the average vanishes for the same reason as for the second term in (7.25). Let R be the contribution to the fourth term in the sums with $n = m$ but $n \neq l$. We introduce $t_+ = (t_1 + t_2)/2$, $\tau = t_2 - t_1$ and denote by a star subscript the conditional averages, constrained by keeping constant the phase $\Psi_{lj}(t'') - \Phi_{lj}(t)$. Then, using the fact that the constraint is mild in view of the many phases of the many waves acting on the two particles,

$$R = \beta_j'^2 \left\langle e^{i[\Psi_{lj}(t'') - \Phi_{lj}(t)]} \int_{t_r}^{t''} \int_{t_r}^{t} \sum_{n=1, n\neq l}^{N} \langle e^{i[\Psi_{nj}(t_2) - \Psi_{nj}(t_1)]} \rangle_* \, dt_2 \, dt_1 \right\rangle$$

$$= \beta_j'^2 \Big\langle e^{i[\Psi_{lj}(t'')-\Phi_{lj}(t)]} \int_{t_r}^{(t+t'')/2} \int\int \int_{2\max(t_r-t_+,t_++t-t'')}^{2\min(t_+-t_r,t-t_+)}$$

$$\sum_{n=1,n\neq l}^{N} P_n(p,x,t_+) \langle e^{i[\Psi_{nj}(t_++\tau/2)-\Psi_{nj}(t_+-\tau/2)]}\rangle_* \, d\tau \, dp \, dx \, dt_+ \Big\rangle$$

$$\simeq \beta_j'^2 \Big\langle e^{i[\Psi_{lj}(t'')-\Phi_{lj}(t)]} \int_{t_r}^{(t+t'')/2} \int\int \int_{2\max(t_r-t_+,t_++t-t'')}^{2\min(t_+-t_r,t-t_+)}$$

$$\sum_{n=1,n\neq l}^{N} P_n(p,x,t_+) \langle e^{i\Omega_j\tau}\rangle_* \, d\tau \, dp \, dx \, dt_+ \Big\rangle \qquad (7.39)$$

where in the last expression we have used the fact that the sum over n yields a narrow support to the integral over τ, as we did in (7.29). The expression inside the integral over t_+ has exactly the same structure as that inside the first integral over t in (7.33). With S_j defined by (7.17) this yields

$$R \simeq S_j \langle e^{i[\Psi_{lj}(t'')-\Phi_{lj}(t)]}\rangle (t+t''-2t_r) \qquad (7.40)$$

which has a structure similar to the first term in (7.27). Therefore, $2\Re(iR)$ adds to (7.28) a term

$$\beta_j'^2 k_j^2 \int_{t_r}^{t_r+\Delta t} \int_{t_r}^{t} \int_{t_r}^{t'} (t+t''-2t_r)S_j \sum_{l=1}^{N} \langle \sin[\Psi_{lj}(t'')-\Phi_{lj}(t)]\rangle \, dt'' \, dt' \, dt$$

$$\simeq S_j \beta_j'^2 k_j^2 \int_0^{\Delta t} \int_0^{t} \sum_{l=1}^{N} \int\int P_l(p,x,t) \langle \sin\Omega_j(t-t_1)\rangle (t^2-t_1^2) \, dx \, dp \, dt_1 \, dt$$

$$\simeq 2S_j\alpha_j\Delta t^2 \partial_p f(v_j,t). \qquad (7.41)$$

This adds to (7.15) a term $\simeq S_j\gamma_{jL}\Delta t \simeq S_j\Delta t/\tau_w \ll |S_j|$ in the limit of interest.

Finally, the element with $l=n=m$ of the fourth term in (7.27) is negligible, since it is of order $1/N$ because of the $\beta_j'^2$ coefficient. Therefore, the fourth term in (7.27) is, indeed, negligible. As a result, the wave amplitude fluctuations have no impact on the final result. Since our estimates for $\langle\Delta|z_j^2(t_r,\Delta t)|\rangle$ follow from expanding integral expressions such as (7.19) and (7.22), we note that 'integrals smoothen estimates'—which is also why integral equations are so central to the study of dynamical systems.

Thus, we find that Landau's formula for the wave intensity growth also holds in the strongly nonlinear regime if the unstable spectrum is broad enough. However, Landau's formula does not always describe the waves' evolution, either for the linear regime, as was shown in section 3.7, or for the nonlinear one, as section 8.2.1 will show for a narrow unstable spectrum. In the regime where $\tau_w \sim \tau_{\text{spread}}$, the wave amplitudes change over the chaotic decorrelation time and the previous technique does not work.

In this derivation we assumed that Δt is small enough for $|k_j \delta x_{lj}(\Delta t)| < \eta$ where η is small. The quantity $|\delta x_{lj}(\Delta t)|$ may be estimated in the case $\tau_w \gg \tau_{\text{spread}}$ by analogy with the technique used in the non-self-consistent case, which yields formulae analogous to (6.118) and to (6.120). The equivalent for the second case is

$$|k_j \delta x_{lj}| \simeq [60(\tau_{\text{spread}} \Delta t)^{1/2}/\tau_{\text{discr}}]^{1/2} \simeq 25 N_0^{1/4} s^{-2/3} \qquad (7.42)$$

where s is the typical resonance overlap parameter, where $N_0 = \Delta t/\tau_{\text{spread}}$ is a large number and where τ_{spread} is defined as in the non-self-consistent case by (6.79) with D_{QL} defined at the end of section 7.1. Therefore $k_j \delta x_{lj}(t)$ may be kept small for any fixed t by having a strong enough resonance overlap. For waves outside the initial resonance box of particle l the estimate is lower. For the case $\tau_w \ll \tau_{\text{spread}}$, $\delta x_l(t)$ and thus $\delta x_{lj}(t)$ are small in the time interval of interest.

Exercise 7.1. Show that (7.35) can be recovered by removing the integrals $\int_{t_r}^{t_r + \Delta t} dt$ of this section, then setting $t_r = 0$ and substituting Δt with t in all expressions. Then Δt must be substituted with t in (7.42), which is a lot more stringent.

Note that, in this section, we used explicitly only the randomness of the particle positions but also the randomness of the wave phases was used as our estimates rely on the results of chapter 6.

7.4 Drag on particles*

The drag on particles with velocity p may be evaluated by an approach similar to the one used in the non-self-consistent case (section 6.9) and to the one used for computing $\langle \dot{I}_j(t) \rangle$ in section 4.1.2. We note that (2.112) is similar to (7.23) where the right-hand side is multiplied by $-k_j/2$ and the sum over l is substituted with a sum over j. Then the calculation of the evolution of waves may be easily transposed. The analogue of (7.28) is

$$\langle p_l(t_r + \Delta t) - p_l(t_r) \rangle_*$$

$$= -\sum_{j=1}^{M} \beta_j'^2 k_j \int_{t_r}^{t_r + \Delta t} \int_{t_r}^{t} \langle \cos(\Psi_{lj}(t') - \Psi_{lj}(t)) \rangle \, dt' \, dt$$

$$+ \sum_{j=1}^{M} \beta_j'^2 k_j^3 \int_{t_r}^{t_r + \Delta t} \int_{t_r}^{t} \int_{t_r}^{t'} \langle |z_j(t)|^2 \rangle \langle \sin[\Psi_{lj}(t'') - \Psi_{lj}(t)] \rangle \, dt'' \, dt' \, dt$$

$$(7.43)$$

where $\langle \bullet \rangle_*$ is the average over all phases restricted by the constraint $p_l(t_r) = p$. $\langle \bullet \rangle_*$ was identified with $\langle \bullet \rangle$ in the right-hand side of (7.43) thanks to the large number of independent phases. From this point the calculation closely follows that for the non-self-consistent case of section 6.10. It yields

$$\langle p_l(t_r + \Delta t) - p_l(t_r) \rangle_*$$

$$= - \sum_{j=1}^{M} k_j (\varepsilon \beta_j)^2 \int_{t_r}^{t_r + \Delta t} \langle I_j(t) \rangle \int_{t_r - t}^{0} \tau \langle \cos(\Omega_j \tau + \delta \Phi_{lj}(\tau | p, t)) \rangle \, d\tau \, dt$$

$$- \sum_{j=1}^{M} \frac{(\varepsilon \beta_j)^2}{k_j} \int_{t_r}^{t_r + \Delta t} \int_{t_r - t}^{0} \langle \cos(\Omega_j \tau + \delta \Phi_{lj}(\tau | p, t)) \rangle \, d\tau \, dt \qquad (7.44)$$

where

$$\delta \Phi_{lj}(\tau | p, t) = k_j \sum_{j'=1}^{M} \int_{t_r}^{t_r + \tau} \int_{t_r}^{t_1} \Re(i\beta'_j k_{j'} z_{j'}(t_2) e^{i[k_{j'} x_l(t_2) - \omega_{j'0} t_2]}) \, dt_2 \, dt_1 \qquad (7.45)$$

with the constraint $p(t) = p$.

Exercise 7.2. Show that, for $\tau_w \gg \tau_{\text{spread}}$, the second term in (7.44) is negligible and this equation is identical to (4.30).

If $\langle I_j(t) \rangle$ has a negligible evolution over the scale Δt, in both time limits, (7.44) has the same structure as (6.113) with one more term corresponding to the spontaneous emission. The two terms in (7.44) may be estimated as the term of (6.113) by taking into account the fact that the integral over τ involves an integrand with a support of order τ_c. This finally yields

$$\langle \dot{p}_l(t) \rangle_* \equiv \frac{\langle p_l(t_r + \Delta t) - p_l(t_r) \rangle_*}{\Delta t} = \frac{\partial D_{\text{QL}}(p)}{\partial p} + F_c(p) \qquad (7.46)$$

where $D_{\text{QL}}(p)$ is defined by (7.13) and where $F_c(p)$ is the smooth interpolation associated with the function

$$F_c(v_j) = -\frac{2\alpha_j}{N \Delta v_j}. \qquad (7.47)$$

In this expression, the kinetic limit $N \to \infty$ and the continuous wave-spectrum limit $\Delta v_j \to 0$ balance each other. If one starts with a given spectrum and lets the number of particles go to infinity, the spontaneous wave emission vanishes, due to destructive interference between the individual particle contributions. Alternatively, if one first takes the limit of a continuous spectrum, the force $F_c(v_j)$ diverges, which means that one cannot neglect the granular nature of the plasma. In other words, in taking the continuous-spectrum limit, one must ensure that there are enough particles per velocity bin Δv_j (their number is $\sim N \Delta v_j / \Delta u$) to guarantee that the limit is well behaved. In the limit of a continuous spectrum, one recovers $F_c(p) = -(L\varepsilon^2\beta^2)/[2\omega(p)]$, i.e. (4.36). As in chapter 6 the derivation of (7.46) uses the randomness of the wave phases.

As shown in section 4.1.3, the system (7.35)–(7.46) satisfies the conservation of momentum; this is also true separately for their collective and spontaneous emission parts (section 4.1.4).

Exercise 7.3. What are the dimensions of $I(p)$, $f(p)$, $D(p)$?

Exercise 7.4. Show that (7.46) can be recovered by removing the integrals $\int_{t_r}^{t_r+\Delta t} dt$ of this section, then setting $t_r = 0$ and substituting Δt with t in all expressions. Then in (7.42), Δt must be substituted with t, which is more stringent.

Combining the results about the drag and the diffusion, and extending these arguments about the quasi-Gaussianity of the velocity distribution function to the self-consistent case, we see that the evolution of particles is ruled in both limits $\tau_w \gg \tau_{\text{spread}}$ and $\tau_w \ll \tau_{\text{spread}}$ by the Fokker–Planck equation (7.14). In contrast, when trying to compute the moments of $p_l(t_r+\Delta t) - p_l(t_r)$ in the regime $\tau_w \sim \tau_{\text{spread}}$, we find correlations which do not exist in the extreme regimes. Therefore it is not even certain that a Fokker–Planck equation is correct in this regime.

7.5   Joint evolution of particles and waves

Coupling (7.14) with (7.15) enables the description of the self-consistent evolution of the wave–particle interaction in both time limits. These equations take a simpler form on introducing $\psi(v)$, the density of wave momentum per particle, which is a smooth function going through the values

$$\psi(v_j) = \frac{k_j I_j}{N \Delta v_j} \tag{7.48}$$

and which is dimensionless (assuming particle masses $m = 1$) as we define it in terms of the wave momentum $k_j I_j$. Then the quasilinear equations for the wave–particle system read:

$$\partial_t f = \partial_p \partial_t \psi \tag{7.49}$$
$$\partial_t \psi = 2\alpha \psi \partial_p f - F_c f. \tag{7.50}$$

These equations are established for both time limits $\tau_w \gg \tau_{\text{spread}}$ and $\tau_w \ll \tau_{\text{spread}}$. The first equation is remarkable as it implies an infinity of conservation laws for the quasilinear regime of plasma turbulence. Let

$$v(p) \equiv f(p) - \partial_p \psi(p). \tag{7.51}$$

Exercise 7.5. Show that, for any $p \in [p_{\min}, p_{\max}]$ where $p_{\min}$ and $p_{\max}$ are arbitrary, $v(p)$ is conserved by system (7.49), (7.50) and that, for any n,

$$h^{(n)}(t) \equiv \int_{p_{\min}}^{p_{\max}} (p^n f(p, t) + np^{n-1} \psi(p, t)) \, dp \tag{7.52}$$

obeys a balance equation of the form

$$\frac{dh^{(n)}}{dt} = j^{(n)}(p_{\max}, t) - j^{(n)}(p_{\min}, t) \tag{7.53}$$

where the functions $j^{(n)}(p, t)$ need to be computed only at the boundary of the spectrum and of the particle velocity distribution. In particular, what are $h^{(n)}(t)$ for $n = 0$, $n = 1$ and $n = 2$? How do these constants of motion relate to the general constants of motion, such as H_{sc}, P and N from self-consistent dynamics?

The conservation law $\partial_t v(p, t) = 0$ may be understood as expressing the detailed balance for momentum between particles and waves at velocity p. This detailed balance is a manifestation of the locality (in velocity) of the wave–particle interaction.

Given a constant $f_0 > 0$, consider the entropy

$$s(t) = - \int_{p_{\min}}^{p_{\max}} \left(f(p, t) \ln \frac{f(p, t)}{f_0} + \frac{F_c(p)}{2\alpha(p)} \ln \psi(p, t) \right) dp. \tag{7.54}$$

This expression may be interpreted in the frame of classical equilibrium statistical mechanics (see, e.g., Balescu (1975) and chapter 9) as the sum of the Boltzmann entropy of the particles (their total number being fixed) and of the populations of classical radiative degrees of freedom (which can be emitted or absorbed) corresponding to the waves.

Exercise 7.6. Show that

$$\frac{ds}{dt} \geq j_{(s)}(p_{\max}, t) - j_{(s)}(p_{\min}, t) \tag{7.55}$$

where the functions $j_{(s)}(p, t)$ need be computed only at the boundary of the spectrum and of the particle population. Moreover, show that equality holds in (7.55) if and only if $\partial_t \psi = 0$ on $[p_{\min}, p_{\max}]$. Show that the right-hand side of (7.55) vanishes if $\delta_t \psi(p_{\min}) = \delta_t \psi(p_{\max}) = 0$ (which implies an H-theorem). Compare with exercise E.6.

7.6   Saturation of the weak-warm-beam instability

Equations (7.14) and (7.15) are a system of coupled equations describing the wave–particle interaction in the case of a broad wave spectrum. We now apply

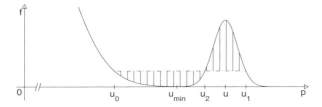

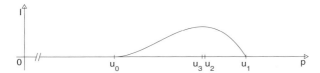

Figure 7.1. Velocity distribution function of the particles at $t = 0$ (continuous line) and at saturation (dotted line) before the final evolution due to spontaneous emission. At $t = 0$ the left-hand side of the distribution corresponds to the decreasing part of the Maxwellian distribution of the thermal plasma and the bump corresponds to the beam. The two hatched domains have the same area.

them to the saturation of the instability due to a weak warm electron beam present in a plasma at thermal equilibrium at $t = 0$. Let u_{min} be the minimum velocity in the beam, let u be the maximum of the beam distribution function as shown in figure 7.1 and assume that the distribution is monotonically increasing between u_{min} and u. We assume that the beam is initially warm enough for the Landau growth rate given by (7.16) to verify $\gamma_{jL} \ll \omega_p$ for all modes made unstable by the beam. Thus, all unstable modes grow with their respective Landau rate.

The time evolution of the waves is given by the largest Landau growth rate, so that $\tau_w = \gamma_{jLmax}^{-1}$. Initially the amplitude of the Langmuir waves is thermal and thus small. Therefore, the condition $\tau_c \ll \tau_w \ll \tau_{spread}$ applies[2] and equations (7.14) and (7.15) may be used to describe the system evolution. This predicts the growth of a broad spectrum of Langmuir waves which are resonant with the beam particles in the region where $f(p, 0)$ has a positive slope. Their spectrum is basically given by the Bohm–Gross dispersion relation (2.1). Simultaneously with the waves' growth, the particles are being diffused by the waves.

Both effects weaken the condition $\tau_w \ll \tau_{spread}$, since γ_{jLmax} decreases due to diffusion and the amplitude of the waves increases. Therefore, after some time, the system enters the regime $\tau_w \simeq \tau_{spread}$ which cannot be described by the system (7.14), (7.15) on the basis of our previous analytical calculations. Let $\mathcal{V}$ be the range of velocities where the initial slope of the distribution function

[2] Condition $\gamma_{jL} \ll \omega_p$ corresponds to $\tau_c \ll \tau_w$ if the beam has a broad enough distribution function.

is positive. Since L is large, particles with a velocity in $\mathcal{V}$ are resonating with strongly overlapping Langmuir waves. Therefore, $[0, L] \times \mathcal{V}$ is a chaotic domain of (x, p) space and the distribution function tends to flatten. This decreases the particle mean momentum and, subsequently, increases the wave amplitude. Both effects drive the system into the regime $\tau_w \gg \tau_{spread}$ which is again described by (7.14) and (7.15).

As the initial unstable waves grow, the domain where they produce a diffusion extends somewhat beyond the initial domain $[u_{min}, u]$. Therefore, the distribution function is eroded there as well, which extends the domain with positive slope of the distribution and further broadens the spectrum of unstable waves[3]. The evolution of the system can be understood as the transfer of momentum from particles to the waves, which translates into the growth of their amplitude: indeed, due to the chaotic flattening of the distribution function, particles lose momentum on average. This process slows down when the chaotic domain starts capturing particles from the zone with an initially negative slope. Indeed, on becoming chaotic, these particles, on average, gain momentum at the expense of the waves' momentum.

To find an equilibrium, one sets the time derivatives to zero in (7.14) and (7.15). To make the definition of this equilibrium simpler, it is interesting to consider first the case where spontaneous emission is negligible in the regime $\tau_w \gg \tau_{spread}$ (this occurs for a sufficiently intense beam).

For a negligible spontaneous emission, (7.14) and (7.15) tell us that the equilibrium corresponds to $\partial_p f = 0$ in the chaotic domain. Indeed, then the Landau growth stops and the first two derivatives of f vanish. This means that a plateau $\mathcal{P}$ develops in the distribution function. The features of $\mathcal{P}$ are easily found graphically in figure 7.1. Let u_0 and u_1 be the minimum and maximum velocity in $\mathcal{P}$ and denote by $f_{\mathcal{P}}$ the value of f in $\mathcal{P}$. As $\mathcal{P}$ is connected to the initial distribution at u_0 and u_1, $f_{\mathcal{P}} = f(u_0, 0) = f(u_1, 0)$. The conservation of the particle number yields

$$\int_{u_0}^{u_1} [f_{\mathcal{P}} - f(p, 0)] \, \mathrm{d}p = 0. \tag{7.56}$$

In figure 7.1 this translates into the equality of areas of the two hatched domains.

To obtain a more quantitative description of the wave spectrum at saturation, we assume that the system is described during the whole relaxation process by (7.14) and (7.15), where γ_{jL} and $D_{QL}(p)$ are possibly modified by a multiplicative factor when $\tau_w \simeq \tau_{spread}$. This factor must be identical, for the evolution of particles and waves is constrained by total momentum conservation[4], as we noted in section 4.1.4, so that (7.49) still holds.

---

[3] At the edge of the domain, the wave amplitudes are small and the $\tau_w \ll \tau_{spread}$ regime may be the dominant one there.

[4] Such a conservation is also present in the Vlasovian derivation of quasilinear theory (Kaufman 1972).

Let us start from (7.49), (7.50) neglecting spontaneous emission. The conservation of (7.51) implies

$$f(p, t) - f(p, 0) = \frac{\partial \psi(p, t)}{\partial p} \tag{7.57}$$

where $\psi(p, 0)$ was neglected. Thus

$$\psi(p, +\infty) = \int_{u_0}^{p} [f_{\mathcal{P}} - f(v, 0)] \, dv \tag{7.58}$$

when the plateau state is reached. $\psi(p, +\infty)$ goes through a maximum for $p = u_2$ where u_2 is the value of p where $\mathcal{P}$ cuts the initial distribution (figure 7.1). As a result of the conservation of the total momentum (exercise 7.5), $\int_{u_0}^{u_1} \psi(p) \, dp = - \int_{u_0}^{u_1} p[f_{\mathcal{P}} - f(p, 0)] \, dp$. The total energy of Langmuir waves in the plateau state is

$$W = \int_{u_0}^{u_1} \langle I(v, +\infty) \rangle \omega(v) \, dv = N \int_{u_0}^{u_1} p \psi(p) \, dp$$

$$= - N \int_{u_0}^{u_1} \frac{p^2}{2} [f_{\mathcal{P}} - f(p, 0)] \, dp \tag{7.59}$$

also in agreement with exercise 7.5. This shows that W is equal to the energy lost by the beam. This result does not follow trivially from the conservation of energy, because of the presence of a coupling energy in H_{sc}. However, the average value of this coupling energy turns out to vanish both in the initial and in the final state, since the waves are weakly coupled to any particle, and since the particle positions are random—initially by assumption and finally because of chaos.

Exercise 7.7. Consider the simplest case of interest, i.e. an initially fast, monokinetic beam, as sketched in figure 7.2, so that $u \gg |u - u_{\min}|$ and $u \gg v_T$. Then $f(p, 0) \simeq \delta(p - u)$. As $u \gg u_{\min} > u_0$, show that when the plateau has been reached, $f_{\mathcal{P}}(p) \simeq 1/u$ and $\psi_{\mathcal{P}}(p) \simeq p/u$ for $p \gg u_0$. Using (7.59), show that $W = Nu^2/3$, which means that two-thirds of the beam energy has gone to the waves.

As already explained in section 4.1.4, the spontaneous emission terms erode the plateau on a longer time-scale, as a prelude to a new final thermal equilibrium of the whole beam–plasma system (the complete walk toward this equilibrium needs to be described by the initial model (1.3)–(1.5)).

Numerical simulations of the dynamics of the self-consistent Hamiltonian (2.109) were performed to check the nonlinear behaviour of the system and to compare it with the predictions of equations (7.14) and (7.15) (Cary *et al* 1992, Doxas and Cary 1997). It is convenient to rescale our parameter μ_{ws} to

$$\mu \equiv \frac{\tau_w}{\tau_D} \simeq 4\mu_{ws} \tag{7.60}$$

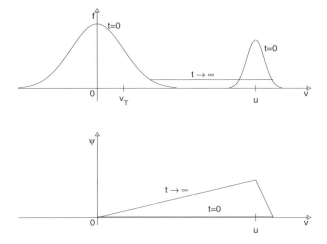

Figure 7.2. Sketch of velocity distribution function and wave intensity spectrum for $t = 0$ and $t \to \infty$ in the case of an initially almost monokinetic beam.

where the estimate uses (6.79), assuming that all wavenumbers have comparable magnitudes.

In the regime $\tau_w \ll \tau_{\text{spread}}$ the system (7.14), (7.15) was found to be correct. However, from $\mu \simeq 0.6$ on when μ increased, the wave spectrum of individual realizations appeared to be jagged with respect to the average one, each wave having a temporal behaviour strongly marked by nonlinear wave coupling. This confirms the behaviour found by Laval and Pesme (1983a), Theilhaber *et al* (1987) and Berndtson *et al* (1994).

In the simulation of Doxas and Cary (1997), the resonance overlap of the neighbouring waves is strong enough for the spectrum to be considered as continuous from the viewpoint of defining the diffusion coefficient. This simulation is run for $\mu \leq 40$. When $\mu \geq 3$ the growth rate is seen to be enhanced over the quasilinear estimate by a factor which increases steadily up to 1.3 for $\mu = 25$. The next and last point of the simulation at $\mu = 40$ stays at 1.3. This might correspond to a saturation of the renormalization before going back to the quasilinear estimate predicted by this chapter. However, the wave spectrum of the calculation is not dense enough for the estimate in (7.42) to be small, even if one takes $\Delta t = \tau_{\text{spread}}$.

An analytical calculation, performed in the frame of the Vlasov–Poisson description for μ small, predicts that the growth rate and the diffusion coefficient should be multiplied by a factor

$$\chi = 1 + 0.02\mu^6 \tag{7.61}$$

with respect to the quasilinear estimate (Laval and Pesme 1983a). According to this estimate, the renormalization effect becomes $O(1)$ for $\mu \simeq 0.02^{-1/6} \simeq 2$

and the renormalization (7.61) is large for larger values of μ. Yet the simulation by Doxas and Cary (1997) agrees only qualitatively with this prediction since the renormalization is hardly visible for $\mu = 3$, when (7.61) predicts the factor χ to be already 16 for this value.

7.7   Historical background and further comments

The system (7.14), (7.15) was first introduced in the framework of the Vlasov–Poisson description of the plasma by Vedenov *et al* (1962) and by Drummond and Pines (1962). These two works dealt with the plasma in a quasilinear way as they neglected mode coupling for the wave growth and considered the particle evolution to be close to a ballistic one[5]. This enabled the introduction of the quasilinear diffusion coefficient long before chaos became fashionable in physics. Furthermore, this introduction was first done for the self-consistent problem.

Nine years later, Roberson and Gentle (1971) published experimental results about the bump-on-tail instability which agreed with the quasilinear predictions. In particular, the plateau formation was observed. However, the precision of the experiments ruled out a precise quantitative check of the theory in the strongly nonlinear regime.

From a theoretical viewpoint, the validity of quasilinear theory was questioned by Adam *et al* (1979) when accounting for nonlinear wave coupling. The importance of this coupling was denied by Galeev *et al* (1980). Then came the analytical proof in dimension one, mentioned in the previous section, of a renormalization of the growth rate and of the diffusion coefficient when μ grows (Laval and Pesme 1983a) and of the inconsistency of quasilinear theory due to mode coupling (Laval and Pesme 1983b). However, Laval and Pesme (1983a) showed the absence of inconsistency in the quasilinear equations in dimensions two and three. The same authors proposed a model predicting a renormalization by a factor 2.2 when the parameter μ (more precisely 10μ) is large (Laval and Pesme 1984). Later, Liang and Diamond (1993a, b) stated this model to be inconsistent and proposed a derivation of the quasilinear equations in the $\mu \gg 1$ regime by extending a technique initially proposed by Boutros-Ghali and Dupree (1981). Shapiro and Sagdeev (1997), by investigating the four-wave coupling, asserted that quasilinear theory works if the particle distribution function and the wave spectrum are averaged over a resonance box width.

More information about this controversy can be found in Doxas and Cary (1997) and Laval and Pesme (1999). This analytical controversy was developed on the basis of the Vlasov–Poisson description of the plasma. It was complemented by numerical simulations with the same description. Such

---

[5] Rigorously speaking, the quasilinear equations had already been mentioned by Romanov and Filippov (1961). This reference makes the ansatz of a Fokker–Planck equation for particle evolution and computes the corresponding diffusion coefficient; it describes the evolution of the Langmuir wave amplitude as the result of spontaneous and stimulated emission of quanta and estimates the corresponding coefficients.

simulations were made for intermediate values of the resonance overlap parameter (Theilhaber *et al* 1987, Berndtson *et al* 1994). They showed a strong nonlinear wave coupling but did not enter the $\mu > 4$ regime. The difficulty of dealing analytically with the strongly nonlinear regime of the Vlasov–Poisson system led from 1989 to the development of the finite-number-of-degrees-of-freedom approach of this book. This made possible the numerical simulations reported in the preceding section.

The controversy about quasilinear theory also motivated a series of experimental works. One was a beam–plasma experiment (Krivoruchko *et al* 1981). However, in order to control the wave dynamics better and to limit noise, experimentalists left beam–plasma experiments for travelling wave-tube experiments where slow waves are amplified by a warm electron beam (Tsunoda *et al* 1991, Hartmann *et al* 1995). All these experimental works confirmed the spectrum jagging seen numerically and the related strong nonlinear wave coupling. The most recent of these experiments reached up to $\mu = 1.4$, where no significant difference with respect to the quasilinear predictions was observed.

Predicting a renormalization by a factor larger than one for μ large sounds natural *a priori*, since analytical results and numerical simulations indicate that such a renormalization occurs when μ increases from small values (without becoming large). Furthermore, recovering the quasilinear estimate for μ large is surprising because the nature of the dynamics is then quite different from that for small μ. Anyhow the quasilinear equations also hold for μ large for several reasons: the weak correlation of any wave and any particle, the even weaker correlation between any two waves or any two particles, the local conservation of wave–particle momentum and the quasilinear character of diffusion in the non-self-consistent case for both the chaotic and non-chaotic regimes in the limit of a continuous spectrum (strong resonance overlap).

Chapter 8

Time evolution of the single-wave–particle system

In this chapter we discuss the results of numerical simulations of the self-consistent system, focusing on the single-wave model ($M = 1$) with N large. *A priori*, this problem sounds simpler than the one dealt with in the previous chapter. Indeed it deals with the self-consistent analogue to the motion of one particle in one wave, whose phase portrait is much simpler than that corresponding to chaotic diffusion. However, its present analytical approach uses Gibbs statistical mechanics, knowledge of which is not a prerequisite to studying this book and which will be discussed in the next chapter. Besides, the previous chapter extended to the self-consistent case techniques and results of chapter 6, whereas we now turn to different approaches.

In order to describe the long-time evolution properly, before turning to numerical simulations, we first recall the relevant time-scales and basic estimates for the nonlinear regime of the system evolution, in particular those related to particle trapping (section 8.1). In section 8.2 we discuss the long-time evolution of the unstable wave, for the case of an initially cold beam and for an initially warm beam. Section 8.3 considers the long-time evolution of the damping wave. For both the unstable and the damped cases, section 8.2.3 compares the finite-N dynamics with the results of kinetic theory sketched in appendix G. This introduces the important fact that the $N \to \infty$ and the $t \to \infty$ limits may not commute in a chaotic system, a fact with possibly far-reaching consequences.

8.1 Self-consistent nonlinear regime with a single wave

The single-wave case is relevant to two kinds of situation. First, in the unstable beam–plasma system, one wave may have a larger growth rate than the others, so that it soon dominates over these and the approximation $M = 1$ is reasonable: this is the case for a cold beam. Second, the physical device of interest, because of its finite length, may have a single wave being resonant with a beam, even a warm one.

If a single wave interacts with all particles, the locality in velocity of the wave–particle interaction is particularly important. In this section we discuss its implications for the nonlinear regime of the beam–wave system. Again, it will appear that only particles with velocities close to the wave phase velocity, up to the resonance width, interact significantly with the wave—and that the relevant velocities thus depend on the current wave intensity.

8.1.1   Instability saturation

It is convenient to introduce the rescaled wave intensity

$$\psi = I/N \tag{8.1}$$

which has the advantage of being closer to the physical quantities (like F_n in (2.41)) as it does not depend on N. The corresponding coupling constant is

$$\varepsilon' = \varepsilon \beta_1 k^{-1} N^{1/2}. \tag{8.2}$$

We drop indices j for the wave-related quantities, except for β_1 to avoid confusion with $\beta = 1/T$ introduced in the Gibbsian approach of the next chapter.

Exercise 8.1. Express ε' in terms of microscopic constants such as V_n and properties of the bulk plasma of chapter 2. Recall that N in chapter 2 is the total number of particles whereas here it denotes the number of 'tail' particles.

With a single wave, initial data are typically a factorized particle distribution $f_n(p_1, \ldots p_n, x_1, \ldots x_n) = \prod_{r=1}^{n} [L^{-1} f_p(p_r)]$ for $n \geq 1$, and a wave $(I_0, \theta_0) = (N\psi_0, \theta_0)$. We discuss several initial data: two elementary velocity distribution functions are the monokinetic beam $\delta(p - p_0)$ and the 'warm' beam. In this second case we take

$$f_p(p) = a + b(p - v_0) \tag{8.3}$$

over a range $v_0 - \Delta v \leq v \leq v_0 + \Delta v$ for some $\Delta v > 0$, to be made more precise later, and $f_p(p) = 0$ outside ($a = 1/(2\Delta v)$ and $b = f_p'(v_0)$).

The instability is investigated with initial conditions for which $\psi \to 0$ in the limit $N \to \infty$. In this limit, the scaled constants of the motion are

$$\sigma \equiv \frac{P}{N} = p_0 = \int_{\mathbb{R}} p f_p(p) \, dp \tag{8.4}$$

$$h \equiv \frac{H}{N} = \frac{p_0^2}{2} + \frac{T_0}{2} \tag{8.5}$$

where p_0 is the beam mean velocity and $T_0 = \int_{\mathbb{R}} (p - p_0)^2 f_p(p) \, dp$ is the beam temperature.

Dynamically, the wave launched at infinitesimal intensity in the warm beam grows at the Landau rate and saturates at a level in agreement with the following simple estimate. Assume that, when the wave intensity saturates, the particle distribution is uniform over the wave cat's eye and that it has not changed outside the cat's eye. Neglecting the wave velocity change due to the interaction, conservation of momentum implies that

$$k\psi = \frac{1}{L} \int_{-L/2}^{L/2} \int_{p_-(x)}^{p_+(x)} (v - \omega_0/k) f(v) \, dv \, dx$$

$$\simeq \frac{1}{2\pi} \int_{-\pi}^{\pi} 2 \int_{\omega_0/k}^{p_+(x)} (v - \omega_0/k)^2 b \, dv \, d(kx) = \frac{64b}{9\pi} \varepsilon'^{3/2} (2\psi)^{3/4} \quad (8.6)$$

where $p_\pm(x)$ correspond to the two branches of the separatrix. Then for the single wave with $k = 1$, the bouncing frequency at saturation is

$$\omega_{sat} = \varepsilon'^{1/2} (2\psi_{sat})^{1/4} \simeq \frac{128b}{9\pi} \varepsilon'^2 = \frac{256}{9\pi^2} \gamma_L \simeq 3\gamma_L. \quad (8.7)$$

The last two forms correspond to the warm beam only. A similar estimate is obtained by using the conservation of energy instead of momentum. It must be considered as an order of magnitude evaluation only, as it is exaggerated to assume that particles outside the 'final' cat's eye will remain distributed as if there were no wave. One merit of (8.7) is that it provides an *a priori* estimate for Δv, say twice the trapping width at this saturation, i.e. $\Delta v \simeq 4(2\varepsilon'^2\psi)^{1/4}$.

8.1.2 Nonlinear regime of wave damping

If the wave damps (if initially $f'_p(\omega_0/k) < 0$ for a broad distribution of velocities), the evolution is more delicate. Indeed, if the wave intensity were constant, trapped particles would undergo bouncing oscillations, with a period of the order of $2\pi/\omega_b = 2\pi \varepsilon'^{-1/2} (2\psi)^{-1/4}$. Then, after half a period (i.e. for $t \simeq \pi/\omega_b$) the trapped particles would have pumped the wave intensity down by a factor $e^{-2|\gamma_L|t}$, so that the bounce frequency is reduced by a factor $e^{-|\gamma_L|t/2}$. At this time, the smaller cat's eye has released some of the initially trapped particles and the remaining ones in the current, smaller, cat's eye slightly pump the wave up, but to a lower intensity than the initial one. Thus, if the initial wave intensity is small enough for $e^{-|\gamma_L|\pi/(2\omega_b)} \ll 1$, i.e. in a strong damping regime, the wave intensity vanishes quickly enough for most particles to be released and Landau's estimate applies. This is called the Landau damping regime.

However, if $e^{-|\gamma_L|\pi/(2\omega_b)} \simeq 1$, a significant fraction of initially trapped particles return to the wave the momentum they borrowed from it. At $t \simeq 2\pi/\omega_b$, most particles have recovered their initial velocity but the momentum balance leaves the wave with a lower intensity. Thus, the bounce frequency is smaller and the decay is accompanied by oscillations of increasing 'period'.

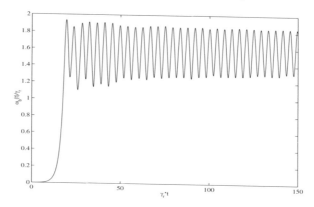

Figure 8.1. Time evolution of $\omega_b(t)/\gamma_L$ for the cold-beam instability. The initial velocity of the beam was chosen to maximize γ_L and the beam is initially spatially homogeneous. (After Firpo *et al* (2001).)

This nonlinear, oscillatory damping regime (see also Mazitov 1965) was discovered by O'Neil (1965), who introduces a current decay rate $\gamma(t)$ such that $dI(t)/dt = 2\gamma(t)I(t)$, and shows that, if the initial ratio $\gamma_L/\omega_b \ll 1$, then $\int_0^\infty \gamma(t)dt = O(\gamma_L/\omega_b)$, which is finite. This implies that, for $t \to \infty$, $\gamma(t)$ vanishes and the wave intensity will not decay to zero.

We now turn to the simulation results in the damping and unstable cases.

8.2   Unstable beam–wave system

8.2.1   Cold-beam–wave instability

For the cold-beam case, $T_0 = 0$. The dynamics of the wave–particle interaction leads to two long-time behaviours.

If the initial beam velocity measured in the wave rest frame is large compared with the expected equilibrium resonance width of the wave, the beam–wave system has no linearly unstable modes and particles keep moving rapidly with respect to the wave (section 3.7.1). Dynamically, if the particles are far from the wave resonance, their motion is in the KAM-like domain of the particle (x, p)-space.

If the beam–wave system has linearly unstable modes, the wave grows but then saturates and displays an oscillatory behaviour (figure 8.1). Indeed the wave traps a significant fraction of particles, but the first trapped particles form a 'macro-particle' moving coherently in the resonance (figure 8.2). To a good approximation, this macro-particle periodically pumps the wave, as its own momentum oscillates due to its bounces. The other particles, crossing the wave separatrix closer to its extremum, are subsequently trapped and detrapped, and

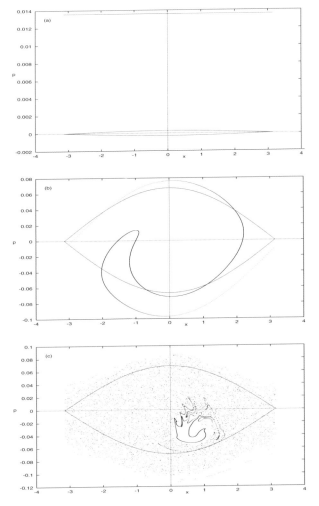

Figure 8.2. Snapshots of the (x, p)-space at (a) $\gamma_L t = 6.24$, (b) $\gamma_L t = 17.5$, (c) $\gamma_L t = 100$ for the cold-beam simulation of figure 8.1. The dots are the $N = 10\,000$ particles, which were initially distributed on a monokinetic beam, faster than the wave (with very small initial intensity). The instantaneous wave resonance 'cat's eye' is drawn to help visualizing the instantaneous force on particles. (After Firpo *et al* (2001).)

their distribution rapidly approaches a nearly uniform distribution over the domain swept by the pulsating separatrix, as seen in figure 8.2. Such behaviour was originally found by O'Neil *et al* (1971) by introducing the set of self-consistent equations which are at the very heart of this book. We note that the separatrix pulsation occurs on the same time-scale as the particle bouncing, which implies that the particle dynamics cannot be considered as similar to the slow chaos of

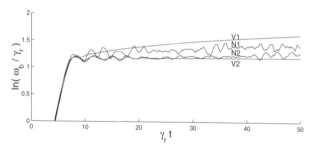

Figure 8.3. Time evolution of $\ln(\omega_b(t)/\gamma_L)$ for CG initial distribution (8.8): kinetic scheme with (V1) 32×128, (V2) 256×1024 (x, p) grid; N-particle system with (N1) $N = 128\,000$, (N2) $N = 512\,000$. (After Doveil *et al* (2001) and Firpo *et al* (2001).)

section 5.5.2.

An efficient simplified Hamiltonian model for the latter behaviour was formulated by Tennyson *et al* (1994), with only four degrees of freedom: the wave, the macro-particle and the two boundaries of the 'chaotic' domain in the long-time regime. In a first approximation, after a few trappings and detrappings, one may assume that the distribution of particles in the domain swept by the wave-pulsating separatrix are quite uniformly distributed in (x, p), so that only the geometry of the domain they occupy affects the wave evolution: this geometry is determined by the boundaries, which one may approximate as KAM boundaries. The density of particles in the chaotic domain is determined by the conservation of their total number.

The other particles remain trapped. As the bounce period is almost the same for all trapped particles (the period diverges only logarithmically near the separatrix), they move quite coherently and behave like a single macro-particle, the mass of which is an unknown parameter in the Tennyson–Meiss–Morrison model: it is determined by the number of trapped particles, which is predicted from their trapping during the initial growth until the onset of nonlinear oscillations (Firpo 1999).

8.2.2   Landau instability

Consider now a warm beam (T_0 is finite). Figures 8.3 and 8.4 display the time evolution of the wave for various initial beams. We run the simulations in the reference frame such that $\omega_0 = 0$ and take $k = 1$. For figure 8.3 the initial distribution f_p is the CG distribution[1] of figure 8.5,

$$f_{CG}(p) = c - \frac{ac}{p + p_1} \qquad (8.8)$$

---

[1] This 'constant growth' distribution is such that the Landau growth rate is the same for all velocities.

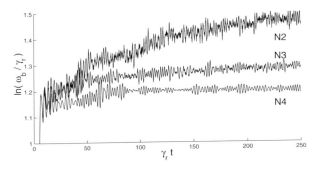

Figure 8.4. For the same conditions as figure 8.3, comparison of CG (N2) with TL initial distribution for (N3) $N = 64\,000$, (N4) $N = 2\,048\,000$. (After Doveil *et al* (2001) and Firpo *et al* (2001).)

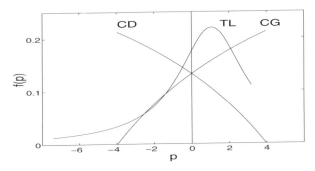

Figure 8.5. Initial warm-beam velocity distributions defined by (8.8), (8.9) and (8.12). (After Doveil *et al* (2001) and Firpo *et al* (2001).)

for $-3.96 \leq p \leq 3.96$ and $f_{CG}(p) = 0$ otherwise. Here, $a = 11.89$ and $p_1 = 15.85$. In figure 8.4 we compare the evolution from the CG initial distribution to the evolution from the TL (truncated Lorentz) distribution

$$f_{TL}(p) = \frac{1}{\pi} \frac{c}{(p - p_1)^2 + a^2} \tag{8.9}$$

for $-7.42 \leq p \leq 3.18$ and $f_{TL} = 0$ otherwise. Here $a = 2.12$ and $p_1 = 1.06$. In (8.8) and (8.9) the normalization constant c ensures that $\int_{\mathbb{R}} f \, dp = 1$.

The runs for figure 8.3 were performed with different numbers of particles: $N_1 = 128\,000$ and $N_2 = 512\,000$, using a symplectic numerical integration scheme (Cary and Doxas 1993). The trapping of particles is illustrated in figure 8.6, which displays the velocity distribution function for line N2 at $\gamma_L t = 200$ (Guyomarc'h *et al* 1996). Figure 8.6(*a*) shows the total distribution, $f_p(p, t) = \int f(x, p, t) \, dx$, and figure 8.6(*b*) shows the conditional distribution given that the particles' initial velocities were in 50 initial 'beamlets'. The latter

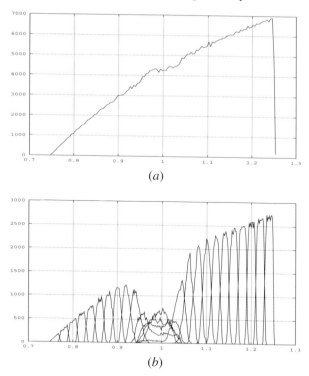

Figure 8.6. Velocity distribution function in the second-growth regime; ordinates are numbers of particles per bin ($N = 512\,000$, $\gamma_L = 1/200$, $t = 40\,000$). Here $\omega_0 = 1$, $k = 1$. (*a*) Total distribution $f(v, t)$ for bins $\Delta v = 0.5/128$. (*b*) Conditional distributions $f(v, t | v_0)$ for 50 initial beamlets ($\Delta v_0 = 0.02$) in bins $\Delta v = 0.2/128$. (After Guyomarc'h *et al* (1996).)

figure confirms that particles away from the wave cat's eye have been weakly affected by the wave.

Figure 8.7 shows snapshots at $\gamma_L t = 200$ for the particles which were initially in a given velocity range. The particles which interacted with the wave from the start are well scattered in the trapped region, whereas particles which interacted later form a thick beam, mainly circulating, but some of them were trapped and some even released by detrapping at a velocity slower than the wave.

Although there appears no 'compact' macro-particle as in the cold-beam case, the trapped particles are not uniformly distributed, and the non-uniformity of the initial velocity distribution also generates a trapped bunch, which pumps the wave. The bunch oscillates approximately at the bounce frequency, which is thus easily observed in figures 8.3 and 8.4.

These oscillations hint at the microscopic process causing the saturation. Initially, for small I, the bouncing period for trapped particles is long compared

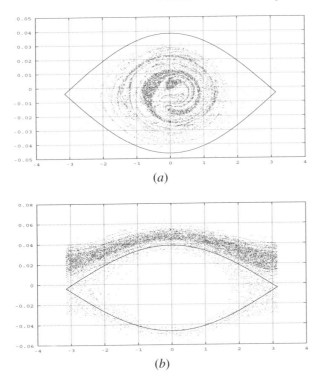

Figure 8.7. Distribution of particles with initial velocities in given ranges. Velocity in wave frame: $v - \omega_0/k = v - 1$ ($N = 512\,000$, $\gamma_L = 1/200$, $t = 40\,000$). (a) $0.99 \leq v_0 \leq 1.0$. (b) $1.03 \leq v_0 \leq 1.04$. (After Guyomarc'h *et al* (1996).)

with the e-folding growth time of the wave. Then the particle motion is nearly ballistic and the growth process obeys the linear equations ensuring exponential growth. But as the wave grows, the bounce period decreases and, when (8.7) holds, the bounce period is close to $2/\gamma_L$. Then the characteristic time-scales for particle motion and for wave evolution are comparable and their evolutions resonate. As the trapped particles are somewhat bunched (because particles trapped at the same time remain together), their periodic motion in the wave implies a periodic source (or loss) for the wave momentum, i.e. for its intensity.

The previous argument holds for the cold-beam case as for the warm case. However, with a warm beam, there are more particles outside the resonance, which may get trapped later by the pulsating-wave separatrix. The pulsations are more chaotic than in the cold-beam case and the trapping–detrapping process is more random. As long as the 'untrapped' particle distribution is unbalanced near the boundary of the resonance, i.e. as long as the distribution function is larger for faster particles than for the slower ones, the wave gains momentum when trapping particles. This process drives a second growth of the wave to still larger amplitudes.

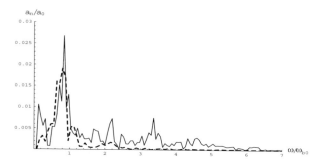

Figure 8.8. Coefficients a_n/a_0 ($n \geq 1$) of the Fourier decomposition of amplitude $\omega_b^2(t)$ during the first time window $38 \leq \gamma_L t \leq 65$ for $N = 48\,000$ (full curve) and $N = 768\,000$ (dashed curve) particles as a function of normalized frequency. (After Firpo *et al* (2001).)

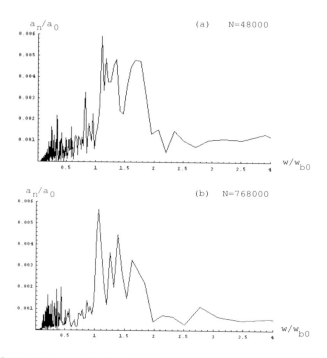

Figure 8.9. Coefficients a_n/a_0 ($n \geq 1$) of the Fourier decomposition of amplitude $\omega_b^2(t)$ during the second time window $250 \leq \gamma_L t \leq 302$ as a function of normalized frequency, for (*a*) $N = 48\,000$ and for (*b*) $N = 768\,000$ particles. (After Firpo *et al* (2001).)

The second growth is much slower than the initial Landau growth and, in contrast with the first one, it is sensitive to the number of particles and to the detailed shape of the initial distribution function. One can monitor the oscillations by Fourier-analysing the wave amplitude. Figures 8.8 and 8.9 display the Fourier amplitudes of the wave intensity for two different time windows, as a function of the frequency normalized to the current bounce frequency ω_b. The dominant oscillations occur indeed at the bounce frequency, as they did already in the cold-beam case but the wave intensity spectrum has significant additional components. These components are more numerous and noisy if N is smaller, and also if one observes the wave at later times.

8.2.3 Comparison with the Vlasov-wave description

The evolutions in the preceding section are now compared with the evolution of the Vlasov-wave system (G.6), (G.7), which describes the limit $N \rightarrow \infty$. It is proved that, given a time $t_* > 0$ and initial data with a number of particles $N \rightarrow \infty$ which approach a smooth initial distribution function $f(x, p, 0)$, the time evolution of the N-particle system coupled with the wave generates a distribution of particles at all $0 \leq t \leq t_*$ close to the solution of the kinetic Vlasov equation (section G.1). However, the larger t_* is, the more particles are needed to keep small the discrepancy between the finite-N evolution and the smooth kinetic evolution.

The Vlasov-wave simulations are performed using a semi-Lagrangian code, which aims at keeping numerical noise to a minimum (Bertrand *et al* 1992, Sonnendrücker *et al* 1999): its accuracy is most sensitive to the size of a mesh $\Delta x \times \Delta p$ over which the Vlasov partial differential equation (G.6) is discretized. The semi-Lagrangian scheme was run on two grids with different meshes, from the corresponding initial data.

The evolution of the wave intensity is displayed in figure 8.3. The Landau instability regime is reproduced accurately by all simulations, in agreement with the kinetic limit theorem. However, the evolutions after saturation diverge. Whereas oscillations persist for the finite-N simulations, they are damped for the kinetic simulations. As the coarser-grid simulation also diverges from the finer one, one suspects a numerical inaccuracy of the kinetic solver. A good accuracy test is provided by the numerical entropy $S_2[f(t)] = \int \int (1-f)f \, dx \, dp$, which is a constant through the evolution according to the Vlasov equation (exercise G.1). As seen in figure 8.10, it is indeed well conserved during the initial wave growth but for all reasonable grid sizes it starts increasing sooner or later.

The finer grid kinetic simulation and the finite-N runs exhibit a few trapping oscillations but they rapidly damp for the kinetic scheme. For the finite-N simulations, the trapping oscillations persist in figure 8.3 and over longer time-scales the wave intensity is seen to drift upward slowly with sustained trapping oscillations in figure 8.4. This second growth of the wave is understood as resulting from the trapping–detrapping of particles by the wave-pulsating separatrix, and persists as long as the particle populations faster and slower than the wave are unbalanced. To check this argument, one may compare the second growth process for the CG and TL initial distributions. The CG distribution can

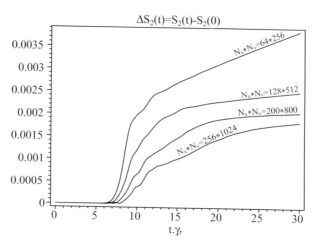

Figure 8.10. Evolution of the 2-entropy $S_2 = \iint (1 - f)f \, \mathrm{d}x \, \mathrm{d}p$ as a function of time $\gamma_L t$ in the Vlasov simulations for several $N_x \times N_v$ grids. The departure from zero indicates that the damping of trapping oscillations is spurious. (After Firpo *et al* (2001).)

maintain the imbalance during the whole time evolution until the cat's eye covers the whole velocity range, whereas the TL distribution has a smaller reservoir of fast particles; indeed the second growth is definitely slower for the TL case than for the CG case.

It is also seen that the simulations with more particles (N4 versus N3 for TL, N2 versus N1 for CG) yield a slower second growth. This suggests that noise is the driving process. If so, this second growth is not described by the kinetic model, because it requires a finite number of particles but the large values of N relevant to physics do not suppress the process, because the time-scale over which the kinetic equation describes the evolution is limited for $N < \infty$, typically with a scaling as $t_{max} \simeq \lambda_{Lyap}^{-1} \ln N$ and $\lambda_{Lyap} \simeq \omega_b$.

All these observations confirm that the late evolution of the wave intensity is controlled by trapping–detrapping of particles, crossing the pulsating separatrix. The separatrix pulsations are, themselves, controlled by the oscillations of the trapped particle momenta. In the kinetic limit $N \to \infty$, there are no individual particles but the mathematical object of interest is the density $f(x, p, t)$, which may be considered as advected (transported) by the particles through (x, p)-space. Then assume that, initially, the level lines of f are smooth curves, so that f is easily interpolated on a finite grid $\Delta x \times \Delta p$. After some time, nearby particles have moved apart, so that these contour lines have deformed in a way similar to the curve in figure 8.2(*b*) and later on to the wrinkled curve of figure 8.2(*c*). At such times, the sampling of f on the grid is poor and an interpolation from grid values is significantly smoother than the Vlasovian ideal f. The poorness of the interpolation is indeed measured by the numerical entropy in figure 8.10. From the kinetic viewpoint, the second growth is missed because the inevitable smoothing loses the microscopic dynamics, which is described by the 'filamentation' of f-level lines in (x, p)-space. This filamentation is the counterpart of (i) the fast separation of particle trajectories in the vicinity of the (local, moving) X-point of the wave and (ii) the secular (linear in t) separation of trajectories in both the trapped and circulating domains[2].

8.3   Linear and nonlinear wave damping

Let us now turn to the damping case with a warm beam. It can be initiated with a finite intensity of the wave, or with an infinitesimal initial intensity to focus on the linear Landau–van Kampen–Case regime (section 3.8.3).

Given a wave launched with intensity $I = N\psi_0$ in a warm beam of N particles with average momentum p_0 and beam temperature T_0, the scaled constants of the motion are

$$\sigma = p_0 + k\psi_0 \tag{8.10}$$

$$h = \frac{p_0^2 + T_0}{2} + \omega_0\psi_0. \tag{8.11}$$

Numerical simulations are performed for an initial distribution[3] CD (figure 8.5),

$$f_{CD}(p) = c - \frac{ac}{p_1 - p} \tag{8.12}$$

for $-\Delta v \le p - \omega_0/k \le \Delta v$ and $f_{CD} = 0$ otherwise.

8.3.1   Evolution of the wave intensity

Firpo (1999) ran simulations for several values of ψ_0 and N, with $\omega_0 = k = 1$, $\Delta v = 0.25$ and constants ε', $a > 0$ and $c > 0$ fixed so that the Landau damping rate is $\gamma_L = -1/200$ and $\int f_p(p) \, dp = 1$; then $T_0 = 1.50 \times 10^{-2}$.

In this case, figure 8.11 displays the scaled intensity ψ after reaching a (hopefully) asymptotic regime. Three ranges of initial data ψ_0 can be recognized and will be further discussed in section 9.7.

- For $\psi_0 \gtrsim 0.1$, the long-time value is the initial value.
- For $\psi_0 \lesssim 10^{-5}$, the long-time value is independent of ψ_0. For N large it is small, but finite.
- A transition domain is present for ψ_0 in between 0.1 and 10^{-5}.

The two extreme regimes are consistent with the trapping and Landau-damping regimes of section 8.1.2.

---

[2] For nonlinear oscillators, such as the pendulum, except for low-amplitude bounce motion, each trajectory has its own period (exercise 5.8). Hence the distance between points belonging to different trajectories increases with time.

[3] This 'constant decay' distribution is symmetrical to the CG distribution (8.12).

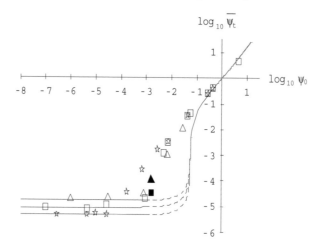

Figure 8.11. Time averages of normalized intensity ψ reached for long times versus ψ_0 for $N = 16\,000$ (triangles), $N = 32\,000$ (squares) and $N = 64\,000$ (stars): filled symbols for simulation times of $2 \times 10^6 \omega_0^{-1}$; open symbols for run times of $5 \times 10^5 \omega_0^{-1}$. Canonical estimations are drawn in continuous lines for $N = 16\,000$, $32\,000$ and $64\,000$ particles. Lines refer to Gibbsian equilibrium estimates of section 9.7: above the transition a single line with $\psi = O(1)$ describes the high-temperature phase, while below the transition the equilibrium estimates for ψ scales like N^{-1}. Near the transition, dashes indicate that estimate (9.26) is not expected to be accurate. (After Firpo and Elskens (2000).)

The time evolution for initial data in the transition domain, below the transition value ψ_{0c}, is displayed in figure 8.12. After a brief linear Landau damping stage (see inset), trapping effects are seen to dominate the slow decay of the wave intensity.

8.3.2   Finite-N effects competing with Landau damping

The previous discussion shows that the effect of a finite N on thermal equilibrium may lead to an increase in wave intensity even in a situation for which kinetic theory predicts Landau damping. This occurs provided that the initial wave amplitude is smaller than the time asymptotic one ('thermal level'). The departure of the finite-N evolution from the kinetic model can also be tested numerically by comparing dynamical simulations with semi-Lagrangian kinetic simulations. Here $\omega_0 = 0$, $k = 1$, $\Delta v = 3.96$, $a = 11.89$ and $p_1 = 15.85$ and c normalizes $\int f_{CD} \, dp = 1$.

In this case, figure 8.13 compares the evolution of the wave intensity coupled to $N = 32\,000$ particles (line N) with its evolution coupled to the kinetic distribution function by the Vlasov–wave system (line V), for an initial intensity such that $\omega_b \simeq 3\gamma_L$. Both simulations start with exponential decay of the wave

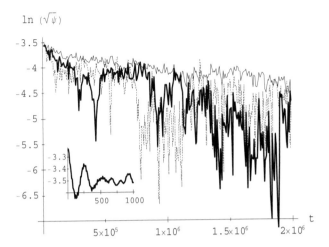

Figure 8.12. Time evolution of normalized amplitude $\sqrt{\psi} = \sqrt{I/N}$ for $N = 16\,000$ (dots), $N = 32\,000$ (bold line) and $N = 64\,000$ (faint line), for $\psi_0 = 1.56 \times 10^{-3} < \psi_{0c}$. Inset: initial evolution, including brief initial linear damping stage. At longer times, for $N = 32\,000$, ψ wanders around half the value associated with $N = 16\,000$, in agreement with canonical predictions. Relaxation towards equilibrium is much longer for $N = 64\,000$ due to critical slowing-down near the phase transition. (After Firpo and Elskens (2000).)

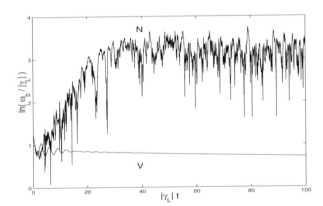

Figure 8.13. Time evolution of $\ln(\omega_b(t)/|\gamma_L|)$ for a CD velocity distribution and initial wave amplitude below thermal level: (N) N-particle system with $N = 32\,000$, (V) kinetic scheme with 32×512 (x, p) grid. The short-time evolution was displayed in figure 3.8. (After Doveil *et al* (2001) and Firpo *et al* (2001).)

at Landau's rate γ_L but the trapped particles bounce after $t \simeq \pi/\omega_b \simeq 1/\gamma_L$ and let the wave intensity grow again to a value close to the initial one, as explained in section 8.1.2. For the kinetic model, the bouncing oscillations of the

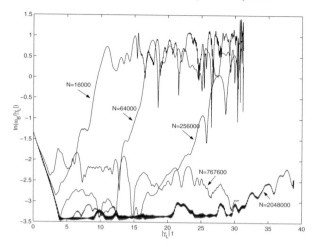

Figure 8.14. Time evolution of $\ln(\omega_b(t)/|\gamma_L|)$ for an initial CD velocity distribution and different values of N. When necessary for readability, curves were smoothed through a sliding window average. (After Firpo *et al* (2001).)

intensity continue and damp and, after a dozen Landau e-folding times, the wave intensity seems to have reached a slowly decaying regime—but this simulation cannot assess whether towards a finite or vanishing asymptotic value. The finite-N case was run with a quiet start initial condition where particles are exactly on a multibeam array, so that spontaneous emission vanishes at $t = 0$. The wave first grossly follows the kinetic prediction, but after a short transient time it grows to its 'thermal' equilibrium defined by (4.19). This growth, which corresponds to the relaxation to the 'thermal' level, results from the spontaneous emission by the particles which eventually sets in, as explained in section 4.3.2.

As shown in figure 8.14, increasing the number of particles delays the time at which spontaneous emission starts dominating the initial damping (for the same quiet types of initial datum) but emission ultimately drives the wave intensity to its non-zero thermal equilibrium level.

8.4   Historical background and further comments

The instability of spatially homogeneous solutions of the Vlasov equation has been discussed in numerous papers and books, with emphasis on the Vlasov–Poisson system, which we introduce in section G.4. A fruitful stability criterion was established by Penrose (1960).

Bernstein, Greene and Kruskal (BGK) (1957) proved the existence of finite-amplitude travelling plasma waves (propagating undeformed at constant velocity) as solutions to the Vlasov–Poisson system; these BGK modes, which have obvious analogues for the Vlasov-wave model (section G.3), are described in the framework of kinetic theory. Their study is subtle, as the Vlasov equation nonlinearly couples the distribution function with the fields. Analogues to these waves in hydrodynamics or charged fluid models lead to the theory of solitons.

One difficulty in the study of the Vlasov–Poisson system for electrostatic waves in a plasma is that the field is generally a superposition of many sinusoidal waves and that the linear combination of solutions to the Vlasov–Poisson or Vlasov-wave systems is not a solution (the term $E(x, t) \partial_p f(x, p, t)$ is nonlinear). However, for a proper understanding, one first focuses on the fate of a single wave. In this respect, BGK modes are good candidates for asymptotic states of a plasma. However, not all BGK modes are stable. In particular, periodic BGK modes with several cat's eyes per period are unstable (Guo and Strauss 1995) and even BGK modes with a single cat's eye but with a 'complicated' distribution function are unstable (Manfredi and Bertrand 2000).

The damping of a single wave in the linear regime was discussed in chapters 3 and 4. The feedback on the wave from the nonlinear motion of particles was described by O'Neil (1965), whose work initiated a debate on the asymptotic evolution of the beam–wave system within its usual formulation—the Vlasov–Poisson system. Holloway and Dorning (1989) proved the existence of waves which linear theory predicts Landau damp but which, in fact, do not damp thanks to nonlinear effects.

Analogues of BGK modes with several waves, with distinct phase velocities, are more complicated to describe. They cannot be reduced to a stationary wavetrain by a Galileo transformation and the nonlinearity of wave–particle interaction implies that a mere linear combination of BGK waves is not a solution to the plasma evolution equations. However, nonlinear superposition methods have been developed for finite-amplitude waves with non-overlapping resonance chains (Buchanan and Dorning 1994, and references therein). A fairly detailed picture of the dynamics of several coupled modes was obtained by Lancellotti and Dorning (1999, 2000), who showed that initial data for the particle distribution function and the electric field evolve to asymptotic states with a finite number of field components. Introducing a decomposition in terms of 'asymptotic' and 'transient' components holding uniformly in time[4], they identify the diversity of behaviours of plasmas in kinetic theory, to which this chapter is a short introduction. In particular, they identify families of initial data such that the transient component vanishes (à la Mazitov–O'Neil) so that the waves do not damp and families of initial data such that the asymptotic component vanishes (à la Landau). See also Lancellotti and Dorning (2000) for an account of the debate on the long-term fate of waves which would Landau damp according to linear theory and the determination of critical initial amplitudes separating the different regimes.

Because of time reversibility, the existence of stable asymptotic solutions, for example, with vanishing field, also implies that the symmetric states (with

---

[4] Uniformity in time is crucial to deal correctly with asymptotic behaviours, in contrast to the elementary perturbative methods of this chapter which allow for secular errors.

reversed velocities) are unstable against some perturbations (Caglioti and Maffei 1998).

The evolution with a linearly unstable wave also has a rich history. The saturation level estimate (8.7) follows the arguments (using energy rather than momentum) by Onishchenko *et al* (1970) and Fried *et al* (1971). However, the longer-time evolution of the plasma due to the trapped particle oscillations is also debated. An early attempt at understanding this evolution in the case of a cold beam interacting with a bulk plasma led O'Neil *et al* (1971) to introduce the self-consistent model on which we have focused. For the weak-warm-beam instability, numerical simulations first revealed the qualitative features of the saturation (Levin *et al* 1972). Then Simon and Rosenbluth (1976) and Janssen and Rasmussen (1981) applied a multiple-time-scale method, in the case $\gamma_L \ll \omega_{j0}$ where the wave envelope evolves slowly in comparison with the wave eigenoscillations, and obtained an evolution equation for the wave intensity such that this intensity may be driven to a limit cycle. Further development of envelope equation analysis showed that the prediction of the asymptotic scaling of the wave amplitude in the limit of weak instability is sensitive to the Hamiltonian nature of the Vlasov equation and to its being a partial differential equation (which leads to continuous spectrum problems à la van Kampen and Case). Moreover, the relevant scalings change significantly if the system involves more than one particle species (Crawford 1994, Crawford and Jayaraman 1996).

Finally, numerical simulations by Doveil *et al* (2001) and Firpo *et al* (2001) have shown that the dynamics with a finite number N of particles leads to a long-time evolution qualitatively distinct from the numerical integration of the kinetic Vlasov model. This occurs despite the fact that the limit $N \to \infty$ formally reduces the many-particle evolution equation to the Vlasov kinetic equation (section G.1.2) but there is no contradiction. Kinetic limit theorems (Firpo and Elskens 1998) show that the limit $N \to \infty$ commutes with the dynamics over any finite time interval $0 \le t \le t_*$ but the discrepancy between both evolutions may well diverge exponentially as $t_* \to \infty$, especially because the wave–particle system has dynamical instabilities. Then, in the long term, given a finite number of particles, there is a time beyond which the smooth solution to the kinetic model (approximating the finite-N system initially) may evolve significantly differently from the physical finite-N system: the limits $t \to \infty$ and $N \to \infty$ need not commute (exercise G.3). In short, in its evolution, the plasma eventually reminds the physicist of its microscopic granular nature, so that the Vlasov equation cannot be used to describe its behaviour accurately over long times (Escande 1989).

The physical mechanism driving the separation of finite-N dynamics from the Vlasov model is the pulsating separatrix crossing, which ensures a good mixing of particles in (x, p)-space—ironically, the basic assumption underlying the statistical description motivating the kinetic model. One way to recover a statistical description is to 'expand' the evolution equation with respect to the small parameter $1/N$: the zeroth-order dynamics is the Vlasov equation, which is reversible and admits many conservation laws (G.12). The first-order correction is known as the Balescu–Lenard equation (Balescu 1960, Lenard 1960, Balescu 1997) for plasmas of particles interacting pairwise by the Coulomb force, and was investigated in accurate simulations by Rouet and Feix (1991, 1996). For our reduced plasma model with particles interacting with waves only, the quasilinear equations of chapter 7 play a similar role. It must be stressed that the rigorous foundation of Balescu–Lenard or Landau equations (Spohn 1991) and quasilinear equations requires further progress in mathematical physics.

Chapter 9

Gibbsian equilibrium of the single-wave–particle system*

In the preceding chapter we considered the numerical time evolution of the single-wave–particle system, and found hints to its asymptotic states. In this chapter we compute analytically the statistical equilibrium states of this system. Indeed these states are generally considered as representative of the typical states of the many-body system and, as such, as typical of the asymptotic states to which the system evolves. The kind of statistics of interest in this chapter is quite different from those used in the remainder of the book. Indeed, it involves averages over the total energy of the system, instead of averages over initial phases. We have already noted that the chaotic motion of particles coupled to one or several waves rapidly forces one to describe the typical state of this system statistically. We now note that there is some freedom in the choice of statistics.

In the many-particle limit $N \to \infty$, the system is described by an invariant distribution over phase space and explicit forms can be found for several physical observables. The case of a single wave is solved analytically and one distinguishes two regimes separated by a phase transition. The two thermodynamic phases are linked to the different dynamical evolutions of the system. In particular, the two regimes of wave damping are recovered in connection with the two phases.

For a reader acquainted with the Gibbsian approach, the most important calculations for section 9.2 are quite straightforward. However, the detailed study of the probability distributions for the high-temperature (section 9.3) and low-temperature phases (section 9.4) is more demanding. Section 9.5 summarizes these results in phase diagrams and section 9.6 relates the statistical equilibrium states with dynamical Bernstein–Greene–Kruskal solutions to the Vlasov-wave kinetic model. In section 9.7 we comment on the numerical simulations of chapter 8 in the light of this chapter's results.

9.1 Gibbs microcanonical and canonical ensembles

The natural ensemble for a self-consistent model of interacting waves and particles is Gibbs' microcanonical ensemble. It is a probability distribution μ_{mic} on the phase space $\Lambda_{N,M}$ for N particles and M waves. Specifically, as positions are on the interval of length L with periodic boundary conditions (denoted by $S_L = \mathbb{R}/(L\mathbb{Z})$), momenta are real numbers and the waves are characterized by their complex amplitudes $Z_j = X_j + iY_j$, the phase space is $\Lambda_{N,M} = S_L^N \times \mathbb{R}^N \times \mathbb{R}^{2M}$. The microcanonical probability distribution on $\Lambda_{N,M}$ is characterized by the functions $\mathcal{H}$ and $\mathcal{P}$, defined by (2.54) and (2.55), which are constants of the motion. Given the values E and P of these constants, the microcanonical distribution is uniform over the accessible domain:

$$d\mu_{\text{mic}}(\boldsymbol{x}, \boldsymbol{p}, \boldsymbol{X}, \boldsymbol{Y}|E, P)$$
$$= \Omega^{-1}\delta(\mathcal{H} - E)\delta(\mathcal{P} - P)\, d^N\boldsymbol{x}\, d^N\boldsymbol{p}\, d^M\boldsymbol{X}\, d^M\boldsymbol{Y}/C_{N,M}. \quad (9.1)$$

The normalizing constant[1] $\Omega(E, P)$, such that $\int_{\Lambda_{N,M}} d\mu_{\text{mic}}(\boldsymbol{x}, \boldsymbol{p}, \boldsymbol{X}, \boldsymbol{Y}) = 1$, is the number of complexions (or microstates).

Gibbs' microcanonical ensemble has a dual use. On the one hand, the microcanonical probability measure (9.1) determines the probability distribution for any observables in phase space $\Lambda_{N,M}$; on the other hand, the number of complexions also links the statistical distribution to the thermodynamic properties of the wave–particle system. In particular, the microcanonical entropy is

$$S = k_{\text{B}} \ln \Omega \quad (9.2)$$

where k_{B} is Boltzmann's constant. In the following, we set $k_{\text{B}} = 1$ for conciseness. The major reason for introducing the Gibbs ensemble is that, while μ_{mic} or S may generally depend smoothly on the parameters (E, P), this dependence may sometimes be dramatic. These remarkable values of (E, P) mark the boundary between 'phases' of the system: in distinct phases, one expects the dynamics of the particles and waves to be significantly different (as it is for the direction of magnetic moments in a ferromagnet above compared with below its Curie point).

The fundamental assumption in assigning the microcanonical measure to describe equilibrium is that all phase-space points $(\boldsymbol{x}, \boldsymbol{p}, \boldsymbol{X}, \boldsymbol{Y})$ with the same energy E and total momentum P are accessible to the microscopic state of the wave–particle system. This is not quite true in practice and one should restrict the microcanonical measure to the actually accessible domain of phase space—which is estimated following the arguments of sections 5.2 and 5.4 by Firpo and Doveil

---

[1] The constant $C_{N,M} = N! h_{\text{P}}^{N+M}$ is necessary to a correct formulation of the statistical theory but it is irrelevant to our present calculations. Here, formally, $h_{\text{P}} = 2\pi\hbar/S$ where $\hbar$ is Planck's constant and S is the cross section such that the three-dimensional volume of our system is LS. Note that $C_{0,M} = C_{0,1}^M = h_{\text{P}}^M$.

(2002). Considering our system as closed and non-isolated, the equilibrium measure is Gibbs' canonical measure:

$$
\begin{aligned}
d\mu_{\text{can}}(x, p, X, Y | T, \sigma) \\
= \mathcal{Z}^{-1} \exp[-\mathcal{H}/T] \delta(\mathcal{P} - N\sigma) \, d^N x \, d^N p \, d^M X \, d^M Y / C_{N,M}
\end{aligned}
\quad (9.3)
$$

where T is the temperature of the bath and $\sigma = P/N$ is the total momentum per particle. The normalizing constant $\mathcal{Z}(T, \sigma, N, M)$, ensuring $\int_{\Lambda_{N,M}} d\mu_{\text{can}} = 1$, is the partition function. It determines the thermodynamic free energy

$$
F = -T \ln \mathcal{Z}. \quad (9.4)
$$

The connection of the canonical partition function $\mathcal{Z}(T)$ with the microcanonical number of complexions $\Omega(E)$ is the two-sided Laplace transform, with $\beta = 1/T$ conjugate to E,

$$
\mathcal{Z}(T, \sigma, N, M) = \int_{-\infty}^{+\infty} \Omega(E, N\sigma, N, M) e^{-\beta E} \, dE = \int_{-\infty}^{+\infty} e^{S - \beta E} \, dE. \quad (9.5)
$$

The integral is well behaved, as the energy per particle H/N of the self-consistent system is bounded below and as Ω does not grow faster than exponentially for $E \to +\infty$. The Laplace transform is invertible using the analytic continuation of $\mathcal{Z}$ to complex values of β.

While the physical motivation for the microcanonical ensemble was that the microscopic state of the system would visit over a 'reasonably short' time lapse a 'reasonably large' part of phase space, the physical motivation for describing a system by the canonical ensemble is that it can exchange energy with its surroundings. This is a different requirement, generally not difficult to implement, but in practice one prefers proving or assuming that the equations of state for the microcanonical ensemble and for the canonical ensemble are equivalent—which is true in typical states but occasionally fails, in particular at phase transitions (Gallavotti 1999, Balescu 1975, 1997). Indeed the calculation of the equation of state and statistical properties of the system is generally easier in the canonical ensemble. This entails that our analytical calculations will not apply directly to the numerical simulations of chapter 8, as we discuss further in section 9.7.

Note that one can also introduce a (bi-)canonical distribution with parameters temperature T and velocity v

$$
d\mu_{T,v}(x, p, X, Y) = \mathcal{Z}_{T,u}^{-1} \exp[-(\mathcal{H} - \mathcal{P}v)/T] d^N x \, d^N p \, d^M X \, d^M Y / C_{N,M} \quad (9.6)
$$

to describe a beam–wave system exchanging both momentum and energy with its environment.

Exercise 9.1. Show that $H/N \geq -\sum_j N\varepsilon^2 \beta_j^2 / (2k_j^2 \omega_{j0})$ and that Ω cannot grow faster than exponentially for $E/N \to \infty$. What does this imply for S/N? Hint: $|\cos a| \leq 1$ and compare with the ideal gas.

Exercise 9.2. After studying the following sections, compare the statistical mechanics of the 'bicanonical' ensemble (9.6) with the canonical one discussed later. As total momentum $\mathcal{P}$ can take any value in $\mathbb{R}$, use the fact that the energy H/N is bounded from below to ensure that integrals of interest converge. Check the equivalence of ensembles, i.e. whether the 'bicanonical' equilibrium properties of the model are equivalent to the canonical ones. Pay special attention to the phase-transition regime.

9.2 Partition function

The case of a single mode is easily amenable to analytic treatment and, as for many systems in statistical physics, the partition function is easier to obtain than the number of complexions. Thus we compute

$$\mathcal{Z}(T, \sigma, N, 1) = \int_{\Lambda_{N,1}} \delta(\mathcal{P} - N\sigma) e^{-\beta H} \, d^N x \, d^N p \, dX \, dY / C_{N,1} \tag{9.7}$$

$$= \int_0^\infty \int_0^{2\pi} \frac{\zeta_p \zeta_x}{C_{N,1}} \exp(-\beta \omega_0 I) \, d\theta \, dI \tag{9.8}$$

where

$$\zeta_p = \int_{\mathbb{R}^N} \delta\left(\sum_r p_r + kI - N\sigma\right) e^{-\beta \sum_r p_r^2/2} \, d^N p \tag{9.9}$$

$$= \frac{(2\pi T)^{N/2}}{\sqrt{2\pi NT}} e^{-\beta(kI - N\sigma)^2/(2N)} \tag{9.10}$$

$$\zeta_x = \int_{S_L^N} e^{-\beta \sum_r \varepsilon k^{-1} \beta_1 \sqrt{2I} \cos(kx_r - \theta)} \, d^N x \tag{9.11}$$

$$= L^N \left(I_0(\beta \varepsilon k^{-1} \beta_1 \sqrt{2I})\right)^N. \tag{9.12}$$

These integrals, where I_n denotes the modified Bessel function of order n, are straightforward (see section B.2), so that

$$\mathcal{Z}(T, \sigma, N, 1) = \frac{2\pi L^N}{C_{N,1}} \frac{(2\pi T)^{N/2}}{\sqrt{2\pi NT}} \int_0^\infty \rho(\sigma, T, \psi) \, d\psi \tag{9.13}$$

where

$$\rho(\sigma, T, \psi) = \exp\left(-\frac{N}{T} f(\sigma, T, \psi)\right) \tag{9.14}$$

with

$$f = \frac{(\sigma - k\psi)^2}{2} + \omega_0 \psi - T \ln\left[I_0\left(\frac{\varepsilon' \sqrt{2\psi}}{T}\right)\right] \tag{9.15}$$

and $\varepsilon' = \varepsilon \beta_1 k^{-1} N^{1/2}$ defined by (8.2). Here $\rho(\sigma, T, \psi)$ is the weight function for the rescaled intensity ψ. The N-dependence in the integral in (9.13) occurs

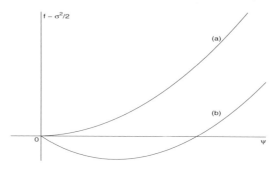

Figure 9.1. Conditional free energy per particle f as a function of wave intensity $\psi = I/N$: (a) high-temperature phase; and (b) low-temperature phase.

only as the factor of f in the exponent. This type of integral is easily estimated by Laplace's method (Bender and Orszag 1978), which reduces the estimation in the limit $N \to \infty$ to the determination of the Taylor expansion of f near its lowest minimum.

Exercise 9.3. For several choices of T and σ, plot each term of (9.15) as a function of ψ. Adding them, obtain the graph of figure 9.1 and discuss the relevance of each term to the location of the minimum.

9.2.1   Thermodynamic discussion

All difficulties in the calculation of $\mathcal{Z}$ lie now in the final integral of (9.13). If there were no wave in the Hamiltonian (i.e. for $M = 0$), the system would reduce to the ideal gas of N independent particles and there would remain no integral in (9.13), yielding the ideal gas free energy $-T \ln(\zeta_p L^N / C_{N,0})$.

Exercise 9.4. Express $e^{-\beta[F(T,\sigma,N,1)-F(T,\sigma,N,0)]} = \mathcal{Z}(T, \sigma, N, 1)/\mathcal{Z}(T, \sigma, N, 0)$ in a form as simple as possible (do not compute it yet).

Therefore, the function f may be considered as an 'excess free energy' for the beam–wave system, conditioned by the intensity $I = N\psi$. When the beam interaction with the wave is taken into account, three contributions, corresponding to the three terms in (9.15), are added to the free energy: the relative kinetic energy of the beam 'centre of mass', the wave energy and the 'interaction free energy'.[2] By inspection, one checks that the first two terms are positive and the interaction term is negative, i.e. that the wave–particle coupling lowers the function f. Thus the uncoupled case ($\varepsilon' = 0$) provides an upper estimate for f, hence a lower estimate for $\mathcal{Z}$ and an upper estimate for $F = -T \ln \mathcal{Z}$.

---

[2] Though the mere notion of energy as a constant of the motion makes it impossible to ascribe a specific contribution to each degree of freedom (lest this contribution be itself a constant of the motion), such decompositions are heuristic as far as we are concerned with additive expressions.

9.2.2 Weight function

The integral of the weight function in (9.13) is dominated by the contribution from the vicinity of the minimum of f if it exists. To find this minimum, it is computationally more convenient to express f as a function of

$$\varphi = \varepsilon' \sqrt{2\psi}/T \tag{9.16}$$

which is a one-to-one increasing function of ψ.

Exercise 9.5. Anticipating the following discussion, rewrite the one-particle contribution to potential energy $\varepsilon k^{-1}\beta_1 \sqrt{2I} \cos(kx - \theta)$ in terms of φ and T, and compare it with the estimate $T/2$ for kinetic energy.

The function $f(\sigma, T, \psi) = g(\sigma, T, \varphi)$ is smooth (C^∞) with respect to $\varphi \in [0, \infty[$ and may reach a minimum at the boundary $\varphi = 0$, at an intermediate point φ^* or tend to an infimum[3] in the limit $\varphi \to \infty$.

We first note that f does not tend to an infimum for $\varphi \to \infty$. Indeed, in this limit the modified Bessel function behaves asymptotically like

$$\ln I_0(\varphi) \simeq \varphi - \frac{1}{2} \ln(2\pi\varphi) + O\left(\frac{1}{\varphi}\right) \tag{9.17}$$

so that $f \simeq \psi^2 k^2/2 + O(\psi)$ increases to infinity as $\psi \to \infty$.

Near $\varphi = 0$, the Taylor expansion of g reads

$$g = \frac{\sigma^2}{2} + (\omega_0 - k\sigma)\psi - \frac{\varepsilon'^2}{2T}\psi + O(\psi^2) \tag{9.18}$$

so that zero is a local minimum for g provided that $\omega_0 > k\sigma + \varepsilon'^2/(2T)$. However, we need a global minimum for g. Thus, we search for extrema of g in $[0, \infty[$. At such points, $\partial_\varphi g = 0$ (and for a regular minimum $\partial_\varphi^2 g > 0$). The derivative reads $\partial_\varphi g = \varphi G(\varphi)$ where

$$G(\varphi) = \left(\frac{T}{\varepsilon'}\right)^2 \left(\omega_0 - k\sigma + \frac{k^2}{2}\left(\frac{T}{\varepsilon'}\right)^2 \varphi^2\right) - T\frac{I_1(\varphi)}{\varphi I_0(\varphi)}. \tag{9.19}$$

The derivative $\partial_\varphi g$ always vanishes at $\varphi = 0$, and vanishes at $\varphi^* > 0$ iff G does. Now note that G is the difference between an increasing function of φ and a decreasing function of φ (see section B.2). Therefore G is an increasing function of φ, which can vanish, at most, once in $[0, \infty[$.

It is clear that $G > 0$ in the limit $\varphi \to \infty$, as $g \to +\infty$ and therefore $\partial_\varphi g \to +\infty$ as well. Asymptotic analysis of (9.19) confirms that G diverges like φ^2. However, $G(0) = (T/\varepsilon')^2(\omega_0 - k\sigma) - T/2$, which may be negative.

---

[3] The concept of infimum generalizes the concept of minimum to cases where it could not be reached over the domain of interest: an infimum may be a minimum, an asymptotic value, or $-\infty$.

Thus, three cases may occur for the minimum of g. When $\sigma < \omega_0/k$, define the critical temperature $T_c(\sigma)$ by

$$T_c(\sigma) = \frac{\varepsilon'^2}{2(\omega_0 - k\sigma)}.$$

(9.20)

- If $G(0) > 0$, g has its absolute minimum at $\varphi = 0$ and the corresponding minimum is quadratic with respect to ψ. This occurs if $\sigma < \omega_0/k$ and $T > T_c(\sigma)$. The corresponding state is called the 'high-temperature phase'.
- If $G(0) = 0$, g has its absolute minimum at $\varphi = 0$ and the minimum of f is quartic with respect to ψ. This occurs if $\sigma < \omega_0/k$ and $T = T_c(\sigma)$. The corresponding state is called the 'phase-transition (critical) state'.
- If $G(0) < 0$, g has its absolute minimum at some $\varphi^* > 0$ solution of (9.19) and the minimum of f is quadratic with respect to $\psi - \psi^*$. This occurs if $\sigma \geq \omega_0/k$ or if $\sigma < \omega_0/k$ and $T < T_c(\sigma)$. The corresponding state is called the 'low-temperature phase'.

For $\sigma < \omega_0/k$, the non-trivial minimum φ^* tends to zero with $\varphi^* \sim \sqrt{\psi^*} \sim (T_c - T)^{1/2}$ as $T \to T_c(\sigma)$. Moreover, when the minimum is non-trivial, the equilibrium state breaks the translation invariance, e.g. by drawing the wave phase according to uniform distribution on the circle. Therefore, the phase transition is of second order (like the ferromagnetic transition in the absence of external field, with the same critical exponent $1/2$).

For each regime, we now discuss the resulting equilibrium estimates for the system's physical properties in the limit $N \to \infty$. As the coupling parameter in (9.20) is ε', the limit $N \to \infty$ implies that $\varepsilon \to 0$ like $N^{-1/2}$. Such an interaction, for which the coupling constant goes to zero as $N \to \infty$, is said to be of the mean-field type: this is the simplest type of system for statistical physics.

9.3   High-temperature phase

9.3.1   Thermodynamics and wave intensity

This phase exists only if $\sigma < \omega_0/k$ and $T > T_c(\sigma)$. Then the excess free energy f is a minimum for $\psi = 0$ and for (9.13)

$$\int_0^\infty e^{-N\beta f} \, d\psi \simeq \frac{\varepsilon'^2 T^2}{2N} \left(\frac{T}{T_c} - 1 \right).$$

(9.21)

The free energy per particle now reads

$$\frac{F}{N} = -\frac{T}{N} \ln \mathcal{Z} \simeq \frac{\sigma^2}{2} - T - T \ln \left[\frac{L}{Nh_P} \sqrt{2\pi T} \right]$$

(9.22)

so that the entropy per particle is

$$\frac{S}{N} = -\frac{1}{N} \left(\frac{\partial F}{\partial T} \right)_{L,N,\sigma} \simeq \frac{3}{2} + \ln \left[\frac{L}{Nh_P} \sqrt{2\pi T} \right]$$

(9.23)

and the pressure is

$$-\left(\frac{\partial F}{\partial L}\right)_{T,N,\sigma} = \frac{TN}{L} \tag{9.24}$$

as for the ideal gas. The expected energy per particle is

$$\left\langle \frac{\mathcal{H}}{N} \right\rangle = \int_{\Lambda_{N,1}} \frac{\mathcal{H}}{N} \, d\mu_{\mathrm{can}} = N^{-1} \partial_\beta \ln \mathcal{Z} \simeq \frac{\sigma^2}{2} + \frac{T}{2} \tag{9.25}$$

and the expected intensity of the wave is

$$\langle I \rangle = -T \partial_{\omega_0} \ln \mathcal{Z} \simeq \frac{1}{\alpha} \tag{9.26}$$

where we define

$$\alpha \equiv \frac{(\omega_0 - k\sigma)(T - T_{\mathrm{c}})}{T^2} = \frac{\varepsilon'^2}{2T}\left(\frac{1}{T_{\mathrm{c}}} - \frac{1}{T}\right). \tag{9.27}$$

This latter quantity does not scale with N in the limit $N \to \infty$. Physically, this means that, if the momentum per particle σ is smaller than ω_0/k, the wave should not have a 'large' amplitude: with an amplitude diverging with N, it would contribute too much to the total momentum. The most likely situation for the system is to have most particles moving like a perfect gas with average velocity σ and temperature T. If the wave intensity scaled like N, it would significantly increase the system's total energy.

More precisely, the canonical ensemble determines the joint probability density for the wave intensity and phase

$$d\mathbb{P}(I, \theta)/(dI \, d\theta) = \int_{\Lambda_{N,1}} \delta(I_1 - I)\delta(\theta_1 - \theta) \, d\mu_{\mathrm{can}}(\boldsymbol{p}, \boldsymbol{x}, I_1, \theta_1)$$

$$= (2\pi N \mathcal{Z})^{-1} \exp(-N\beta f) = (2\pi)^{-1} \alpha e^{-\alpha I}. \tag{9.28}$$

Thus the excess free energy f also determines the probability density, following the general scheme of Gibbs equilibrium measures. The phase θ has uniform distribution on $[0, 2\pi[$ and is independent of the intensity, which has exponential distribution on $[0, \infty[$. This implies that the Cartesian variables of the wave have a joint Gaussian distribution

$$d\mathbb{P}(X, Y)/(dX \, dY) = (2\pi)^{-1} \alpha e^{-\alpha(X^2 + Y^2)/2} \tag{9.29}$$

with independent components X and Y.

9.3.2 Particle distribution and joint probabilities

The probability distribution function for a particle in (x, p)-space, given a wave with phase θ and amplitude I, is

$$d\mathbb{P}(p, x | I, \theta)/(dp \, dx) = \int_{\Lambda_{N,1}} \delta(p_1 - p)\delta(x_1 - x)\delta(I_1 - I)\delta(\theta_1 - \theta) \, d\mu_{\mathrm{can}}$$

$$\times \, [d\mathbb{P}(I_1, \theta_1)/(dI_1 d\theta_1)]^{-1}$$

$$\simeq \frac{e^{-(p-\sigma)^2 \beta/2}}{\sqrt{2\pi T}} \, \frac{e^{-\varepsilon' \beta \sqrt{2I/N} \cos(kx - \theta)}}{L I_0(\varepsilon' \beta \sqrt{2I/N})} \tag{9.30}$$

$$\simeq \frac{e^{-(p-\sigma)^2/(2T)}}{L\sqrt{2\pi T}} \tag{9.31}$$

for $N \to \infty$. Similarly, the n-particle joint probability distributions, given the wave (I, θ), are found to reduce to those of the ideal gas of free independent particles in the limit $N \to \infty$.

The particle distribution at any time determines the instantaneous phase velocity of the wave

$$\dot{\theta} = \omega_0 - \frac{\varepsilon \beta_1 k^{-1}}{\sqrt{2I}} \sum_r \cos(kx_r - \theta) \tag{9.32}$$

which is also a random variable in the canonical ensemble. As the particles are independent and identically distributed, the central limit theorem implies that $\dot{\theta}$ has a normal distribution with parameters

$$\langle \dot{\theta} \rangle = \omega_0 - \frac{\varepsilon'}{\sqrt{2\psi}} \frac{I_1(\varphi)}{I_0(\varphi)} \tag{9.33}$$

$$\langle \dot{\theta}^2 \rangle - \langle \dot{\theta} \rangle^2 = \frac{\varepsilon'^2}{2I} (\langle \cos^2(kx - \theta) \rangle - \langle \cos(kx - \theta) \rangle^2)$$

$$= \frac{\varepsilon'^2}{2I} \left(\frac{1}{2} + \frac{I_2(\varphi) I_0(\varphi) - I_1(\varphi)^2}{I_0(\varphi)^2} \right) \tag{9.34}$$

given the wave intensity I.

Averaging over the values of I according to the distribution (9.28), it is then easily seen that

$$\langle \dot{\theta} \rangle = \omega_0 - \frac{\varepsilon'^2}{2T} = \left(1 - \frac{T_c}{T} \right) \omega_0 + \frac{T_c}{T} k\sigma \tag{9.35}$$

so that the coupling significantly reduces the phase velocity from its nominal value ω_0 (but not in the limit $T \to \infty$) and keeps it larger than the average particle velocity $\langle p \rangle = \sigma$. The variance of $\dot{\theta}$ is evaluated similarly as

$$\langle \dot{\theta}^2 \rangle - \langle \dot{\theta} \rangle^2 \simeq \left\langle \frac{\varepsilon'^2}{4I} \right\rangle. \tag{9.36}$$

For given intensity I, this variance diverges as $I \to 0$. Indeed, when the wave intensity is small, its phase is very sensitive to fluctuations. As a result, the variance of $\dot{\theta}$ (averaged with respect to the distribution of I) is infinite.

Exercise 9.6. Show that the rate of change $(\dot{X}, \dot{Y})$ of the wave in variables (X, Y) has a finite expectation and a finite covariance.

To estimate the probability for a particle to be trapped in the wave's resonance, we first compute this probability for a wave with intensity I and phase velocity $\dot{\theta}$:

$$\mathbb{P}\left(\frac{(p - \dot{\theta}/k)^2}{2} - \frac{\varepsilon\beta_1}{k}\sqrt{2I}(\cos(kx - \theta) + 1) < 0\right)$$

$$= \int_0^L \int_{p_-(x)}^{p_+(x)} d\mathbb{P}(p, x|I, \theta) \tag{9.37}$$

where $p_\pm(x) = \dot{\theta}/k \pm [2A_1(1 + \cos(kx - \theta))]^{1/2}$ and $A_1 = (2I)^{1/2}\varepsilon k^{-1}\beta_1$ (exercise 1.3). As $\varepsilon \to 0$, the integrand in (9.37) is almost constant and the probability reduces to

$$2^{5/2}\pi^{-3/2}\varphi^{1/2}\exp\left(-\frac{(\dot{\theta}/k - \sigma)^2}{2T}\right). \tag{9.38}$$

Thus, in the limit $N \to \infty$,

$$\mathbb{P}(\text{particle 1 is trapped}) \simeq cN^{-1/4}\int_0^\infty \frac{e^{-(\dot{\theta}/k-\sigma)^2/(2T)}}{\sqrt{2\pi T}}\alpha e^{-\alpha I}\, dI$$

$$\simeq c'N^{-1/4}(2\pi T)^{-1/2}\exp\left(-\frac{\varepsilon'^4}{k^2 T}\left(\frac{1}{T_c} - \frac{1}{T}\right)^2\right) \tag{9.39}$$

where c, c' are numerical constants. This probability estimate vanishing for $N \to \infty$ confirms that the particles are basically free in this high-temperature phase.

9.4 Low-temperature phase

This is the more interesting phase, with a wave intensity scaling like N. It is found for $\sigma \geq \omega_0/k$ at any T and for $\sigma < \omega_0/k$ at $T < T_c(\sigma)$.

9.4.1 Thermodynamics and wave intensity

The excess free energy is minimum for $\varphi = \varphi^* > 0$ and, for (9.13),

$$\int_0^\infty e^{-N\beta f}\, d\psi \simeq e^{-N\beta g(\varphi^*)}\left(\frac{2\pi T}{Ng''(\varphi^*)}\right)^{1/2} \tag{9.40}$$

where $g''(\varphi^*) = \varphi^* \partial_\varphi G(\varphi^*)$. Then the free energy per particle and the expected (total) energy per particle read, for $N \to \infty$:

$$\frac{F}{N} = g(\varphi^*) - T - T \ln \left(\frac{L}{Nh_P} \sqrt{2\pi T} \right) \tag{9.41}$$

$$\left\langle \frac{\mathcal{H}}{N} \right\rangle = g(\varphi^*) + \frac{T}{2}. \tag{9.42}$$

The wave intensity is now scaling like N. The probability distribution function for $\psi = I/N$ is

$$d\mathbb{P}(\psi)/d\psi \simeq (2\pi)^{-1/2}(N\beta f''(\psi^*))^{1/2} e^{-N\beta f''(\psi^*)(\psi - \psi^*)^2/2}. \tag{9.43}$$

Thus ψ has a normal distribution, with expectation

$$\left\langle \frac{I}{N} \right\rangle = \langle \psi \rangle = \psi^* = \frac{1}{2} \left(\frac{\varphi^* T}{\varepsilon'} \right)^2 \tag{9.44}$$

and variance $\langle \psi^2 \rangle - \langle \psi \rangle^2 = \mathrm{O}(N^{-1}) \to 0$.

9.4.2 Particle distribution and joint probabilities

Following the invariance of $\mathcal{H}$ with respect to translations of x and θ/k, the distribution of θ for any given ψ is uniform. The one-particle probability distribution function, given the wave phase θ, reads:

$$d\mathbb{P}(p, x|\theta)/(dp\,dx) = f_1(p, x|\theta) = \frac{e^{-(p + k\psi^* - \sigma)^2/(2T)}}{\sqrt{2\pi T}} \frac{e^{-\varphi^* \cos(kx - \theta)}}{L I_0(\varphi^*)} \tag{9.45}$$

and the n-particle joint probability distribution, given the phase θ, factorizes since the n-particle contribution to the energy $\mathcal{H}$ is the sum of n single-particle contributions (1.12). In other words, any n particles are independent in the limit $N \to \infty$, given the phase θ of the wave. However, the phase θ is not independent of the particle positions, as shown by (9.45). This contrasts with the independence between the phase θ and the particle positions x_n in the high-temperature phase: the low-temperature phase breaks the translation symmetry in the microcanonical ensemble and the canonical ensemble restores the symmetry statistically (i.e. with the distribution of θ being uniform).

Exercise 9.7. Consider two particles labelled 1 and 2. Show that $\langle \cos(kx_1 - \theta)\cos(kx_2 - \theta) \rangle = \prod_{j=1}^2 \langle \cos(kx_j - \theta) \rangle$ but that $\langle \cos(kx_1)\cos(kx_2) \rangle \neq \prod_{j=1}^2 \langle \cos(kx_j) \rangle$. More generally, show that $kx_1 - \theta$ and $kx_2 - \theta$ are independent but that x_1 and x_2 are not independent.

The phase velocity $\dot{\theta}$ is given by (9.32). It is again a Gaussian random variable, with expectation

$$\langle \dot{\theta} \rangle = \omega_0 - \frac{\varepsilon'^2}{T\varphi^*} \frac{I_1(\varphi^*)}{I_0(\varphi^*)} = k\sigma - k^2\psi^* = k\langle p \rangle. \tag{9.46}$$

The final expression shows that the average particle velocity equals the average wave phase velocity, in contrast with the high-temperature case (9.35). Note that the variance of $\dot{\theta}$ scales like N^{-1} while the variance of p is independent of N.

The (spatial) probability density for a particle $n(x)$ is maximal at each[4] trough x^* of the wave potential well and minimal at each crest of the potential. The probability for a particle to be trapped can be written as in (9.37) but the scaling of $I \simeq N\psi^*$ implies that this probability is finite in the limit $N \to \infty$,

$\mathbb{P}(\text{particle 1 is trapped})$

$$= \int_{-\pi}^{\pi} \left[\mathrm{erf}\left(\frac{p_+(x) - \sigma}{\sqrt{2T}} \right) - \mathrm{erf}\left(\frac{p_-(x) - \sigma}{\sqrt{2T}} \right) \right] \frac{e^{-\varphi^* \cos y}}{2 I_0(\varphi^*)} \, dy \tag{9.47}$$

with $x = (y + \theta)/k$. This probability scales like φ^* for $\varphi^* \to 0$. More interestingly, it goes exponentially to zero if $|\sigma - \omega_0/k| \gg T^{1/2}$.

Exercise 9.8. Compute the entropy per particle and pressure in the low-temperature phase and compare with the high-temperature results (9.23) and (9.24).

9.5 Phase diagrams

In summary, the wave–particle system in the canonical ensemble undergoes a phase transition for $T = T_c(\sigma)$. This phase transition is characterized by the breaking of invariance of the statistical state with respect to a change in space origin: in the high-temperature phase, the wave phase θ and particle positions x_n are statistically independent (i.e. knowing some of them provides no information on the other variables); in the low-temperature phase, the wave θ and the particle x_n are not independent (knowing one of them provides some information on the other variables).

As the phase transition breaks a continous symmetry of the model, the low-temperature phase is characterized by an order parameter, $\psi = I/N$, which is a continuous function of T at the transition, so that this transition is second order.

Figure 9.2 displays the variation of the order parameter as a function of σ, for various temperatures. The phase transition is better observed by plotting the constant order parameter lines in the (σ, T) plane (figure 9.3). In this plane, the low-temperature phase is mapped by a family of lines $\psi = \psi_0$ with $\psi_0 > 0$, whereas the condition $\psi = 0$ defines a two-dimensional domain, bounded by the phase transition line. The dashed line simply marks the value of σ above which the high-temperature phase does not exist.

---

[4] For $k = 2\pi m/L$, the wave has m equivalent troughs.

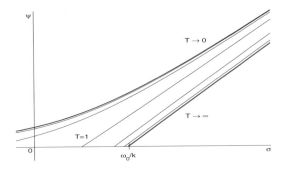

Figure 9.2. Order parameter ψ as a function of σ for various values of T.

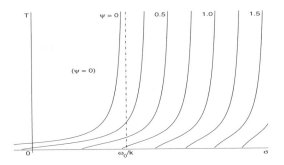

Figure 9.3. Lines ψ =constant in the (T, σ)-plane. The transition line $\psi = 0$ separates the high-temperature phase ($\psi = 0$) from the low-temperature phase ($\psi > 0$).

The equation of state, relating the average energy $\mathcal{H}/N$ to (σ, T), is sketched in figure 9.4. The upper line is always in the low-temperature phase and it bends downward for all values of the energy $\langle h \rangle$ (or of the temperature T). The middle line (dashed) corresponds to the critical value $\sigma = \omega_0/k$. The lower line connects at $T = T_c(\sigma)$ a bending part (low-temperature phase, for $T < T_c(\sigma)$) with a straight part (high-temperature phase, for $T > T_c(\sigma)$).

9.6 Bernstein–Greene–Kruskal state

The canonical equilibrium distribution of the particles and the wave also corresponds to a Galilean solution of the Vlasov-wave equations (G.2), (G.3). This travelling solution is of the Bernstein–Greene–Kruskal type (sections 8.4 and G.3), as the distribution function $f_1(p, x, t)$ depends only on the single-particle

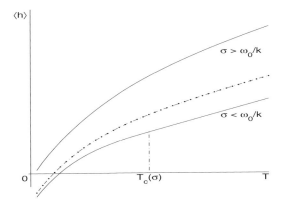

Figure 9.4. Equation of state for $\sigma > \omega_0/k$ (upper line), $\sigma = \omega_0/k$ (dashed line) and $\sigma < \omega_0/k$ (lower line).

energy H_1 given by (1.12) in the comoving frame. Indeed,

$$f_1(p, x, t) = K^{-1} \exp\left(-\frac{H_1(p - v/k, x - \theta/k)}{T}\right) \tag{9.48}$$

where $\theta = \theta(0) + kvt$, the constant K normalizes f_1 to unity over $\Lambda_1 = \mathbb{R} \times S_L$ and the parameters $v \in \mathbb{R}$ and $T > 0$ are to be determined. This distribution function is a source for the wave, by (G.3), which preserves the travelling wave at velocity v if and only if $\dot{\theta} = kv$ and

$$-kv\sqrt{2I} = -\omega_0\sqrt{2I} + \varepsilon\beta_1 k^{-1} N \int_{\Lambda_1} e^{-i(kx-\theta)} f_1 \, dp \, dx \tag{9.49}$$

i.e.

$$(\omega_0 - kv)\sqrt{2\psi} = \varepsilon'\frac{I_1(\varphi)}{I_0(\varphi)} \tag{9.50}$$

which is equivalent to equation $\varphi G(\varphi) = 0$ discussed in section 9.2.2.

Exercise 9.9. Show that (9.50) admits non-trivial solutions ($\psi > 0$) for the values of (v, T) corresponding, through (9.46), to the values of (σ, T) for which $\varphi G = 0$ admits non-trivial solutions. Show also the converse.

Exercise 9.10. Show that travelling solutions to the Vlasov-wave system, with a distribution $f_1(p, x, t) = f_p(p - v)f_x(x - \theta)$ and a wave $\psi = $ constant, $\theta = \theta(0) + kvt$, must be such that f_1 is given by (9.48).

Bernstein–Greene–Kruskal modes may be considered as nonlinear analogues of the linear wavelike modes of chapter 3. However, for finite N, no exact solutions to the nonlinear system (2.111), (2.112), (2.114) with a single wave ($M = 1$) with constant velocity and non-zero constant intensity exist (Elskens 2001). Finite-N analogues of Bernstein–Greene–Kruskal modes must involve time-dependent waves, which may be periodic, along with regular particle motions, just as there exist time-periodic exact solutions to the gravitational N-body problem (these are called choreographies; see, e.g., Montgomery (2001) and Chenciner *et al* (2001)).

9.7 Comparison with numerical simulations

Before comparing with the simulations of chapter 8, we observe that numerical simulations are deterministic runs from fixed initial conditions, whereas the statistical physics framework introduces an ensemble of states (expected to be realized during the simulation). Moreover, simulations apply to an isolated wave–particle system, while the theoretical canonical calculations apply to a closed system. However, both ensembles are equivalent in the thermodynamic limit $N \to \infty$ in the high-temperature phase and they are also equivalent in the low-temperature phase once the wave phase θ is fixed. The second issue for the comparison is the determination of the equilibrium ensemble parameters. Since only particles close to resonance with the wave interact with it and wander in (x, p)-space, the total energy and momentum must be determined by taking into account only those particles. Thus, one should self-consistently estimate the wave's typical amplitude, the chaotic domain and the 'relevant' particles, with their total number. This procedure can be implemented for some types of initial condition.

Because of the locality in velocity of the wave–particle interaction, one should not blindly use the motion constants of the original distribution f_p to discuss the fate of our wave–plasma system. Given the wave resonance's central velocity $v_0 \equiv \dot{\theta}$ and bounce pulsation ω_b, the particles interacting with the wave are those with velocities in a range $v_0 - \Delta v \leq v \leq v_0 + \Delta v$ for some $\Delta v > 0$ of the order of a few times the wave trapping width $2\omega_b/k$. Estimating the boundaries of the chaotic sea for the single-wave model, Firpo and Doveil (2002) find $\Delta v \sim 5\omega_b/k$. In the following we restrict the particle distribution to a similar range and we approximate the particle velocity distribution by (8.3).

Exercise 9.11. Show that for f_p given by (8.3) the particles' average velocity is $v_0 + 2b\Delta v^2/3$, and their variance is $T_0 = \Delta v^2/3 - 4b^2\Delta v^6/9 \geq 2\Delta v^2/9$. The critical temperature is $T_c = -3\varepsilon'^2 b/(4\Delta v^3)$. As f_p is positive, note that $|b| \leq a/\Delta v$.

If the wave amplitude is small, the relevant choice for v_0 is the unperturbed wave phase velocity ω_0/k. This is also a natural choice if the wave is damped. However, if the wave amplitude is large and the relevant thermodynamic state is the low-temperature one, the equilibrium phase velocity is given by (9.46). As the wave amplitude undergoes a significant change, the resonance width varies accordingly and the selection of relevant particles to estimate σ and h too. Since the estimation of v_0 and Δv is meant here only to provide typical initial data, we keep $v_0 = \omega_0/k$ in all choices for simplicity. Once the total energy and momentum are known for the distribution (8.3), equation (9.25) or (9.42) yields the relevant estimate for the temperature of the corresponding thermal equilibrium state.

Section 8.2.1 considered the cold-beam case. If $\sigma = p_0 < \omega_0/k$, the system is in the high-temperature phase and the scaled wave intensity ψ at thermal equilibrium is zero. If $\sigma = p_0 > \omega_0/k$, the energy h lies in the low-temperature phase and the wave intensity may grow, following the cold beam–wave linear instability discussed in section 3.7.1, to saturate at values in the range of (9.44). The numerical test of these scenarios confirms the thermodynamic expectations only partly in figures 8.1 and 8.2. Indeed, in the case $\sigma = p_0 > \omega_0/k$, the instability does not quite lead to the prescribed Bernstein–Greene–Kruskal-like thermal state, as a fraction of the particles have an almost regular motion which makes the ergodic domain smaller than assumed for the comparison with the canonical predictions.

The positive-slope condition for the Landau instability translates into $\sigma > \omega_0/k$ if the initial wave amplitude is small. According to section 9.2.2, the system is then in the low-temperature phase and the wave intensity is expected to grow, a fact confirmed in section 8.2.2, even beyond the linear regime where Landau's prediction holds, towards the thermal equilibrium intensity. Then T_0 is not the equilibrium temperature of the self-consistent system, which is determined by the equation of state (9.42).

Section 8.3 considered the damping case with a warm beam. In the high-temperature phase, the constants of the motion (8.11) determine an equilibrium temperature $T = T_0 + 2(\omega_0 - kp_0)\psi_0 - k^2\psi_0^2$ and a critical temperature (9.20). Elementary algebra reduces the high-temperature criterion $T > T_c$ to the inequality $2\varepsilon'^{-2}(\psi_0 - u/k)(k\psi_0 - \alpha_+)(k\psi_0 - \alpha_-) > 1$, where $u = \omega_0/k - p_0 > \omega_0/k - \sigma > 0$ and $\alpha_\pm = u \pm \sqrt{u^2 + T_0}$. This defines a critical value ψ_{0c} (for which equality holds in the criterion) for the initial amplitude and the high-temperature regime corresponds to $\psi_0 < \psi_{0c}$. For the numerical simulations of section 8.3.1, $\psi_{0c} = 5.59 \times 10^{-2}$ which falls in the domain $[10^{-5}, 0.1]$ of the transition between the two extreme regimes of damping ($|\gamma_L| \ll \omega_b$ and $|\gamma_L| \gg \omega_b$).

In the large-ψ_0 case (i.e. large ω_b), we found that the long-time value is the initial value, as expected from the previous discussion. Indeed, from the thermodynamical viewpoint, a small wave amplitude implies that the total energy is essentially the sum of particles' kinetic energies, whereas a large wave amplitude provides the system with a large (x, p)-space region where particles have a low energy: for the microcanonical ensemble the latter means a larger entropy for a given total energy; and for the canonical ensemble this means a lower free energy for a given temperature—as we computed earlier. This result agrees with the dynamical argument of Mazitov (1965) and O'Neil (1965) after which the asymptotic damping of the wave is found to vanish (section 8.1.2).

In the small-ψ_0 case, the long-time value is independent of ψ_0 but scales with N like $N^{-1/2}$ as predicted for the high-temperature phase (figure 8.11). In the limit $N \to \infty$, the predicted value vanishes, as indicated by the vertical asymptote to the curve on the plot. For ψ_0 in the intermediate regime, the long-time estimate was found to be harder to reach. Then the runs exhibit a type of critical slowing-down as it occurs in the vicinity of a second-order phase transition.

9.8   Further comments

This chapter rests mainly on work by Firpo (1999) and Firpo and Elskens (2000). The thermodynamic approach to the wave–particle system is complementary to the dynamics approach investigated through most of this book. Its validity relies on the applicability of the ergodic hypothesis. The role of chaos to ensure an 'effective ergodicity' was discussed in particular by Escande *et al* (1994) for a model with nearest-neighbour interactions, and for the wave–particle model by Firpo and Doveil (2002); the former paper provides a Gibbsian check of the validity of the Gibbs technique.

Sharp applications of Gibbs equilibrium statistical mechanics to plasma models started with Lenard (1961) and Prager (1962) who determined the equation of state for one-dimensional models. Further results are reviewed by Choquard *et al* (1981) and Brydges and Martin (1999). A striking property of the one-dimensional Coulomb model of section 1.1.2 is that it undergoes no phase transition.

It is thus unexpected that we identify a phase transition in the wave–particle model. But this is no paradox because the basic assumption in the thermodynamic approach is the ergodic hypothesis, and thermodynamics alone gives no clue to the time-scales needed to visit a large enough domain in phase space. The thermodynamic treatment of the binary interaction assumes that all particles reach a 'global' equilibrium, whereas the thermodynamic treatment of the wave–particle system applies to a reduced system, in which only the faster, more efficient interactions are taken into account. Identifying the relevant effective variables is central to physics: in everyday thermodynamics, we do not take the 'thermally dead universe' as the relevant equilibrium state! Moreover, the argument involves the recognition that the plasma is a finite-N many-body system, in which 'thermal noise' induces behaviours which the limiting $N \to \infty$ Vlasovian description discards.

Chapter 10

Conclusion

10.1 Summary

A description of the Langmuir wave–particle interaction only through classical mechanics has been provided. Starting from an N-body model of the plasma, the fundamental aspects of the beam–plasma interaction were studied, encompassing two extreme cases of the saturation process of the beam–plasma instability. To this end, several intermediate steps were taken.

The first step was the construction of a field–particle description of the plasma. This field corresponds to the collective oscillations of electrons close to the plasma frequency. When Fourier analysed in space, these oscillations appear as Langmuir waves. Electrons with a velocity close to the phase velocity of a wave may be trapped into it. This is the paradigm of nonlinear resonance. The particles of the field–particle description are those electrons with a velocity in the range of the velocities of the considered Langmuir waves. A Hamiltonian was built to describe the self-consistent wave–particle dynamics.

The second step of the theory was to study the self-consistent dynamics in the case of a weak field. The cold beam–plasma instability was found as a particular limit of the theory. The slope of the particle distribution function, if weak, was found to be an essential control parameter for the self-consistent dynamics. For a positive slope, a Langmuir wave with a given wavenumber was found to be dominated by unstable eigenmodes growing with the Landau growth rate, which is proportional to this slope. For a negative slope, a given wavenumber was related to a set of marginally stable modes, the analogues of van Kampen–Case modes. Taking the reversibility of the dynamics into account then also required us to introduce van Kampen–Case-like eigenmodes in the description of the Landau growth.

At this point, we turned from the study of one mechanical realization of the system to consider statistics over initial wave phases and particle positions. This led, still for the weak-field case, to the derivation of a wave evolution equation incorporating a spontaneous emission of waves and a stimulated emission or absorption of waves by almost resonant particles; the non-spontaneous effects were ruled by the Landau growth rate. The wave evolution is related to a tendency for the almost resonant particles to synchronize with a single wave. If the particles correspond to the tail of a thermal plasma, the Langmuir waves were found to saturate at a thermal level as a consequence of the phase mixing among the analogues of van Kampen–Case modes. The Vlasovian limit of the description turns out to be highly singular.

The description of the nonlinear stage of the beam–plasma instability required the preliminary introduction of several concepts and tools of Hamiltonian chaos to be recalled later. This made it possible to derive the quasilinear equations for the weak-warm-beam instability by taking advantage of decorrelations resulting from the initial random phases and positions for the weak-field case and from the spreading of the particle positions due to chaos in the strong-field case. The weak-warm-beam instability was found to saturate with the formation of a plateau in the particle distribution function. Such phenomena are measured in a plasma with simple electrostatic tools like Langmuir probes. From this description of the weak-warm-beam–plasma instability, completely based upon classical mechanics, an old dream of the 19th century was realized: the non-trivial evolution of a macroscopic many-body system was described by taking into account the true character of its chaotic motion.

In the case where a unique Langmuir wave resonates with the weak warm beam, the instability was numerically shown to saturate in two stages. The first one corresponds to trapping most of the beam particles in the wave troughs. The second one is reached by trapping a large fraction of the fast particles, after a slow second growth caused by their progressive capture induced by the fluctuations of the wave amplitude due to the finite number of particles. The cold beam–plasma instability was shown to saturate with a Langmuir wave whose large amplitude fluctuates due to the particles bouncing in the wave troughs.

As far as particle dynamics is concerned, Landau growth or damping corresponds to a coherent synchronization mechanism with a single wave or to a diffusive mechanism in the presence of a broad spectrum of waves. This diffusion may be chaotic or not, depending on the wave amplitude.

Finally, the case of a single Langmuir wave interacting with a more general population of particles was dealt with analytically using Gibbs equilibrium ensembles. In the case of a negative slope for the distribution function, this approach enabled us to give evidence of a second-order phase transition, corresponding in the dynamics to the separation between Landau damping and damping with trapping.

Now we recall the concepts of Hamiltonian chaos, on which the description of the nonlinear part of the Langmuir wave–particle interaction relies heavily. These concepts were tackled from the low-dimensional (i.e. few degrees of freedom) viewpoint. Nonlinear resonances were found to be central actors of chaotic motion. They were typically found to form a dense set in phase space but a renormalization technique showed that, in domains where motion is regular, their size vanishes at small scale. An adiabatic approach to slow Hamiltonian chaos helped in evidencing the geometrical structure underlying chaotic transport associated with resonance overlap.

Chaotic diffusion is of central importance in plasma physics as well as in other Hamiltonian dynamics. It was explained by introducing the concept of a resonance box, which corresponds to the set of resonances actually acting for chaotic transport when a particle has a given velocity. Then diffusion was proved in the strongly chaotic regime by a two-step technique evidencing the two mechanisms at its root, namely the small dependence of the motion on the phase of any wave over a finite time and the corresponding spreading of velocities and positions.

With respect to the general problems of statistical physics, it is worth noting first that insisting on mechanical, finite-N pictures sheds light on the meaning of the kinetic models—both the deterministic-like Vlasov equation and the probabilistic-like Fokker–Planck equation. These two models are paradigmatic of many-body physics. Besides, the long-range nature of Coulomb interaction, on which the basic pictures of plasmas rest, turned out to have limited importance.

10.2   Open issues

The part of this book devoted to nonlinear phenomena points to many open issues. First, though the strong resonance overlap is the most relevant regime for physical applications, the description of chaotic diffusion still exhibits a theoretical gap corresponding to the supra-quasilinear regime of diffusion. There, progress is requested both in the intuitive understanding of the process and in its calculation.

For the weak-warm-beam–plasma instability, the intermediate regime, where the time-scales of chaos and wave growth are comparable, is still a mystery. Is it a diffusive regime?

Moreover, in the regimes we considered, the derivation of the quasilinear equations has been debated for a long time and its extensive check through numerical simulations remains desirable; an experimental verdict is even more important. On the mathematical side, the derivation of quasilinear equations could likely be sharpened by formulating it in functional-analytic form, similarly to the kinetic limit discussed in appendix G.

Further insight into the long-term fate of many waves interacting with many particles in the strong-overlap regime should also be gained from applying the Gibbs statistical equilibrium theory to this case. The condition allowing one to use the Gibbs formalism for describing a dynamical property such as equilibrium is microscopic chaos, which holds for the weak-turbulence regime. A hint as to the relevance of the Gibbs equilibrium for this system is that the entropy (7.54) which is irreversibly produced by the quasilinear equations is identical to the Gibbs entropy for a system of particles interacting with an infinite number of radiation degrees of freedom.

Numerical simulations (in the elementary case of a single Langmuir wave interacting with a warm beam) exhibited a second growth of the wave amplitude due to the beam particles trapping and detrapping, which is not seen in the Vlasovian simulation of the plasma. This suggests that the competition between the separation of solutions (exponential in t) and the regular dependence on initial data (discrepancy at t proportional to initial discrepancy) allows for the infinite time limit and the infinite number of particles limit not to commute. It would be important to check this discrepancy between the two descriptions analytically.

Indeed, if real, this discrepancy might be stronger in more turbulent cases like the weak-warm-beam instability. This non-commutation of the two limits might also hold for other continuous media. Mathematically, it relates to the fact that the approach to equilibrium in our system is related to the filamentation of distribution functions over phase space. This process can be rapid thanks to large-scale chaos or slow—it may even occur in integrable dynamics thanks to anisochronism.

Finally, we hope the reader will take hold of our views and challenge them, will extend them to other problems of interest and will find ways of making our theory more compact, more intuitive and more rigorous. Indeed, as explained by Popper, there is no such thing as a final theory. In agreement with Descartes' method, doubt must remain a fundamental item of the scientific adventure.

Appendix A

Continuous and discrete Fourier transforms

A.1 Perturbed lattices and discrete Fourier transforms

One is often interested in the motion of point particles in the vicinity of a lattice state. Thus, the reference positions are

$$x_n = \left(\frac{n}{N} + \frac{\phi}{2\pi} \right) L \tag{A.1}$$

where ϕ is a constant, N the number of lattice points, n the index of each point and L the spatial length of the lattice.

The small perturbation of interest is represented by a vector $y = (y_n) \in \mathbb{R}^N$. It is often useful to change basis in the y-space, thus introducing another representation $z = Ay$ of the vector. The matrix A describing this change of basis must be invertible in order for the mapping $y \mapsto z$ to be one-to-one; for finite N, it is necessary and sufficient that $\det A \neq 0$.

A special case of change of variables is the discrete Fourier transformation, which acts on the complexified vector space $\mathbb{C}^N$. In this case,

$$z_m = N^{-1/2} \sum_{n=1}^{N} e^{2\pi imn/N} y_n. \tag{A.2}$$

The matrix A so defined for $1 \leq m \leq N$ is invertible, as can be seen in two ways.

- Its rows are in the form $(a_n)^m$, with $a_n = e^{2\pi in/N}$. Hence A is a Vandermonde matrix, with determinant $\det A = \pm \prod_{n=2}^{N} \prod_{n'=1}^{n-1} (a_n - a_{n'})$. In our case, all coefficients a_n are distinct Nth roots of unity; however, this invertibility property would still hold if the coefficients were just $a'_n = e^{2\pi i x_n/L}$ for an arbitrary choice of positions $0 \leq x_n < L$, provided all positions are distinct.

- Inverting matrix A explicitly is easy for the discrete Fourier transform, as it is unitary. Indeed, it is clear that

$$N^{-1} \sum_{n=1}^{N} e^{2\pi i m n/N} e^{-2\pi i m' n/N} = \delta_{m,m'} \tag{A.3}$$

where the Kronecker symbol $\delta_{m,m'}$ is 1 for $m = m'$ and 0 otherwise. For $m = m'$, this is obvious. For $m \neq m'$, the number $c = e^{2\pi i(m-m')/N}$ is an Nth root of unity, distinct from 1. Then the left-hand side of (A.3) reduces to $(c^N - 1)c/(c - 1)$, which vanishes.

Thus (A.3) shows that the conjugate transposed matrix A^* is the reciprocal of A. Hence

$$\mathbf{y} = A^* \mathbf{z}. \tag{A.4}$$

Unitarity of A is equivalent to the property that mapping (A.2) preserves the Euclidian (in $\mathbb{R}^N$) or Hermitian (in $\mathbb{C}^N$) norm, $\|\mathbf{y}\|^2 = \mathbf{y}^* \cdot \mathbf{y} = \|\mathbf{z}\|^2$. This is physically interesting when this norm has a direct interpretation, e.g. if y_n is a velocity and $\|\mathbf{y}\|^2$ is a kinetic energy (up to a constant).

In addition to these change-of-representation properties, it is worth noting that (A.2) defines complex numbers z_m for all integers $m \in \mathbb{Z}$. It is easily seen that $z_m = z_{m'}$ if and only if $m \equiv m' (\text{mod } N)$.

Note also that, if $\mathbf{y}$ is a real-valued vector, then vector $\mathbf{z}$ has a symmetry: $z_{N-m} = z_m^*$ (with $z_N = z_0$ as mentioned).

Finally, one easily expresses matrix A in terms of the reference positions x_n and finds that

$$z_m = e^{-im\phi} N^{-1/2} \sum_{n=1}^{N} e^{2\pi i m x_n/L} y_n. \tag{A.5}$$

A.2   Representation of trigonometric functions

Trigonometric functions verify the identities

$$\sin \pi z = \pi z \prod_{k=1}^{\infty} \left(1 - \frac{z^2}{k^2}\right) \tag{A.6}$$

$$\cos \pi z = \prod_{k=1}^{\infty} \left(1 - \frac{z^2}{(k - 1/2)^2}\right) \tag{A.7}$$

where the right-hand side converges for all $z \in \mathbb{C}$. Moreover, for all $z \in \mathbb{C} \setminus \mathbb{Z}$,

$$\frac{\pi \cos \pi z}{\sin \pi z} = \frac{1}{z} + \frac{2z}{\pi} \sum_{k=1}^{\infty} \frac{1}{z^2 - k^2} \tag{A.8}$$

$$\frac{\pi^2}{\sin^2 \pi z} = \sum_{k=-\infty}^{\infty} (z-k)^{-2} \qquad \text{(A.9)}$$

$$\frac{\pi^3 \cos \pi z}{\sin^3 \pi z} = \sum_{k=-\infty}^{\infty} (z-k)^{-3}. \qquad \text{(A.10)}$$

Moreover, for $a = c - iw$ with $-1/2 \le w \le 1/2$ and $c > 0$, let $u = ia = w + ic$ and $b = \exp(-2\pi a)$. Then, for $y \in \mathbb{C}$,

$$\pi^2 \sin^{-2} \pi u - y\pi \frac{\cos \pi u}{\sin \pi u} = \pi i y \left(1 - 4b' - \frac{yb'^2}{16\pi^2}\right)(1-b)^{-2} \qquad \text{(A.11)}$$

where $b = iyb'/(4\pi)$.

A.3   Laplace and Fourier transforms

Given a function $f : \mathbb{R} \to \mathbb{C}$, its Laplace transform (if it exists) is the function $\hat{f}$ defined by

$$\hat{f}(s) = \int_0^{\infty} f(t)e^{-st}\, dt \qquad \text{(A.12)}$$

and its Fourier transform $\tilde{f}$ (if it exists) is defined by

$$\tilde{f}(k) = \int_{-\infty}^{\infty} f(x)e^{ikx}\, dx. \qquad \text{(A.13)}$$

The Fourier transform of $\tilde{f}$ restores $2\pi f$, up to a complex conjugation, as

$$f(x) = (2\pi)^{-1}\int_{-\infty}^{\infty} \tilde{f}(k)e^{-ikx}\, dk. \qquad \text{(A.14)}$$

The Parseval identity reads:

$$\int_{-\infty}^{\infty} |\tilde{f}(k)|^2\, dk = 2\pi \int_{-\infty}^{\infty} |f(x)|^2\, dx. \qquad \text{(A.15)}$$

Given a function $f : \mathbb{R} \to \mathbb{C}$ such that $\int_{\mathbb{R}}(1+x^2)|f(x)|^2\, dx < \infty$, let

$$m(f) = \int |f(x)|^2\, dx \qquad \text{(A.16)}$$

$$c(f) = \frac{1}{m(f)}\int x|f(x)|^2\, dx \qquad \text{(A.17)}$$

$$w(f) = \left[\frac{1}{m(f)}\int (x - c(f))^2|f(x)|^2\, dx\right]^{1/2} \qquad \text{(A.18)}$$

which we call, respectively, the mass, centre and width of f. One shows that

$$m(\tilde{f}) = 2\pi m(f) \tag{A.19}$$
$$w(\tilde{f})w(f) \geq \pi. \tag{A.20}$$

The first relation implies that $(2\pi)^{-1/2}\tilde{f}$ has the same mass as f (which associates an isometry in function space L^2 with the Fourier transform). The second relation shows that a peaked original f must have a transform $\tilde{f}$ with broad support. These observations are fundamental to signal-processing and to quantum mechanics (where (A.20) relates to the uncertainty relations).

Exercise A.1. Given $a \in \mathbb{R}$ and $b > 0$, compute $m(f)$, $c(f)$, $w(f)$ and $w(\tilde{f})$ for $f(x) = (2\pi b^2)^{-1/2}e^{-(x-a)^2/(2b^2)}$. Compare them with the normalization, expectation and standard deviation of the Gaussian probability distribution f (section E.2) and check (A.20). Do the same for the probability density $g(x) = (2b)^{-1}e^{-|x|/b}$ ('two-sided' exponential). Note that the Gaussian distribution is optimal for (A.20), while g is not.

Exercise A.2. Given $a > 0$, let

$$f(t) = e^{-at}H(t \geq 0) \qquad g(t) = f(-t) \qquad h(t) = f(t) + g(t) \tag{A.21}$$

where H denotes the Heaviside function (section B.3). Show that

$$\tilde{f}(\omega) = (a - i\omega)^{-1} \qquad \tilde{g}(\omega) = \tilde{f}^*(\omega) \qquad \tilde{h}(\omega) = \frac{2a}{a^2 + \omega^2}. \tag{A.22}$$

Appendix B

Special functions

We recall elementary properties of special functions, which belong to the physicist's standard toolbox. Useful formulae are collected by Abramowitz and Stegun (1970) and in electronic libraries such as as the digital library of mathematical functions. Throughout this book, we denote the real and imaginary parts of a number $z \in \mathbb{C}$ as $z = \Re z + i\Im z$, and its complex conjugate as $z^* = \Re z - i\Im z$.

B.1 Euler gamma function

The Euler gamma function is defined by the integral representation for any $u > 0$

$$\Gamma(u) = \int_0^\infty e^{-t} t^{u-1} \, dt \tag{B.1}$$

and verifies the identity $\Gamma(u+1) = u\Gamma(u)$. For positive integer u, $\Gamma(u+1) = u!$. In the limit $u \to \infty$, its asymptotic expansion is the Stirling formula (see, e.g., Bender and Orszag 1978, section 5.4):

$$\Gamma(u) = (u-1)! \sim \left(\frac{2\pi}{u}\right)^{1/2} \left(\frac{u}{e}\right)^u \left(1 + \frac{1}{12u} + \cdots\right). \tag{B.2}$$

Note that $\Gamma(\frac{1}{2}) = \sqrt{\pi} = 1.7\ldots$, and that Γ has a simple pole at $u = 0$. The function $1/\Gamma(u)$ is entire on $\mathbb{C}$ and has simple zeros at negative integers.

Note that, for $a, b, c > 0$,

$$\int_0^\infty e^{-ax^b} x^c \, dx = b^{-1} a^{-(c+1)/b} \Gamma\left(\frac{c+1}{b}\right). \tag{B.3}$$

B.2   Bessel functions

Bessel functions of the first kind, of order $n \in \mathbb{N}$, have the integral representation for $z \in \mathbb{C}$

$$J_n(z) = \frac{1}{\pi} \int_0^\pi e^{iz\cos\theta} \cos(n\theta) \, d\theta \tag{B.4}$$

and $J_{-n}(z) = (-1)^n J_n(z)$. Modified Bessel functions of order n satisfy

$$I_n(z) = I_{-n}(z) = e^{-i\pi n/2} J_n(ze^{-i\pi/2}) = \frac{1}{2\pi} \int_0^{2\pi} e^{z\cos\theta} \cos(n\theta) \, d\theta. \tag{B.5}$$

For integer order $n \geq 0$, Bessel and modified Bessel functions are entire functions of z, with Taylor expansion

$$I_n(z) = \left(\frac{z}{2}\right)^n \sum_{k=0}^\infty \frac{(z/2)^{2k}}{k!(n+k)!}. \tag{B.6}$$

They occur in the Fourier expansion of the exponential of trigonometric functions,

$$e^{z\cos\theta} = \sum_{k=-\infty}^\infty I_k(z) \cos(k\theta) = I_0(z) + 2 \sum_{k=1}^\infty I_k(z) \cos(k\theta) \tag{B.7}$$

and have the (Laurent series) generating function $e^{z(t+1/t)/2} = \sum_{k=-\infty}^\infty I_k(z)t^k$ for $t \neq 0$, convergent for all $z \in \mathbb{C}$.

They satisfy the recurrence relations

$$I_{n-1}(z) - I_{n+1}(z) = \frac{2n}{z} I_n(z) \tag{B.8}$$

$$\frac{d}{dz} I_n(z) = \frac{1}{2} I_{n-1}(z) + \frac{1}{2} I_{n+1}(z) = I_{n+1}(z) + \frac{n}{z} I_n(z). \tag{B.9}$$

For n fixed and $|z| \to \infty$ (with $|\operatorname{Arg} z| < \pi/2$), they admit the asymptotic expansion

$$I_n(z) \approx \frac{e^z}{\sqrt{2\pi z}} \left(1 - \frac{\mu - 1}{1!(8z)} + \frac{(\mu - 1)(\mu - 9)}{2!(8z)^2} - \cdots\right) \tag{B.10}$$

where $\mu = 4n^2$.

To show that $I_1(z)/[zI_0(z)]$ is decreasing[1] as claimed in section 9.2.2, we note that $I_0(z)$ and $z^{-1} I_1(z)$ have series converging on the whole complex plane, which satisfy the following lemma.

---

[1] The asymptotics for $z \to \infty$ of the ratio $I_1(z)/[zI_0(z)]$ is directly seen from the asymptotic behaviour of the modified Bessel function.

Lemma. Consider two entire functions $A(z) = \sum_{n=0}^{\infty} \alpha_n z^n$ and $B(z) = \sum_{n=0}^{\infty} \alpha_n \beta_n z^n$, with positive coefficients α_n and β_n (and $\alpha_0 > 0$ with no loss of generality). If the sequence (β_n), $n \in \mathbb{N}$, is decreasing, then the ratio $B(z)/A(z)$ is a real positive decreasing function of z for $0 \leq z < \infty$ and $\lim_{z \to +\infty} B(z)/A(z)$ exists.

Proof. Consider $0 \leq x \leq y < \infty$. As the series are entire, with positive coefficients, $0 < \alpha_0 \leq A(x) \leq A(y) < \infty$. To prove that $B(x)/A(x) \geq B(y)/A(y)$, it suffices to prove that $B(x)A(y) - B(y)A(x) \geq 0$, which is clear as

$$B(x)A(y) - A(x)B(y) = \sum_{k=0}^{\infty}\sum_{l=0}^{\infty} \alpha_k \alpha_l (\beta_k - \beta_l) x^k y^l \qquad \text{(B.11)}$$

is a convergent series with all terms positive. Then $\lim_{z \to +\infty} B(z)/A(z)$ exists since a positive decreasing function of z has a limit for $z \to \infty$.

B.3   Dirac distribution

The Dirac distribution δ is not a function. It is often called a generalized function, because it is the limit (in a natural sense) of well behaved functions. It is mathematically meaningful under integral signs; by definition, it is such that

$$\int g(x)\delta(x)\,dx = g(0) \qquad \text{(B.12)}$$

for bounded continuous functions g. If one views a function g as a vector in abstract space, with an infinity of components indexed by x, then the Dirac distribution δ extracts the component with index 0 from this vector. Similarly, for the 'component with index y', one uses

$$\int g(x)\delta(x-y)\,dx = g(y) \qquad \text{(B.13)}$$

i.e. δ is the unit factor for convolution. Note that, as it appears in (B.12) under an integral sign, it is not dimensionless: $[\delta(x)] = [x^{-1}]$.

The rules of calculus apply to distributions in a similar way to that for functions under integral signs. Using a generalized derivative,

$$dH(x > a) = \delta(x - a)\,dx \qquad \text{(B.14)}$$

where H is the indicator function ($H(C) = 1$ iff C holds and 0 otherwise). Speaking of the value of the Dirac distribution at a single point x is meaningless.

The Dirac distribution is the limit of many functions of interest.

Exercise B.1. Show that

$$\pi\delta(x) = \lim_{\varepsilon \to 0^+} \frac{1 - \cos(x/\varepsilon)}{x^2 \varepsilon} = -\lim_{\varepsilon \to 0^-} \frac{1 - \cos(x/\varepsilon)}{x^2 \varepsilon} \qquad \text{(B.15)}$$

$$= \lim_{T \to +\infty} \int_0^T \cos(xt) \, dt = - \lim_{T \to -\infty} \int_T^0 \cos(xt) \, dt \qquad \text{(B.16)}$$

$$= \lim_{\varepsilon \to 0^+} \frac{\varepsilon}{x^2 + \varepsilon^2} = - \lim_{\varepsilon \to 0^-} \frac{\varepsilon}{x^2 + \varepsilon^2} \qquad \text{(B.17)}$$

in the sense of distributions, i.e. for use in (B.12). Sketch the functions of x for various ε and T in terms of which these limits are defined and comment on the importance of signs in the limits.

The fact that the Fourier transform is involutive is best expressed by the identity

$$\int_{-\infty}^{\infty} e^{ikx} e^{-iky} \, dk = 2\pi \delta(x - y) \qquad \text{(B.18)}$$

which is also a resolution of the identity in the Fourier basis of $L^2(\mathbb{R})$.

The Dirac distribution also occurs in the integral of a function passing through a simple pole. Let $f(x)$ be a real-valued function of $-\infty < x < \infty$, such that $|f(x) - f(y)| \leq A|x - y|^c$ for any $x, y \in \mathbb{R}$ and $|f(x)| \leq A|x|^{c'}$ for $x \to \pm\infty$, with $A > 0, 0 < c \leq 1$ and $c' > 0$. Then, for $z \in \mathbb{C}$, consider

$$F(z) = \int_{-\infty}^{\infty} \frac{f(x)}{x - z} \, dx. \qquad \text{(B.19)}$$

If $\Im z \neq 0$, this integral defines a smooth function F. More precisely, it defines two continuous functions, F_+ for $\Im z > 0$ and F_- for $\Im z < 0$. But the definition (B.19) is singular for $z \in \mathbb{R}$. The principal value of such a singular integral is defined for $b < a < c$,

$$\mathrm{PP} \int_b^c \frac{f(x)}{x - a} \, dx = \lim_{z \to 0^+} \left(\int_b^{a-z} \frac{f(x)}{x - a} \, dx + \int_{a+z}^c \frac{f(x)}{x - a} \, dx \right). \qquad \text{(B.20)}$$

One shows (see, e.g., Balescu (1963) or van Kampen and Felderhof (1967)) for $a \in \mathbb{R}$ that

$$\mathrm{PP} \int_{-\infty}^{\infty} \frac{f(x)}{x - a} \, dx = \lim_{b \to 0^+} F_+(a + ib) - \pi i f(a)$$

$$= \lim_{b \to 0^+} F_-(a - ib) + \pi i f(a). \qquad \text{(B.21)}$$

These Plemelj formulae imply that the two functions $F_\pm$ have different limits on the real axis and that

$$f(a) = 2\pi i (F_+(a) - F_-(a)) \qquad \text{(B.22)}$$

so that the right-hand side of (B.22) provides an integral operator representation of the Dirac distribution.

One may define the period-P Dirac distribution $\delta_P(x)$ in a similar way,

$$\delta_P(x) = \sum_{n=-\infty}^{\infty} \delta(x - nP) = \frac{1}{P} \sum_{k=-\infty}^{\infty} e^{2\pi i k x / P}. \qquad (B.23)$$

Classical approximations to $\delta_{2\pi}(x)$ are the periodic window functions $W^{(M)}(x) = \sum_{n=-\infty}^{\infty} M H(|x - 2k\pi| < 1/2M)$.

Exercise B.2. The Fourier coefficients $W_k^{(M)} = (2\pi)^{-1} \int_{-\pi}^{\pi} W^{(M)}(x) e^{-ikx} \, dx$ of the periodic window function are the unique numbers $W_k^{(M)}$ such that $W^{(M)}(x) = \sum_{k=-\infty}^{\infty} W_k^{(M)} e^{ikx}$. Show that $\lim_{M\to\infty} W_k^{(M)} = 1$ for all k and conclude that $\lim_{M\to\infty} W^{(M)}(x) = \delta_{2\pi}(x)$ in the sense of distributions.

Appendix C

Phase space structure and orbits

Chaos does not occur in continuous-time dynamical systems in the plane. More precisely, consider a system whose state is described by a point $z = (z_1, \ldots, z_N) \in \mathcal{M}$, where $\mathcal{M}$ is the N-dimensional phase space and whose dynamics obeys a first-order autonomous differential equation

$$\dot{z} = f(z) \tag{C.1}$$

where $f : \mathcal{M} \to \mathbb{R}^N$ is called its velocity vector field. For the pendulum of section 1.2, $z = (q, p)$ belongs to the cylinder $\mathcal{M} = S_{2\pi} \times \mathbb{R}$ (where $S_{2\pi}$ is the circle, i.e. the interval of length 2π with periodic boundary conditions) and $f(q, p) = (p, -A \sin q)$.

The following classical theorem (see, e.g., Hirsch and Smale 1974) is crucial for the study of (C.1):

> *Existence and uniqueness of solutions to C^1 dynamics*:
> Let U be an open subset of $\mathcal{M}$ and $x_0 \in U$. If f is continuously differentiable on U, there exist $\tau > 0$ and a unique solution ϕ : $]-\tau, \tau[\to U$ to the differential equation (C.1) with initial condition $\phi(0) = x_0$.

In short, orbits do not cross (this would violate uniqueness) and motion is well behaved (it is a well-defined, continuously differentiable function of time) as long as the system state remains in the 'regular' part of phase space, i.e. the domain U where f is well behaved.

One must allow for various types of phase space $\mathcal{M}$, for topology is central to understanding the dynamics. For instance, the circle for the pendulum (where adding 2π to the angle brings the pendulum back to the same position) is not identical to the line for the electron (where a displacement by $L = 2\pi$ does not bring the electron to the same position but only to a similar place in the next trough of the wave potential).

This no-crossing theorem restricts the evolution of dynamical systems in continuous time much more than the dynamics in discrete time. Abraham and Shaw (1992) provide a rich visual tutorial to such geometric considerations.

In one-dimensional phase space, the continuous-time dynamics are elementary. Indeed, at any point z, either the velocity $f(z)$ vanishes (then the point is fixed) or the velocity is positive (and motion proceeds to the right) or the velocity is negative (and motion proceeds to the left). Thus on an orbit the velocity may change its algebraic value but never its sign. The only ways for a motion to return to its initial point z_0 are:

(i) to have $z(t) = z_0$ for all times, i.e. $f(z_0) = 0$ or
(ii) to have $f(z) \neq 0$ for all $z \in \mathcal{M}$, provided that $\mathcal{M}$ is (equivalent to) the circle.

On the line, it is also clear that, given two orbits $z_1(t)$ and $z_2(t)$, if $z_1(0) < z_2(0)$ then $z_1(t) < z_2(t)$ for all times t.

For a two-dimensional phase space $\mathcal{M}$, the orbits' mutual avoidance also ensures some order. More precisely, if $\mathcal{M}$ is the plane (or a connected piece of it), the cylinder (which is topologically equivalent to an annulus on the plane) or the sphere (which is mapped to the plane by removing a single point, say its north pole), then any closed orbit separates $\mathcal{M}$ in two unconnected domains, say $\mathcal{D}_1$ and $\mathcal{D}_2$. Then any orbit with initial condition in $\mathcal{D}_1$ remains in $\mathcal{D}_1$ for all time.

Another aspect of this order is emphasized by the

Poincaré–Bendixson theorem:
For smooth dynamics (C.1) on the plane, the asymptotic behaviour of any orbit is either stationary (fixed point) or periodic (limit cycle).

In spite of its strength, this theorem still leaves room for important open questions, such as estimating the number and position of limit cycles from the mere knowledge of f (Hilbert's 16th problem).

The mutual avoidance argument applies to both the pendulum and the electron in a wave. More complicated dynamical systems may have topologically more complicated phase spaces.

In a three- or higher-dimensional phase space $\mathcal{M}$, a single orbit is still a line, which cannot separate $\mathcal{M}$ into unconnected domains. Then orbits can braid around each other, in more or less complicated patterns. Typically, it requires an $(N-1)$-parameter family of orbits to separate $\mathcal{M}$ into unconnected domains, but already for $N = 3$ these families often form manifolds with a more complicated topology than the plane, so that even on them the orbits can be weird. Then chaos can occur.

For Hamiltonian systems, no chaos occurs for one-degree-of-freedom systems with $q \in S_L$ or $q \in \mathbb{R}$ while $p \in \mathbb{R}$. It is allowed as soon as there are more degrees of freedom (see also section 5.5.1).

Appendix D

Symplectic structure and numerical integration

In this appendix we first present a distinctive feature of the canonical structure of Hamiltonian dynamics. Then we show how one can respect it in numerical applications.

D.1 Symplectic dynamics*

The canonical structure of (5.1), (5.2) implies that Hamiltonian dynamics preserves an area in phase space. For a single degree of freedom, consider two orbits, $(q(t), p(t))$ and $(q(t) + \delta q(t), p(t) + \delta p(t))$, close to each other. Then decompose the small vector $(\delta q, \delta p)$ in two vectors, parallel to the coordinate system, $\delta \boldsymbol{q} = (\delta q, 0)$ and $\delta \boldsymbol{p} = (0, \delta p)$, and compute the oriented area $\delta \boldsymbol{p} \wedge \delta \boldsymbol{q} = \delta p \delta q$ of their parallelogram. The change of $\delta \boldsymbol{p} \wedge \delta \boldsymbol{q}$ in time is

$$\frac{\mathrm{d}}{\mathrm{d}t} \delta \boldsymbol{p} \wedge \delta \boldsymbol{q} = \delta \dot{\boldsymbol{p}} \wedge \delta \boldsymbol{q} + \delta \boldsymbol{p} \wedge \delta \dot{\boldsymbol{q}}$$

$$= -\frac{\partial^2 H}{\partial q^2} \delta \boldsymbol{q} \wedge \delta \boldsymbol{q} - \frac{\partial^2 H}{\partial q \partial p} \delta \boldsymbol{p} \wedge \delta \boldsymbol{q}$$

$$+ \frac{\partial^2 H}{\partial q \partial p} \delta \boldsymbol{p} \wedge \delta \boldsymbol{q} + \frac{\partial^2 H}{\partial p^2} \delta \boldsymbol{p} \wedge \delta \boldsymbol{p}$$

$$= 0 \qquad\qquad (\mathrm{D.1})$$

where we have used the identity $\delta \boldsymbol{q} \wedge \delta \boldsymbol{q} = \delta \boldsymbol{p} \wedge \delta \boldsymbol{p} = 0$ (a parallelogram defined by two parallel vectors has vanishing area). For N degrees of freedom, using the canonical Poisson brackets, one shows that the Poincaré–Cartan invariant two-form[1] $\mathrm{d}\bar{\omega} \equiv \sum_i \delta p_i \wedge \delta q_i$ is conserved. Its preservation in numerical simulations

---

[1] A two-form (denoted by a wedge or exterior product) is a differential geometric object, defining locally the area of a parallelogram built on two infinitesimal vectors in a many-dimensional space. More generally, an n-form defines the oriented n-volume of the n-parallelepiped (see, e.g., Schutz 1980).

of Hamiltonian dynamics requires the use of symplectic integration schemes.

Note that (D.1) holds as well if H depends on time. In terms of the volume in phase space, it implies that the Hamiltonian flow is divergence free, or incompressible,

$$\text{div } f = \partial_q \partial_p H + \partial_p(-\partial_q H) = 0. \tag{D.2}$$

In terms of orbits, it implies that, given a set of initial data $\mathcal{D}_0 \subset \mathcal{M}$ at time 0, the orbits evolving from it at time t form a set $\mathcal{D}_t \subset \mathcal{M}$ such that

$$\mu(\mathcal{D}_t) = \int_{\mathcal{D}_t} d^N p \, d^N q = \int_{\mathcal{D}_0} d^N p \, d^N q = \mu(\mathcal{D}_0) \tag{D.3}$$

i.e. the Hamiltonian flow preserves the Liouville measure[2] $d\mu \equiv d^N p \, d^N q = \prod_i dp_i \, dq_i$.

D.2   Numerical integration schemes

Integrating numerically differential equations is central to the understanding of dynamics. Given a first-order system

$$\dot{z} = f(z) \tag{D.4}$$

on a phase space $\mathcal{M}$, we want to find the function $\Phi : \mathcal{M} \times \mathbb{R} \to \mathcal{M}$ such that, given $z_0 \in \mathcal{M}$, the function $\phi(t) \equiv \Phi(z_0, t)$ is the solution to (D.4) with initial condition $\phi(0) = z_0$. This function Φ is called the flow on $\mathcal{M}$ associated with (D.4).

A numerical integrator for (D.4) is a map $\Psi : \mathcal{M} \times \mathbb{R} : (z, t) \mapsto (z', t')$ such that $z' \simeq \Phi(z, t' - t)$ within the requested accuracy. A good integrator is fast, accurate (i.e. $\|z' - \Phi(z, t' - t)\|$ is small), robust (i.e. not too sensitive to a poor knowledge of f).... It must also preserve crucial properties of the dynamics (D.4), e.g. conserve known constants of the motion (as energy or momentum) within a given tolerance. If the dynamics has some symmetries, e.g. if it is time-reversible (for the pendulum, equations of motion are invariant for the mapping $(q, p, t) \mapsto (q, -p, -t)$), one expects the integrator to preserve the symmetries too.

For Hamiltonian dynamics, conservation of the Poincaré–Cartan invariant is essential. This requirement leads to the construction of symplectic numerical integrators.

A (poor) straightforward scheme is the Euler integrator

$$z' = z + f(z)\Delta t. \tag{D.5}$$

---

[2] For $N = 1$, the Liouville measure is the Poincaré–Cartan invariant but for $N > 1$ the conservation of $d\bar{\omega}$ is stronger than the conservation of $d\mu = |(d\bar{\omega})^N|/(N!)$.

Exercise D.1. For the pendulum (1.11), let $z = (q, p)$, with $p = \dot{q}$. Given a time step $\Delta t > 0$, write down the Euler integrator $(q, p) \mapsto (q', p')$ explicitly. Denote by $(Q(q, p, \Delta t), P(q, p, \Delta t)) = \Phi(q, p, \Delta t)$ the exact solution of the pendulum equation of motion for the time step Δt.

(i) Accuracy: show that $|Q(q, p) - q'| \leq C_1(q, p)(\Delta t^2)$ and that $|P(q, p) - p'| \leq C_2(q, p)(\Delta t^2)$. Find upper estimates for C_1 and C_2: can you find uniform estimates (i.e. independent of the actual point (q, p))?

(ii) Reversibility: show that
$Q(q, -p, -\Delta t) = Q(q, p, \Delta t)$, $P(q, -p, -\Delta t) = -P(q, p, \Delta t)$,
$q'(q, -p, -\Delta t) = q'(q, p, \Delta t)$ and $p'(q, -p, -\Delta t) = -p'(q, p, \Delta t)$.

(iii) Area conservation: show that

$$\det \frac{\partial(q', p')}{\partial(q, p)} \neq 1 \tag{D.6}$$

while (D.1) implies $\det(\partial(Q, P)/\partial(q, p)) = 1$. Estimate the discrepancy.

These inequalities are understood as stating that, in general, equality does not hold—though equality may hold at some peculiar points (q, p). Hint: use Taylor expansions.

The non-conservation of area rules out the Euler scheme for integrating Hamiltonian dynamics. Similar criticisms apply to many other schemes which prove satisfactory for non-Hamiltonian dynamics.

Exercise D.2. Discuss the Runge–Kutta schemes of order 2 or 4.

It often turns out that conservation of the constants of motion cannot be ensured by the numerical schemes. This may be corrected easily in some cases by monitoring their evolution explicitly. However, for symplectic schemes applied to autonomous Hamiltonian dynamics, it appears that the energy fluctuates around its initial value but does not wander far away (Channell and Scovel 1990). This is a natural consequence of the theorem by Benettin and Fassò (1999) quoted in the next section.

D.3   Symplectic integration of Hamiltonian dynamics

The simplest way to construct a symplectic integrator is to decompose the dynamics in integrable Hamiltonian steps. The chain rule for derivatives ensures that, if both mappings $\Psi_i : \mathcal{M} \to \mathcal{M}$ for $i = 1, 2$ are symplectic, then $\Psi_2 \circ \Psi_1 : \mathcal{M} \to \mathcal{M}$ is also symplectic. This is easily checked if $\mathcal{M}$ has only two dimensions, for symplecticity then reduces to area conservation.

Now consider a Hamiltonian

$$H(q, p) = K(p) + V(q) \tag{D.7}$$

i.e. the sum of two integrable Hamiltonians. Then, for $H(q, p) = K(p)$, the equations of motion reduce to $\dot{p}_i = 0$, $\dot{q}_i = \partial_{p_i} K \equiv u_i(p)$, so that $\Phi(q, p, \tau) = \Psi_K(q, p, \tau) \equiv (q + u(p)\tau, p, \tau)$. Similarly, for $H(q, p) = V(q)$, the equations of motion reduce to $\dot{q}_i = 0$, $\dot{p}_i = -\partial_{q_i} V \equiv a_i(q)$, so that $\Phi(q, p, \tau) = \Psi_V(q, p, \tau) \equiv (q, p + a(q)\tau, \tau)$. Note the complete symmetry between the treatment of K and V, in agreement with the basic symmetry of coordinates and momenta in Hamiltonian dynamics.

For a Hamiltonian (D.7), the integrator $\Psi_K(\Delta t) \circ \Psi_V(\Delta t)$ is symplectic. Besides, it is accurate to $O(\Delta t^2)$.

Exercise D.3. Show that this integrator yields

$$q' = q + u(p')\Delta t \tag{D.8}$$
$$p' = p + a(q)\Delta t \tag{D.9}$$
$$t' = t + \Delta t. \tag{D.10}$$

Check its accuracy and area conservation. How does this scheme differ from the Euler scheme? For the pendulum, show that it is not reversible.

If $K(-p) = K(p)$, the dynamics of (D.7) is reversible. A simple reversible symplectic integrator is the centred leapfrog

$$\Psi_{\text{leapfrog}} = \Psi_K\left(\frac{\Delta t}{2}\right) \circ \Psi_V(\Delta t) \circ \Psi_K\left(\frac{\Delta t}{2}\right). \tag{D.11}$$

Exercise D.4. Write down the (more usual) integrators $\Psi_K(\Delta t) \circ \Psi_V(\Delta t)$ and $\Psi_V(\Delta t) \circ \Psi_K(\Delta t)$ for the pendulum. Compare them with each other and with (D.11). Relate them to the standard map (5.59), (5.60) by changes of variables and parameters.

The relation between the numerical integration of the pendulum and the standard map involves a connection between the map parameter K, the integrator time step Δt and the pendulum parameter A. As the standard map may also be derived from a continuous-time dynamics with an infinity of waves, one may view the effect of time discretization (by Δt) as perturbing the original dynamics by additional terms. While the pendulum is integrable, the standard map approximation to the pendulum is chaotic and one may ascribe its non-integrability to those parasitic resonances. This invites one to choose a small enough time step to keep the parasitic resonances far away from the pendulum's cat's eye in which one is interested.

The leapfrog integrator is second order in Δt. One may construct higher-order symplectic integrators by composing successive maps Ψ_i with different substeps τ_i. The separation into two integrable Hamiltonians may also be replaced by a separation into more terms if this leads to simpler integration. In particular, for the wave–particle system, a fourth-order integrator was developed by Cary and Doxas (1993).

For Hamiltonian dynamics, another way to construct symplectic integrators is to use generating functions, which goes beyond the level of this book. This also leads to analytical insight on the dynamics (somewhat related to the Lagrangian approach to mechanics).

The converse to constructing a discrete-time integrator for a given continuous-time flow is the suspension of a given discrete-time mapping in a continuous-time dynamics—which is more difficult. Yet it is an important question, as the theory of area-preserving maps and, more generally, chaotic mappings is more advanced than the theory of flows (Hamiltonian or not).

In particular, given a symplectic map Ψ_ε, analytic and ε-close to identity, there exists an analytic autonomous Hamiltonian H_ε such that its time-1 mapping differs from Ψ_ε by a quantity exponentially small in $1/\varepsilon$. This implies that, given an analytic Hamiltonian H_0, a symplectic integrator of order p (so that $\varepsilon = \mathrm{O}(\Delta t^{p+1})$) generates orbits 'very' close to the orbits of a Hamiltonian H' which is ε' close to H_0 over 'long' integration times with $\lim_{\varepsilon\to 0} \varepsilon' = 0$ (Benettin and Giorgilli 1994, Benettin and Fassò 1999). The resulting accuracy estimates are important, for example, in simulating the dynamics of many-body systems.

Exercise D.5. Write down the algorithms for the integration schemes and implement them in a genuine computer language. Check our statements numerically with these codes. Beware that a tiny algorithmic or encoding mistake may result in a deep change in the actual flowchart and assignments of variables, and dismiss the dream of a 'black box universal integrator'. Check your code on integrable examples.

Appendix E

Probability and stochastic processes

E.1 General random variables and vectors

The support (of the distribution) of a real random variable X is the smallest closed set $S \subset \mathbb{R}$ such that $\mathbb{P}(X \in S) = 1$. Given a function $h : \mathbb{R} \to \mathbb{R} : x \mapsto h(x)$, its expectation with respect to the distribution of X is

$$\langle h(X) \rangle = \mathcal{E}h(X) = \int_{\mathbb{R}} h(x)\mathbb{P}(x \leq X < x + \mathrm{d}x) \tag{E.1}$$

if this integral exists. The characteristic function of a real random variable X is defined for all $u \in \mathbb{R}$ as

$$\varphi_X(u) = \langle e^{iuX} \rangle \tag{E.2}$$

and, if all moments of X exist,

$$\varphi_X(u) = \sum_{k=0}^{\infty} \frac{(iu)^k \langle X^k \rangle}{k!}. \tag{E.3}$$

A random variable uniformly distributed on $[a, b] \subset \mathbb{R}$ (with $a \leq b$) has expectation $\langle X \rangle = (a + b)/2$ and variance $\langle X^2 \rangle - \langle X \rangle^2 = \sigma_X^2$, with the standard deviation $\sigma_X = (b - a)/(2\sqrt{3})$.

The expectation $m_X \equiv \langle X \rangle$ provides a characteristic value for the random variable. It is the least-squares estimate for X, i.e. the number a_0 for which the function $g(a) \equiv \langle (X - a)^2 \rangle$ reaches its minimum. Moreover $g(m_X) = \sigma_X^2$.

The standard deviation provides a scale for the discrepancy between randomly drawn values of X and its expectation. According to the Bienaymé–Chebyshev inequality, for any $b > 0$, $\mathbb{P}(|X - m_X| \geq b\sigma_X) \leq b^{-2}$.

Similarly, for any random variable X with finite moment of degree $k \geq 1$, the Markov inequality ensures that, for any $b > 0$,

$$\mathbb{P}(|X - m_X|^k \geq b^k \langle |X - m_X|^k \rangle) \leq b^{-k}. \tag{E.4}$$

For any two random variables (X, Y) with finite variances, their covariance $\mathrm{Cov}(X, Y) \equiv \langle XY \rangle - \langle X \rangle \langle Y \rangle$ satisfies the bound $|\mathrm{Cov}(X, Y)| \leq \sigma_X \sigma_Y$.

Random variables X_r, $1 \leq r \leq n$, are independent if and only if their joint probability distribution factorizes, i.e.

$$d\mathbb{P}(x_1 \leq X_1 < x_1 + dx_1, \ldots x_n \leq X_n < x_n + dx_n)$$

$$= \prod_{r=1}^{n} d\mathbb{P}(x_r < X_r \leq x_r + dx_r) \tag{E.5}$$

or, equivalently, their joint characteristic function factorizes,

$$\varphi_X(\boldsymbol{u}) \equiv \langle e^{i\boldsymbol{u}\cdot X} \rangle = \prod_{r=1}^{n} \varphi_{X_r}(u_r). \tag{E.6}$$

Given $n \to \infty$ independent random variables X_r, $1 \leq r \leq n$, with finite second moment, let $c_r \equiv \langle X_r \rangle$ and $\sigma_r \equiv (\langle X_r^2 \rangle - c_r^2)^{1/2}$. Assume there exists an $a > 0$ such that $\langle |X_r^3| \rangle < a$ for all r. Let $c = \lim_{n\to\infty} n^{-1} \sum_{r=1}^{n} c_r$ and $\sigma^2 = \lim_{n\to\infty} n^{-1} \sum_{r=1}^{n} \sigma_r^2$. Consider the sequence of random variables $M_n = n^{-1} \sum_{r=1}^{n} X_r$ and $Z_n = n^{-1/2} \sum_{r=1}^{n} (X_r - c_r)$. Then, the law of large numbers and the central limit theorem, respectively, ensure that

$$\lim_{n\to\infty} \mathbb{P}(y \leq M_n < y + dy) = \delta(y - c) dy \tag{E.7}$$

$$\lim_{n\to\infty} \varphi_{Z_n}(u) = e^{-\sigma^2 u^2 / 2}. \tag{E.8}$$

Thus the average M_n, which is a random variable, converges in law to the expectation c, which is a deterministic data. Moreover, the properly scaled fluctuations Z_n, which are random variables whose law depends on the specific laws of the original X_r, converge in law to a Laplace–Gauss random variable.

Exercise E.1. Let X_r, $1 \leq r \leq n$, be independent random variables with identical law, such that $\langle X \rangle = 0$ and $\langle X^2 \rangle < \infty$. Let $Y_n = n^{-1/2} \sum_{r=1}^{n} X_r$ and $V_n = (n-1)^{-1/2} \sum_{r=2}^{n} X_r$. Show that X_1 and V_n are independent but X_1 and Y_n are not independent. To what law does the joint law of (X_1, V_n) converge as $n \to \infty$? To what law does the joint law of (X_1, Y_n) converge as $n \to \infty$? Let Q be independent of all the X_r: are $Q + X_1$ and $Q + V_n$ independent?

E.2   Gaussian distributions

A random variable X following the Laplace–Gauss, or normal, law with expectation $\langle X \rangle = a$ and standard deviation $\langle (X - a)^2 \rangle^{1/2} = \sigma > 0$, has the probability density on $\mathbb{R}$

$$f_X(x) = d\mathbb{P}(x \leq X < x + dx)/dx = (2\pi)^{-1/2} \sigma^{-1} \exp\left(-\frac{(x - a)^2}{2\sigma^2}\right). \tag{E.9}$$

Its characteristic function is

$$\langle e^{iuX} \rangle = \int_{\mathbb{R}} e^{iux} f_X(x)\, dx = \exp(iua - \tfrac{1}{2}u^2\sigma^2). \tag{E.10}$$

For $a = 0$, one easily finds that $\langle X^n \rangle = 0$ for odd $n > 0$ and that, for any $r \geq 0$,

$$\langle |X|^r \rangle = 2^{r/2}\pi^{-1/2}\Gamma\left(-\frac{r+1}{2}\right)\sigma^r \tag{E.11}$$

where $\Gamma(s) = (s - 1)!$ is Euler's gamma function.

The cumulative probability function for X is related to the error function erf:

$$\mathbb{P}(X < b) = \int_{-\infty}^{b} f_X(x)\, dx = \frac{1}{2} + \frac{1}{2}\,\mathrm{erf}\left(\frac{b - a}{\sqrt{2}\sigma}\right). \tag{E.12}$$

One shows that $1 - \mathrm{erf}\, x \approx \pi^{1/2}e^{-x^2}x^{-1}$ as $x \to \infty$.

In $\mathbb{R}^n$, the random vector X with expectation $a = (a_j) = (\langle X_j \rangle)$ and positive covariance matrix $C = (C_{ij}) = (\langle X_i X_j - a_i a_j \rangle)$ follows the multivariate normal law if its characteristic function is

$$\langle e^{i\boldsymbol{u}\cdot\boldsymbol{X}} \rangle = \int_{\mathbb{R}} e^{i\boldsymbol{u}\cdot\boldsymbol{x}} f_X(\boldsymbol{x})d^n\boldsymbol{x} = \exp(i\boldsymbol{u}\cdot\boldsymbol{a} - \tfrac{1}{2}\boldsymbol{u}\cdot\boldsymbol{C}\cdot\boldsymbol{u}) \tag{E.13}$$

where the dot denotes the scalar product. Its probability density reads:

$$f_X(\boldsymbol{x}) = (2\pi)^{-n/2}(\det C)^{-1/2}\exp(-\tfrac{1}{2}(\boldsymbol{x} - \boldsymbol{a})\cdot C^{-1}\cdot(\boldsymbol{x} - \boldsymbol{a})). \tag{E.14}$$

The multivariate normal X has independent components if its covariance matrix is diagonal.

Exercise E.2. Let X be a Gaussian random variable with expectation a and standard deviation σ. Show that $\langle \cos(bX + c) \rangle = e^{-b^2\sigma^2/2}\cos(ba + c)$ for any real b, c.

E.3   Fokker–Planck equation

A stochastic process is a 'function-valued random variable'. Among stochastic processes, Markov processes play a prominent role in physics (see van Kampen (1992) for a thorough discussion).

The Fokker–Planck equation is a fundamental equation in the theory of Markov processes. We consider a process $X(t)$ in $\mathbb{R}$ such that, for $t' \leq t$, the increment

$$Y(t, t') = X(t) - X(t') \tag{E.15}$$

is a random variable whose distribution depends only on $t - t'$ and on $X(t')$. If we denote by $f_X(x, t)$ the probability density for $X(t)$ and by $f_Y(y, t|x)$ the probability density for $Y(t, 0)$ given the initial value $X(0) = x$, they fulfill the Chapman–Kolmogorov equation

$$f_X(x, t) = \int_{\mathbb{R}} f_X(x - y, t') f_Y(y, t - t' | x - y) \, dy. \tag{E.16}$$

It is clear that $Y(t, t) = 0$ for any x and t. Let us now assume that $t' \to t$ and that the distribution of $Y(t, t')$ is regular enough for $\langle |Y(t, 0)|^2 \rangle$ to be finite and to be a smooth function of t in the limit $t \to 0$. Finally assume that $f_X(x, t)$ is also a smooth function of x and t. We Taylor expand the left-hand side of (E.16) with respect to $t - t'$:

$$f_X(x, t) = f_X(x, t') + (t - t') \partial_t f_X(x, t') + \cdots \tag{E.17}$$

and the integrand in the right-hand side with respect to y:

$$\int_{\mathbb{R}} f_X(x - y, t') f_Y(y, t - t' | x - y) \, dy$$

$$= \int_{\mathbb{R}} f_X(x, t') f_Y(y, t - t' | x - y) \, dy$$

$$\quad - \int_{\mathbb{R}} (\partial_x f_X(x, t')) y f_Y(y, t - t' | x - y) \, dy$$

$$\quad + \int_{\mathbb{R}} (\partial_x^2 f_X(x, t')) \frac{y^2}{2} f_Y(y, t - t' | x - y) \, dy + \cdots \tag{E.18}$$

$$= \int_{\mathbb{R}} f_X(x, t') f_Y(y, t - t' | x) \, dy - \int_{\mathbb{R}} f_X(x, t') y \partial_x f_Y(y, t - t' | x) \, dy$$

$$\quad + \int_{\mathbb{R}} f_X(x, t') \frac{y^2}{2} \partial_x^2 f_Y(y, t - t' | x) \, dy + \cdots$$

$$\quad - \int_{\mathbb{R}} (\partial_x f_X(x, t')) y f_Y(y, t - t' | x) \, dy$$

$$\quad + \int_{\mathbb{R}} (\partial_x f_X(x, t')) y^2 \partial_x f_Y(y, t - t' | x) \, dy + \cdots$$

$$\quad + \int_{\mathbb{R}} (\partial_x^2 f_X(x, t')) \frac{y^2}{2} f_Y(y, t - t' | x) \, dy + \cdots \tag{E.19}$$

In each term of (E.19) the first factor in the integrand is independent of y, leaving only a moment of the transition probability density f_Y, namely

$$\int_{\mathbb{R}} f_X(x - y, t') f_Y(y, t - t' | x - y) \, dy = f_X(x, t') - \partial_x (f_X(x, t') \langle Y(t, t') \rangle_{|x})$$

$$+ \tfrac{1}{2} \partial_x^2 (f_X(x, t') \langle Y(t, t')^2 \rangle_{|x}) + \cdots. \tag{E.20}$$

The subscript to the averages recalls that f_Y depends on the arrival point x. The Taylor expansion has translated the dependence of the transition probability density on the departure point $x - y$ into a dependence on the arrival point x. Comparing (E.20) with (E.17) in the limit $t' \to t$ yields the Fokker–Planck equation

$$\partial_t f_X(x, t) = -\partial_x \left(f_X(x, t') \frac{\langle Y(t, t') \rangle_{|x}}{t - t'} \right)$$

$$+ \partial_x^2 \left(f_X(x, t') \frac{\langle Y(t, t')^2 \rangle_{|x}}{2(t - t')} \right) + \cdots. \tag{E.21}$$

The simplest instance[1] of a Fokker–Planck equation describes random walks with independent identically distributed increments, with expectation b and variance σ^2 over a single time step δt, such that $\langle Y(\delta t, 0) \rangle_{|x} = b$ and $\langle Y(\delta t, 0)^2 \rangle_{|x} = b^2 + \sigma^2$. Then the independence of successive increments implies that

$$\langle Y(t, t') \rangle_{|x} = b \frac{t - t'}{\delta t} \tag{E.22}$$

$$\langle (Y(t, t'))^2 \rangle_{|x} = \langle Y(t, t') \rangle_{|x}^2 + \sigma^2 \frac{t - t'}{\delta t}. \tag{E.23}$$

In the limit $(t - t')/\delta t \to \infty$, the central limit theorem[2] ensures that the distribution of $X(t)$ converges to a normal distribution, which is indeed the Green function of the diffusion equation (to which the Fokker–Planck equation reduces in this 'homogeneous diffusion' case).

More generally, the coefficients $\langle Y(t, t') \rangle_{|x}$ and $\langle Y(t, t')^2 \rangle_{|x}$ in (E.21) may depend on x and t. The microscopic description by a random walk will also have (x, t)-dependent coefficients, with the natural scaling in the limit $(t - t')/\delta t \to \infty$,

$$b(x, t) = v(x, t)\delta t \tag{E.24}$$

$$\sigma(x, t)^2 = 2D(x, t)\delta t \tag{E.25}$$

so that

$$\langle Y(t, t') \rangle_{|x} = (t - t')v(x, t) \tag{E.26}$$

$$\langle Y(t, t')^2 \rangle_{|x} = (t - t')^2 v(x, t)^2 + 2(t - t')D(x, t). \tag{E.27}$$

Recalling that (E.21) was derived in the limit $t' \to t$, and assuming that v and D are macroscopic quantities ('of the order of unity' on the scale of x and t), we may neglect the square term in (E.27) with respect to the linear term. Then (E.21)

---

[1] Random walks for which single steps have infinite variance lead similarly to Lévy processes (see, e.g., Balescu 1997).

[2] Hence higher moments of $(\delta t)^{-1/2}[Y(t' + \delta t, t') - \langle Y(t' + \delta t, t') \rangle]$ are also Gaussian in this limit. The nth-order moment would contribute to (E.23) in the ∂_x^n term but this contribution is negligible for $n \geq 3$ in the limit $\delta t \to 0$ as it scales at least like $\delta t^{1/2}$.

reduces to

$$\partial_t f_X(x,t) = -\partial_x(v(x,t)f_X(x,t)) + \partial_x^2(D(x,t)f_X(x,t)) \qquad \text{(E.28)}$$
$$= -\partial_x(w(x,t)f_X(x,t)) + \partial_x(D(x,t)\partial_x f_X(x,t)) \qquad \text{(E.29)}$$

where

$$w = v - \partial_x D. \qquad \text{(E.30)}$$

Exercise E.3. On the macroscopic scale, assume that all moments of f_X are defined. Show (using integration by parts) that the evolution equations of the first moments of f_X are:

$$\frac{d}{dt} 1 = 0 \qquad \text{(E.31)}$$

$$\frac{d}{dt}\langle X \rangle = \langle v(X) \rangle \qquad \text{(E.32)}$$

$$\frac{d}{dt}\langle X^2 \rangle = 2\langle X v(X) \rangle + 2\langle D(X) \rangle \qquad \text{(E.33)}$$

$$\frac{d}{dt}\langle (X - \langle X \rangle)^2 \rangle = 2\,\mathrm{Cov}(X, v(X)) + 2\langle D(X) \rangle. \qquad \text{(E.34)}$$

The actual average velocity is thus not $v(\langle X \rangle)$. The dependence of the diffusion coefficient D on the position x also contributes to the average transport of particles. Moreover, the rate of change of the variance of X is not $D(\langle X \rangle)$ either, in general.

Exercise E.4. Show that the variance $\langle \dot{X}^2 \rangle - \langle \dot{X} \rangle^2$ of the velocity is typically infinite in the limit of infinitesimal time steps $\delta t \to 0$. Estimate the higher moments of $Y(t, t + \delta t)$ in the limit $\delta t \to 0$.

Exercise E.5. Show that, if w and D are constant, the solution to (E.28) with initial condition $f(x, 0) = \delta(x - x_0)$ and vanishing boundary conditions as $x \to \pm\infty$ is a normal distribution. Compute its moments.

Exercise E.6. Assuming that D does not vanish in the spatial domain $[0, L]$, compute the stationary solution f_0 to (E.28) in the case $\partial_t w = 0$, $\partial_t D = 0$ on $[0, L]$ with boundary conditions $\partial_x f(0, t) = \partial_x f(L, t) = 0$, normalized to unity ($\int_0^L f\,dx = 1$). For any solution f to (E.28) normalized to unity, show that the relative entropy

$$S(t) \equiv S[f|f_0] = -\int_0^L \left(\frac{f(x,t)}{f_0(x)} \ln \frac{f(x,t)}{f_0(x)} \right) f_0(x)\,dx \qquad \text{(E.35)}$$

can only increase with time, and that it is constant if and only if f is the stationary solution f_0.

This entropy production result is central to the analysis of the approach to the stationary solution, hence to the physical understanding of the Fokker–Planck equation. It emphasizes the stochastic nature of the microscopic process $X(t)$. This contrasts with the deterministic nature of the processes underlying the Vlasov equation, as discussed in appendix G.

Exercise E.7. Consider the Fokker–Planck equation on $\mathbb{R}$, with coefficients $w(x,t)$ and $D(x,t)$ vanishing outside the spatial interval $]0, L[$. Assume that the coefficients are continuous functions of (x,t). Show that, for all times $t \geq 0$, $\int_0^L f(x,t)\,dx = \int_0^L f(x,0)\,dx$ and that $f(x,t)$ inside $[0, L]$ is independent of initial data $f(y,0)$ with $y \notin]0, L[$. Conclude that the vanishing coefficients isolate the interior of the interval from the exterior. Extend your argument to higher dimension.

Exercise E.8. Consider a system such that $w(x) = 0$ and $D(x) = D_0 + D_1 x$. Over what time-scale does the non-uniformity of $D(x)$ manifest itself? For instance, given Dirac-distributed initial data $f(x,0) = \delta(x - x_0)$, estimate a time such that $|\langle X(t) - x_0 \rangle|^2 \geq \langle |X(t) - \langle X(t) \rangle|^2 \rangle$. What does your estimate become if $w(x) = w_0$?

Appendix F

Estimates for chapter 6

F.1 n-time correlation function of $C(\tau)$ and $S(\tau)$

The n-time correlation function of $C(\tau)$ as defined by (6.41) is

$$\langle C(\tau_1)\dots C(\tau_n)\rangle = \frac{1}{2^n A^{n/3}} \sum_{\epsilon_1}\dots\sum_{\epsilon_n} \sum_{m_1=-\lambda_\beta}^{\lambda_\beta} \dots$$

$$\dots \sum_{m_n=-\lambda_\beta}^{\lambda_\beta} \left\langle \cos\left[\sum_{i=1}^{n}\epsilon_i(m_i\tau_i A^{-2/3} - \varphi_{m_i})\right]\right\rangle \quad \text{(F.1)}$$

where $\epsilon_i = \pm 1$ for all $1 \le i \le n$. We compute it following closely Bénisti (1995). The randomness of the φ_m implies that the terms of (F.1) with $\sum_{i=1}^n \epsilon_i\varphi_{m_i} \ne 0$ vanish. Condition $\sum_{i=1}^n \epsilon_i\varphi_{m_i} = 0$ implies $\sum_{i=1}^n \epsilon_i = 0$ and the existence of a one-to-one mapping σ of the indices i such that $\epsilon_i = 1$ onto the indices j such that $\epsilon_j = -1$ such that $m_{\sigma(i)} = m_i$. For n odd, this second condition cannot be satisfied, and $\langle C(\tau_1)\dots C(\tau_n)\rangle$ vanishes.

If n is even, let $l = n/2$. There are only l independent ϵ_i in (F.1). Let $\mathcal{I}_+$ be the set of the i such that $\epsilon_i = 1$, and $\mathcal{M}_+$ the set of l-uplets $(m_{i_1}, \dots, m_{i_l}) \in \mathbb{Z}^l$ such that $i_j \in \mathcal{I}_+$ and $-\lambda_\beta \le m_{i_j} \le \lambda_\beta$ for all j such that $1 \le j \le l$. With this notation we may write

$$\langle C(\tau_1)\dots C(\tau_{2l})\rangle = \frac{1}{2^{2l} A^{2l/3}} \sum_{\mathcal{I}_+}\sum_{\sigma}\sum_{m\in\mathcal{M}_+} \cos\left[\sum_{i\in\mathcal{I}_+} x_i(\tau_i - \tau_{\sigma(i)})\right]$$

$$- C'(\tau_1, \dots, \tau_{2l}) \quad \text{(F.2)}$$

where $C'(\tau_1, \dots, \tau_{2l})$ is defined to cancel the terms unduly repeated in the first term and $x_i = A^{-2/3}m_i$. We estimate C' after the first term.

The first term in (F.2) is expressed in terms of Riemann sums for integrals of the type

$$I_{2l} = \int_{-\beta}^{\beta} \ldots \int_{-\beta}^{\beta} \cos\left[\sum_{i\in\mathcal{I}_+} (\tau_i - \tau_{\sigma(i)})x_i\right] dx_1 \ldots dx_p$$

$$= 2^l \prod_{i\in\mathcal{I}_+} \frac{\sin[\beta(\tau_i - \tau_{\sigma(i)})]}{\tau_i - \tau_{\sigma(i)}} \tag{F.3}$$

whose value does not depend on A. Replacing the Riemann sums by integrals is valid, provided that each τ_i verifies (6.43).

We now focus on C'. It contains sums which are similar to those of (F.2) where at least two among the indices m_i such that $\epsilon_i = 1$ are always equal. Possibly by relabelling them, let the independent indices be the first p ones ($p \le l - 1$). Then C' is a sum of a finite number of terms of the type

$$S_p = \frac{1}{A^{2l/3}} \sum_{m_1} \ldots \sum_{m_p} \cos\left[\sum_{i=1}^{p} x_i \sum_{j\in J(i)} \epsilon_i \tau_j\right] \tag{F.4}$$

where $J(i)$ is the set of indices corresponding to the same m_i. When each τ_i satisfies (6.43), S_p is estimated as

$$S_p \simeq J_p = \frac{1}{A^{2(l-p)/3}} \int_{-\beta}^{\beta} \ldots \int_{-\beta}^{\beta} \cos\left[\sum_{i=1}^{p} x_i \sum_{j\in J(i)} \epsilon_i \tau_j\right] dx_1 \ldots dx_p. \tag{F.5}$$

As $|\cos u| \le 1$, it is clear that $|S_p| \le (2\beta)^p A^{-2(l-p)/3}$, which vanishes for $A \to \infty$.

The way indices are repeated in the sequence $(m_1, \ldots, m_l)$ is determined by the mapping from integers $(1, \ldots, p)$ to the integers $(1, \ldots, l)$. Given the finite number R_{lp} of such mappings, $|C'| \le R_{lp}(2\beta)^p A^{-2(l-p)/3}$ and

$$\langle C(\tau_1) \ldots C(\tau_{2l}) \rangle \simeq \frac{1}{2^{2l-1}} \sum_{\mathcal{I}_+} \sum_{\sigma} \prod_{i\in\mathcal{I}_+} \frac{\sin[\beta(\tau_i - \tau_{\sigma(i)})]}{\tau_i - \tau_{\sigma(i)}}. \tag{F.6}$$

This justifies (6.45), as can be checked by computing the n-time correlation function of this expression. A similar calculation holds for the multiple-time correlation functions $\langle C(\tau_1) \ldots C(\tau_n) S(\tau_{n+1}) \ldots S(\tau_{n+n'}) \rangle$.

The core of this argument is geometrical: in l-dimensional space with coordinates $(x_1, \ldots, x_l)$, each constraint $m_i = m_j$ corresponds to a domain of width $A^{-2/3}$ around the plane $x_i = x_j$. As $A \to \infty$, the volume of these domains vanishes. But there is competition between large l and large A: the number of planes $x_i = x_j$ grows factorially with l, while volumes scale geometrically with A. Therefore, for large l, i.e. 'many-times correlation functions', one needs a large A to ensure the accuracy of approximations.

F.2   Estimate of the mean number of visited boxes

According to the definition of section 6.5, the average number of resonance boxes visited at time t for A large is

$$N_b(t) = \frac{\langle p_{max}(t) - p_{min}(t) \rangle}{2 \Delta v_{box}} \tag{F.7}$$

where $p_{max}(t)$ and $p_{min}(t)$ are, respectively, the maximum and minimum value of $p(t')$ for $0 \le t' \le t$. $N_b(t)$ can be estimated by considering that over this time interval $p(t')$ is a diffusive process (Bénisti 1995). Then p may be considered as a random walk. Consider the random walk

$$S_N = \sum_{i=1}^{N} \epsilon_i \tag{F.8}$$

where the ϵ_i are random variables, independent of each other, and taking values 1 or -1 with probability $1/2$ each. The probability for $S_N = r \le N$ is $(1/2^N)\binom{N}{(N+r+\eta)/2}$, where $\eta = 0$ if N and r have the same parity and $\eta = 1$ if they do not (Feller 1970, volume I, section III.7). Assume N is even and let $n = N/2$. Let $S_N^{max} \equiv \max_{0 \le t \le N} S_t$. Then the expectation of the maximum of S_t for $0 \le t \le N$ is

$$\langle S_N^{max} \rangle = \sum_{l=1}^{n} \frac{2l}{2^N} \binom{N}{(N+2l)/2} + \sum_{l=1}^{n} \frac{2l-1}{2^N} \binom{N}{(N+2l)/2}$$

$$= \sum_{l=1}^{n} \frac{4l-1}{2^{2n}} \binom{2n}{n+l} \simeq \frac{1}{\sqrt{\pi n}} \int_1^\infty (4l-1) \exp(-l^2/n) \, dl \tag{F.9}$$

where the final form follows from noting that $n + l$ has a binomial distribution with expectation n and variance $n/2$, and applying the central limit theorem (this amounts to using Stirling formula (B.2) and approximating the sum by an integral). Since $\langle S_N^2 \rangle = N$,

$$\langle S_N^{max} \rangle \simeq \sqrt{\frac{2 \langle S_N^2 \rangle}{\pi}}. \tag{F.10}$$

By symmetry, $\langle S_N^{min} \rangle = -\langle S_N^{max} \rangle$, and

$$\langle S_N^{max} - S_N^{min} \rangle \simeq \sqrt{\frac{8 \langle S_N^2 \rangle}{\pi}}. \tag{F.11}$$

In this approach $N_b(t)$ is $\langle S_N^{max} - S_N^{min} \rangle$ and $\langle S_N^2 \rangle = 2 D_{QL} t / (2 \Delta v_{box})^2$, which yields (6.47).

Exercise F.1. Show that (F.11) also holds for N odd, either by a direct estimate or by sandwiching N between $N - 1$ and $N + 1$.

Exercise F.2. Show that the probability for S_N to be at the origin is estimated for $N \gg 1$ by $\mathbb{P}(S_N = 0) \approx (\pi N/8)^{-1/2}$. For what values of N is the approximation accurate within 5%? Within 1%? Within ε for $\varepsilon \ll 1$ given?

F.3   Initial Brownian motion

We now compute the order n moment of $\Delta p(t)$ as defined in section 6.8.1 (Bénisti 1995)

$$\langle \Delta p^n(t) \rangle = \frac{1}{2^n} \int_0^t \cdots \int_0^t \sum_{\epsilon_1,\ldots,\epsilon_n} \sum_{m_1,\ldots m_n} \left(\prod_{i=1}^n (k_{m_i} A_{m_i}) \right)$$

$$\times \left\langle \cos \sum_{i=1}^n [\epsilon_i (k_{m_i} q_0 + \Omega_{m_i} t_i + \varphi_{m_i})] \right\rangle dt_1 \ldots dt_n \quad \text{(F.12)}$$

where Ω_m is defined by (6.53), $\epsilon_i = \pm 1$ for $1 \le i \le n$, and each m_i runs from 1 to M. The average over the phases is non-zero if and only if for each $\epsilon_i = +1$ there is an $\epsilon_j = -1$ such that $m_i = m_j$. This implies

$$\langle \Delta p^{2l+1}(t) \rangle = 0 \quad \text{(F.13)}$$

for any l. For $n = 2l$, the rule implies that only l ϵ_i may be chosen arbitrarily. There are only $\binom{2l}{l}$ choices of this kind. When such a choice is made, the only freedom left is the ordering of the set of j to couple with the ordered set of i. The number of such orderings is $l!$. As integral (F.12) is unchanged on permuting the i, we find

$$\langle \Delta p^{2l}(t) \rangle = G(t) - G'(t) \quad \text{(F.14)}$$

where

$$G(t) = \frac{1}{2^{2l}} \frac{(2l)!}{l!} \int_0^t \cdots \int_0^t \sum_{m_1,\ldots m_l} \left(\prod_{i=1}^l (k_{m_i} A_{m_i}) \right)^2$$

$$\times \cos \left[\sum_{i=1}^l \Omega_{m_i} (t_i - t_{i+l}) \right] dt_1 \ldots dt_{2l} \quad \text{(F.15)}$$

and $G'(t)$ stands to avoid counting twice in (F.12) terms with at least two identical indices m_i with $\epsilon_i = +1$. Using the identity $\cos(a+b) = \cos a \cos b - \sin a \sin b$ and noting that sines do not contribute to the sum by symmetry, we reduce (F.15) to

$$G(t) = \frac{(2l)!}{l! 2^{2l}} \prod_{i=1}^l \int_0^t \int_0^t \sum_{m_i} k_{m_i}^2 A_{m_i}^2 \cos \Omega_{m_i} (t_i - t_{i+l}) \, dt_i \, dt_{i+l}$$

$$= \frac{(2l)!}{l! 2^{2l}} (2\langle \Delta p(t)^2 \rangle)^l. \tag{F.16}$$

To estimate $G'(t)$ one computes the number N_2 of $2l$-uplets $(m_1, \ldots, m_{2l})$ where the set of indices m_i with $\epsilon_{m_i} = 1$ is identical to that with $\epsilon_{m_i} = -1$ and only $\nu = 2$ indices m_i with $\epsilon_{m_i} = 1$ are equal. $N_2 = \binom{2l}{l} \binom{l}{2}^2 (l-2)!$ as there are $\binom{l}{2}$ ways of choosing a pair of indices i such that $\epsilon_{m_i} = 1$; there are $\binom{l}{2}$ ways of relating each of these pairs to a pair of indices j such that $\epsilon_{m_j} = -1$ and there are $(l-2)!$ ways of pairing the remaining m_i with the remaining m_j. Thus $\nu = 2$ gives to $G'(t)$ a contribution

$$G_2'(t) = \frac{1}{2^{2l}} \frac{(2l)!}{2! 2! (l-2)!} \int_0^t \cdots \int_0^t \sum_{m_1, \ldots, m_{l-1}} \left(\prod_{i=1}^{l-2} (k_{m_i} A_{m_i})^2 \right) (k_{m_{l-1}} A_{m_{l-1}})^4$$

$$\times \cos \left[\Omega_{m_{l-1}} (t_l + t_{l-1} - t_{2l} - t_{2l-1}) + \sum_{i=1}^{l-2} \Omega_{m_i} (t_i - t_{i+l}) \right] dt_1 \ldots dt_{2l}$$

$$- G_2''(t) \tag{F.17}$$

where $G_2''(t)$ stands to avoid counting several times in (F.17) terms with at least three identical indices m_i with $\epsilon_i = +1$ or with two distinct m_i repeated. Rewriting the sums as was done for (F.15) yields

$$G_2'(t) = \frac{(2l)! (D_{QL}(p_0)t)^l \Delta\Omega(p_0)t}{6\pi (l-2)!} - G_2''(t) = \frac{G(t)l(l-1)\Delta\Omega(p_0)t}{12\pi} - G_2''(t) \tag{F.18}$$

where $\Delta\Omega(p_0)$ is the typical value of $\Delta\Omega_m$ as defined by (6.54). Like $G_2'(t)$, $G_2''(t)$ is positive, but smaller than $G_2'(t)$. Therefore, $G_2'(t)$ overestimates the correction to $G(t)$ due to $\nu = 2$. The larger ν yield similar, but smaller, contributions. Hence $G'(t)/G(t) \leq l(l-1)t/(12\pi)$. Thus, if

$$t\Delta\Omega(p_0) \ll 12\pi/l^2 \tag{F.19}$$

the correction $G'(t)$ is negligible in (F.14). This result implies that the distribution of $\Delta p(t)$ may be considered as Gaussian for t small enough. Though the higher moments $(l \to \infty)$ may be non-Gaussian for fixed M and t (and they should be non-Gaussian in view of the construction of $\Delta p(t)$ from the random phases $\{\varphi_m\}$), one is usually interested only in lower moments for which the restriction (F.19) is mild.

Exercise F.3. Check that the moments of $p(t)$ imply that its characteristic function is $\langle e^{iup(t)} \rangle \simeq \exp(iup_0 - u^2 D_{QL}t)$ formally, i.e. neglecting the restriction (F.19).

To show that the velocity $p(t)$ evolves according to a Brownian motion, it is not sufficient to prove that the single-time higher moments follow the

Gaussian distribution. It is also necessary to show that the n-time moment $\langle \Delta p(t_1) \ldots \Delta p(t_n) \rangle$ has the same distribution as for a Brownian motion. This is proved by (6.77) for the second moment. It is clear that

$$\langle \Delta p(t_1) \ldots \Delta p(t_n) \rangle = 0 \qquad \text{(F.20)}$$

for odd n, and that for $n = 2l$ the combinatorial analysis and the way of treating sums in (F.12) applies again. This yields

$$\langle \Delta p(t_1) \ldots \Delta p(t_{2l}) \rangle \simeq [2D_{QL}(p_0)]^l \sum_{(s)} \prod_{i=1}^{l} \min(t_{s_1(i)}, t_{s_2(i)}) \qquad \text{(F.21)}$$

$$= \sum_{(s)} \prod_{i=1}^{l} \langle \Delta p(t_{s_1(i)}) \Delta p(t_{s_2(i)}) \rangle \qquad \text{(F.22)}$$

where the sum runs over all possible partitions, denoted by (s), of the set $\{1, 2, \ldots, 2l\}$ into l pairs $\{s_1(i), s_2(i)\}$, with $s_1(i) < s_2(i)$. This result is exactly the decomposition of the nth moment characterizing a multi-Gaussian random variable, ensuring that $\Delta p(t)$ is a Brownian random walk. The corrections like G' were neglected, as was done for the single-time nth moment $\langle \Delta p^n(t) \rangle$.

Then (F.20) and (F.21) show that $\Delta p(t)$ has a Brownian behaviour (moments follow Gaussian statistics) in the scaling range $\tau_c \ll t \ll \tau_{\text{spread}}$ with (F.19). Therefore $\Delta p(t)$ diffuses with a quasilinear diffusion coefficient. Note that this initial diffusion is non-chaotic: it is merely due to the breadth of the wave spectrum and to the nearly ballistic motion.

Exercise F.4. The Brownian behaviour of $\Delta p(t)$ makes the process $\Delta q(t) = \int_0^t \Delta p(t') dt'$ rather simple. In particular, show that $\langle \Delta q(t_1) \ldots \Delta q(t_n) \rangle = 0$ if n is odd, and that

$$\langle \Delta q(t_1) \ldots \Delta q(t_{2l}) \rangle = \sum_{(s)} \prod_{i=1}^{l} \langle \Delta q(t_{s_1(i)}) \Delta q(t_{s_2(i)}) \rangle \qquad \text{(F.23)}$$

which implies that $\Delta q(t)$ is also a Gaussian process. To fully characterize it, one only needs to compute from (6.77), for $0 \leq t_1 \leq t_2$,

$$\langle \Delta q(t_1) \Delta q(t_2) \rangle = D_{QL} \left(t_1^2 t_2 - \frac{t_1^3}{3} \right). \qquad \text{(F.24)}$$

Using (E.3), deduce from this result the one-time characteristic function for $t \geq 0$

$$\langle e^{iu \Delta q(t)} \rangle = e^{-u^2 D_{QL}(t^3/3)} \qquad \text{(F.25)}$$

and the characteristic function of position increments for $t_1 \geq 0, t_2 \geq 0$

$$\langle e^{iu(\Delta q(t_2) - \Delta q(t_1))} \rangle = e^{-\frac{u^2}{3} D_{QL}|t_2-t_1|^3 - u^2 D_{QL}|t_2-t_1|^2 \min(t_1,t_2)}. \qquad \text{(F.26)}$$

F.4   Dependence of an orbit on one phase

As argued in section 6.8.2, we show that the initial quasilinear regime extends over a time longer than τ_{spread} for E large. To this end, we take advantage of the fact that in (6.81) the average cosine is small if $q(t)$ weakly changes when the two phases φ_{m_1} and φ_{m_2} are varied, even though it may strongly change when all phases are modified. To avoid cumbersome formulae, we give the explicit derivation for the spreading due to one phase and we extend the result to two phases afterwards. To estimate this spreading, we study how the orbit which is at $p = p_0, q = q_0$ at $t = 0$ is modified when the phase φ_n is varied from zero to a finite value. Let $(q_{n'}(t), p_{n'}(t))$ be the orbit for $\varphi_n = 0$. For any φ_n let

$$\delta q_n(t) = q(t) - q_{n'}(t) \tag{F.27}$$

$$\delta p_n(t) = p(t) - p_{n'}(t). \tag{F.28}$$

These two functions vanish at $t = 0$. We assume t to be small enough so that $k_{\max}|\delta q_n(t)| \ll \pi$. As $\delta q_n(t)$ is small, its evolution is governed by the second-order linear inhomogeneous equation (6.85). Such an equation may be solved, in general, by first computing the Green function (solving with only the first term in the right-hand side with a non-zero initial condition), and then by convoluting this Green function with the source term, i.e. the second term in the right-hand side of (6.85). However, if one is interested in short-time behaviour, it is also valuable to compute $\delta q_n(t)$ iteratively. This is the standard Picard procedure, used also to prove the existence and uniqueness of the solutions to differential equations. Picard's algorithm is explicit, whereas the Green function is not known explicitly for this differential equation. However, we actually focus on upper estimates for covariances.

Thus we consider the orbit separations $(\delta p_n(t), \delta q_n(t))$ due to φ_n, with all phases φ_m random for $m \neq n$, and denote expectations by $\langle \ldots \rangle_{m \neq n}$. We start from (6.85) and initial conditions $(\delta p_n(0), \delta q_n(0)) = (0, 0)$, so that $\delta q_n(t)$ is given by (6.89). To estimate the size of $\delta q_n(t)$, we compute successively $\langle \delta q_n(t) \rangle_{m \neq n}$ and $\langle \delta q_n^2(t) \rangle_{m \neq n}$ for the random distribution of phases φ_m (all $m \neq n$). We use an iterative procedure whose first step approximates δq_n by δq_{n0}.

The expectation of (6.89) for the random distribution of phases φ_m (all $m \neq n$) reads:

$$\langle \delta q_n(t) \rangle_{m \neq n} = \int_0^t \int_0^{t'} \langle F(t'') \delta q_n(t'') \rangle_{m \neq n} \, dt'' \, dt' + \langle \delta q_{n0}(t) \rangle_{m \neq n}. \tag{F.29}$$

Then

$$\langle \delta q_{n0}(t) \rangle_{m \neq n} = A_n k_n (\cos \varphi_n - 1) \int_0^t \int_0^{t'} \langle \sin(k_n q_{n'}(t'') - \omega_n t'') \rangle_{m \neq n} \, dt'' \, dt'$$

$$+ (A_n k_n \sin \varphi_n) \int_0^t \int_0^{t'} \langle \cos(k_n q_{n'}(t'') - \omega_n t'') \rangle_{m \neq n} \, dt'' \, dt'.$$

$$\tag{F.30}$$

As many waves are acting on the orbit, its statistics is not much modified if one of them has its phase kept fixed. Thus, we may approximate the averages in the integrals by averages over all phases, which yields

$$\langle e^{i[k_n q_{\eta}(t'') - \omega_n t'']} \rangle_{m \neq n} \simeq e^{i[k_n q_0 + (k_n p_0 - \omega_n)t'']} \langle e^{ik_n \Delta q(t'')} \rangle \qquad (F.31)$$

where $\Delta q(t'')$ has the same distribution as if the initial quasilinear regime extended up to t''. Then (F.25) ensures that the integrand in (F.30) is small as soon as $t'' \gg \tau_{\text{spread}}$. Moreover, as the expression in (F.25) is real positive, the modulus of (F.31) is maximal for $(q_0, p_0) = (0, \omega_n/k_n)$, and[1]

$$|\langle \delta q_{n0}(t) \rangle_{m \neq n}| \leq 3 A_n k_n \int_0^t \int_0^{t'} e^{-k_n^2 D_{\text{QL}} t''^3 / 3} \, dt'' \, dt'$$

$$\leq 3 A_n k_n t \int_0^t e^{-k_n^2 D_{\text{QL}} t''^3 / 3} \, dt''$$

$$\leq \Gamma\left(\frac{1}{3}\right) \left(\frac{k_n^2 D_{\text{QL}}}{3}\right)^{-1/3} A_n k_n t \qquad (F.32)$$

where $\Gamma(1/3) \approx 2.7$ (see section B.1). Thus, for any φ_n,

$$k_n |\langle \delta q_{n0}(t) \rangle_{m \neq n}| \leq 3.2 \sqrt{\gamma_{\text{D}n} k_n |\Delta v_n|} t \qquad (F.33)$$

which is O(1) for $t \approx (\tau_{\text{discr}} \tau_{\text{spread}})^{1/2} \sim \mathcal{B}^{1/2} \tau_{\text{discr}} \sim E^{-1/3} k^{-1/3} \tau_{\text{discr}}^{1/3}$.

However, (F.33) estimates only the expectation of $\delta q_{n0}(t)$. For many processes of interest, the expectation is small or vanishes identically while the standard deviation grows. Thus we now consider the variance, which we first estimate with respect to the distribution of φ_n for fixed values of the other phases,

$$C_0(t) = \langle \delta q_{n0}(t) \delta q_{n0}(t) \rangle_{\varphi_n} - \langle \delta q_{n0}(t) \rangle_{\varphi_n} \langle \delta q_{n0}(t) \rangle_{\varphi_n}$$

$$= \frac{(k_n A_n)^2}{2} \int_0^t \int_0^{t_1'} \int_0^t \int_0^{t_2'} \cos[k_n q_{\eta}(t_1'') - k_n q_{\eta}(t_2'')$$

$$- \omega_n(t_1'' - t_2'')] \, dt_2'' \, dt_2' \, dt_1'' \, dt_1'. \qquad (F.34)$$

We average this expression with respect to the distribution of all other phases φ_m, using (F.26). Thus

$$|\langle \cos(k_n[q_{\eta}(t_2'') - q_{\eta}(t_1'')] - p_0(t_2'' - t_1''))] \rangle_{m \neq n}|$$

$$\lesssim |\cos[(k_n p_0 - \omega_n)(t_2'' - t_1'')]| \exp[-\frac{1}{3} k_n^2 D_{\text{QL}} |t_2'' - t_1''|^3]$$

$$\leq \exp(-\frac{1}{3} k_n^2 D_{\text{QL}} |t_2'' - t_1''|^3) \qquad (F.35)$$

---

[1] This estimate is very conservative. If one estimates the sine average by zero (which holds for $(q_0, p_0) = (0, \omega_n/k_n)$), then one may divide estimates in (F.32) and (F.33) by three.

where we used again the argument on a single phase fixed among many. The largest estimate occurs again for $\omega_n \simeq k_n p_0$. Integrating (F.35) yields

$$C_0(t) \leq (k_n A_n)^2 \Gamma\left(\frac{4}{3}\right) \left(\frac{k_n^2 D_{\mathrm{QL}}}{3}\right)^{-1/3} \int_0^t \int_0^t \min(t_1', t_2') \, dt_1' \, dt_2'$$

$$\leq C_{0\mathrm{M}}(t) \equiv k_n^{-2} C_n' \gamma_{\mathrm{D}n}^3 t^3 \tag{F.36}$$

where

$$C_n' \equiv 3^{1/3} \Gamma\left(\frac{1}{3}\right) \frac{2}{\pi} \frac{k_n |\Delta v_n|}{\gamma_{\mathrm{D}n}} \simeq 0.28 \mathcal{B}_n. \tag{F.37}$$

This estimate ends the first step of our iterative procedure. We note that $\langle \Delta q^2(t) \rangle / C_{0\mathrm{M}}(t) \simeq 2.4/\mathcal{B}$ as $\langle \Delta q^2(t) \rangle$ is given by (6.78); thus is large for large values of E. Moreover $k_n^2 C_0(t)$ becomes of the order of unity for $t \simeq (\gamma_{\mathrm{D}n}^2 k_n |\Delta v_n|)^{-1/3} = (k_n |\Delta v_n|)^{-1} \mathcal{B}^{2/3} \sim E^{-4/9}$, which is shorter than the time-scale $(k_n |\Delta v_n|)^{-1} \mathcal{B}^{1/2} \sim E^{-1/3}$ at which $k_n \langle \delta q_0(t) \rangle \sim 1$ but larger than $\gamma_{\mathrm{D}n}^{-1} = (k_n |\Delta v_n|)^{-1} \mathcal{B} \sim E^{-2/3}$.

The second step of the iterative procedure takes into account the first term in the right-hand side of (6.85). For E large enough, this term yields the exponential growth of the distance between two nearby orbits in the case $\delta q_n(0) \neq 0$ and $\varphi_n = 0$. Here we have $\delta q_n(0) = 0$ and $\varphi_n \neq 0$. This non-zero φ_n yields a non-zero $\delta q_n(t)$ when time goes on because of the second term in (6.85), which is, in turn, subject to exponential growth. We, therefore, expect our second estimate of δq_n to be larger than the first one.

For this second step of the iterative procedure, as δq_n is small, we may treat $F(t)$ as a Gaussian process with moments

$$\langle F(t) \rangle = 0 \tag{F.38}$$

$$\langle F(t_1) F(t_2) \rangle \simeq 2 \gamma_{\mathrm{D}n}^3 \delta(t_1 - t_2) \tag{F.39}$$

where $\delta(t)$ is the Dirac distribution. Indeed $q_\eta(t)$ has a weak dependence on any phase, which makes $\langle F(t_1) F(t_2) \rangle$ a Bragg-like function with the small width τ_c in $t_1 - t_2$. Higher moments of F are assumed to factorize, i.e. F is treated as a white noise, which is consistent with approximating $\dot{q}_\eta(t) - p_0$ by a Brownian motion.

This assumption implies that

$$\langle \delta q_n(t) \rangle \simeq \langle \delta q_{n0}(t) \rangle \simeq 0 \tag{F.40}$$

as the integrand in (F.29) now vanishes (because the values of $F(\tau)$ which may influence $\delta q_n(t'')$ correspond to $\tau < t''$, hence are independent of $F(t'')$), and using the previous estimate for $\langle \delta q_{n0}(t) \rangle$.

We estimate the spreading of $\delta q_n(t)$ by computing

$$C(t) \equiv \langle \delta q_n(t)^2 \rangle$$

$$\simeq \int_0^t \int_0^{t_1'} \int_0^t \int_0^{t_2'} \langle F(t_1'') F(t_2'') \rangle \langle \delta q_n(t_1'') \delta q_n(t_2'') \rangle \, dt_2'' \, dt_2' \, dt_1'' \, dt_1' + C_0(t)$$

$$= 2\gamma_{Dn}^3 \int_0^t \int_0^t \int_0^{\min(t_1', t_2')} C(t'') \, dt'' \, dt_2' \, dt_1' + C_0(t). \tag{F.41}$$

It follows from (6.85) and our assumptions on F that

$$C(t) = C_0(t) + LC(t) \tag{F.42}$$

with

$$Lf(t) \equiv 2\gamma_{Dn}^3 \int_0^t \int_0^t \int_0^{\min(t_1', t_2')} f(t'') \, dt'' \, dt_2' \, dt_1'. \tag{F.43}$$

With inequality (F.36), the integral equation (F.42) implies an explicit upper estimate for $C(t)$. We prove it later using a form of the classical Gronwall lemma (see, e.g., Hirsch and Smale 1974).

For any $\lambda > 0$, consider the norm $\| f \|_\lambda = \lambda \int_0^\infty e^{-\lambda t} |f(t)| dt$ on the space of functions on $[0, \infty[$. Then, provided that $\lambda > 4^{1/3} \gamma_{Dn}$, L is a contraction operator, i.e. $\| Lf \|_\lambda \leq \| f \|_\lambda$. This ensures that the operator $(1 - L)^{-1}$ is represented on the space of functions with finite norm for λ by the convergent series $(1 - L)^{-1} = \sum_{l=0}^\infty L^l$.

Moreover, L preserves positivity: if $f(t) \geq 0$ for all $t \geq 0$, then $Lf(t) \geq 0$ for all $t \geq 0$. Then $(1 - L)^{-1}$ also preserves positivity, which implies

$$C = (1 - L)^{-1} C_0 \leq (1 - L)^{-1} C_{0M} \equiv C_M \tag{F.44}$$

where C_{0M} was defined in (F.36).

At this point, it is convenient to introduce the Laplace transform (A.12). In particular, (F.36) implies $\hat{C}_0(s) \leq \hat{C}_{0M}(s) = 6k_n^{-2} C_n' \gamma_{Dn}^3 s^{-4}$. However,

$$\widehat{LC}(s) = 4\gamma_{Dn}^3 s^{-4} \int_0^\infty \int_0^u \int_0^{u_1} \int_0^{u_2} e^{-u} C(u_3/s) \, du_3 \, du_2 \, du_1 \, du$$

$$= 4\gamma_{Dn}^3 s^{-4} \int_0^\infty \int_0^\infty \int_0^\infty \int_0^\infty e^{-(v_3 + v_2 + v_1 + v_0)} C(v_3/s) \, dv_3 \, dv_2 \, dv_1 \, dv_0$$

$$= 4\gamma_{Dn}^3 s^{-3} \hat{C}(s) \tag{F.45}$$

where $u_3 = v_3$, $u_2 = v_3 + v_2$, $u_1 = v_3 + v_2 + v_1$, $u = v_3 + v_2 + v_1 + v_0$. Applying the Laplace transform to both sides of $C_M = C_{0M} + LC_M$, which results from the definition of C_M in (F.44), enables computing C_M. On introducing $\gamma_n' \equiv 4^{1/3} \gamma_{Dn}$, this yields

$$\hat{C}_M(s) = \frac{s^3}{s^3 - 4\gamma_{Dn}^3} \hat{C}_{0M}(s)$$

$$= \frac{C_n'}{2k_n^2} \left(-\frac{3}{s} + \frac{1}{s - \gamma_n'} + \frac{1}{s + \gamma_n'(1 - i\sqrt{3})/2} + \frac{1}{s + \gamma_n'(1 + i\sqrt{3})/2} \right). \tag{F.46}$$

Inverting the Laplace transform and using (F.44) then yields

$$k_n^2 C(t) \le \frac{C_n'}{2} \left(e^{\gamma_n' t} + 2 e^{-\gamma_n' t/2} \cos\left(\frac{\sqrt{3}}{2} \gamma_n' t\right) - 3 \right) \qquad \text{(F.47)}$$

with C_n' defined by (F.37). This corresponds to (6.91).

F.5   Estimate of non-quasilinear terms

Finally, we show that the non-quasilinear terms $\Delta_\pm$ of section 6.8.2 are negligible over time τ_{QL} by using the fact that q has a small dependence on any given pair of phases over time τ_{QL}. We rewrite (6.81) for $j = \epsilon = \pm$ in the form

$$\Delta_j = -\frac{\epsilon}{2} \int_0^t \int_0^t \Theta_j(t_1, t_2) \, dt_1 \, dt_2 \qquad \text{(F.48)}$$

where

$$\Theta_j(t_1, t_2) = \sum_{m_1=1}^{M} \sum_{m_2=1}^{M} A_{m_1} k_{m_1} A_{m_2} k_{m_2} \langle \cos[\Psi(t_1, t_2)] \rangle. \qquad \text{(F.49)}$$

The argument of the cosine in (F.49) is

$$\Psi(t_1, t_2) = \Psi_\pm(t_1, t_2) + k_{m_1} \delta q_1(t_1) \pm k_{m_2} \delta q_2(t_2) \qquad \text{(F.50)}$$

where

$$\Psi_\pm(t_1, t_2) = k_{m_1} q_{\not m_1, \not m_2}(t_1) \pm k_{m_2} q_{\not m_1, \not m_2}(t_2) - \omega_{m_1} t_1 \mp \omega_{m_2} t_2 + \varphi_{m_1} \pm \varphi_{m_2} \quad \text{(F.51)}$$

$q_{\not m_1, \not m_2}(t)$ is $q(t)$ computed with $\varphi_{m_1} = \varphi_{m_2} = 0$ and

$$\delta q_i(t) \simeq k_{m_i} A_{m_i} \int_0^t \int_0^{t'} (\sin[q_{\not m_1, \not m_2}(t'') - \omega_{m_i} t'' + \varphi_{m_i}]$$
$$- \sin[q_{\not m_1, \not m_2}(t'') - \omega_{m_i} t'']) \, dt'' \, dt' \qquad \text{(F.52)}$$

with $i = 1, 2$. For t small, the decomposition (F.50) of Ψ is interesting because it makes its main dependences on φ_{m_1} and φ_{m_2} explicit. Approximation (F.52) for $\delta q_i(t)$ by its estimate given by the first step of the iteration procedure of section F.4 is justified for $t < \alpha \tau_{QL}$ for $\alpha < 1$ for E large enough. For such times, $\delta q_i(t)$ is small and expanding $\cos \Psi$ in the vicinity of $\Psi_\pm$ yields

$$\cos \Psi = \cos \Psi_\pm - (k_{m_1} \delta q_1 \pm k_{m_2} \delta q_2) \sin \Psi_\pm$$
$$- \tfrac{1}{2} (k_{m_1}^2 \delta q_1^2 \pm 2 k_{m_1} k_{m_2} \delta q_1 \delta q_2 + k_{m_2}^2 \delta q_2^2) \cos \Psi_\pm \ldots \qquad \text{(F.53)}$$

where the first term to contribute in (6.81) with a non-zero average (with respect to φ_{m_1} and φ_{m_2}) is

$$\langle \delta q_1 \delta q_2 \cos \Psi_{\pm} \rangle_{\varphi_{m_1}, \varphi_{m_2}}$$

$$= \mp \kappa \int_0^{t_1} \int_0^{t_1'} \int_0^{t_2} \int_0^{t_2'} \langle \cos[\Psi_{\pm}(t_1, t_2) - \Psi_{\pm}(t_1'', t_2'')] \rangle \, dt_2'' \, dt_2' \, dt_1'' \, dt_1'$$

$$= \mp \kappa \int_0^{t_1} \int_{t_1''}^{t_1} \int_0^{t_2} \int_{t_2''}^{t_2} \langle \cos[\Psi_{\pm}(t_1, t_2) - \Psi_{\pm}(t_1'', t_2'')] \rangle \, dt_2' \, dt_2'' \, dt_1' \, dt_1''$$

$$= \mp \kappa \int_0^{t_1} \int_0^{t_2} \langle \cos(k_{m_1} q_{\hat{m}_1, \hat{m}_2}(t_1) \pm k_{m_2} q_{\hat{m}_1, \hat{m}_2}(t_2) - k_{m_1} q_{\hat{m}_1, \hat{m}_2}(t_1'')$$

$$\mp k_{m_2} q_{\hat{m}_1, \hat{m}_2}(t_2'') - \omega_{m_1}(t_1 - t_1'') \mp \omega_{m_2}(t_2 - t_2'')) \rangle$$

$$\times (t_2 - t_2'')(t_1 - t_1'') \, dt_2'' \, dt_1'' \tag{F.54}$$

where we write $\kappa \equiv k_{m_1} A_{m_1} k_{m_2} A_{m_2}/4$ and, as in section F.4, we approximate the statistics of $q_{\hat{m}_1, \hat{m}_2}(t)$ with respect to all phases by the statistics of $q_0 + p_0 t + \Delta q(t)$. Then, to lowest order in the δq_i, the integrand in (F.48) is

$$\Theta_{\pm}(t_1, t_2) = \sum_{m_1=1}^{M} \sum_{m_2=1}^{M} k_{m_1}^2 A_{m_1} k_{m_2}^2 A_{m_2} \langle \delta q_1 \delta q_2 \cos \Psi_{\pm} \rangle$$

$$= \Re \left\langle \prod_{j=1}^{2} \sum_{m_j=1}^{M} \frac{k_{m_j}^3 A_{m_j}^2}{2} \int_0^{t_j} e^{\epsilon_j i [\Omega_{m_j}(t_j - t_j'') + k_j \Delta q(t_j, t_j'')]} (t_j - t_j'') \, dt_j'' \right\rangle \tag{F.55}$$

with $\epsilon_1 = 1$ and $\epsilon_2 = \pm 1$ according to the subscript of Θ and $\Delta q(t, t') = \Delta q(t) - \Delta q(t - t')$. If we do not exclude $m_1 = m_2$ in the sum, we find the product of two expressions with a structure reminiscent of that of Δ_0 in (6.81), with the expectation computed for the product of two velocity and time integrals.

Before performing phase averages, we define, in the dense spectrum limit $\Delta v_m \to 0$,

$$\Xi(t, t') \equiv t' \int_{v_{\min}}^{v_{\max}} \gamma^3 e^{-ikt'(v-p) + ik \Delta q(t, t')} \, dv \tag{F.56}$$

where $\gamma = \gamma_D(v)$ and $k = k(v)$ are smooth functions of the phase velocity v. Then (F.55) reads

$$\Theta_{\pm}(t_1, t_2) = \Re \left\langle \prod_{j=1}^{2} \int_0^{t_j} \pi^{-1} \Xi_{\epsilon_j}(t_j, t_j') \, dt_j' \right\rangle \tag{F.57}$$

where the sign $\epsilon = \pm$ selects $\Xi_{+} = \Xi$ or its complex conjugate $\Xi_{-} = \Xi^*$. This Ξ vanishes for $t' = 0$. For $t' \to \infty$ it is an oscillating integral, which remains

O(1), and we note that in the exponent $\langle \Delta q(t, t') \rangle = 0$ and $\langle (\Delta q(t, t'))^2 \rangle = \frac{2}{3} D_{QL} t'^3 + 2 D_{QL} t'^2 (t - t')$. Then

$$
\Xi(t, t') = t' \gamma(p)^3 e^{ik(p)\Delta q(t,t')} \int_{v_{\min}}^{v_{\max}} \left(1 + \frac{\gamma(v)^3 - \gamma(p)^3}{\gamma(p)^3} \right)
$$

$$
\times e^{i(v-p)[-k(p)t' + \partial_p k(p)\Delta q(t,t')] + O(v-p)^2} \, dv \tag{F.58}
$$

$$
\simeq i \frac{t' \gamma(p)^3 e^{ik(p)\Delta q(t,t')}}{k(p)t' - \partial_v k(p)\Delta q(t, t')}
$$

$$
\times [e^{i(v-p)[-k(p)t' + \partial_p k(p)\Delta q(t,t')] + O(v-p)^2}]_{v_{\min}}^{v_{\max}}. \tag{F.59}
$$

Define $c_{\min} = (v_{\min} - p)k(p)\tau_c$ and $c_{\max}$ similarly, which are of the order of unity assuming $k(p)$ is of the order of $k(v_{\min})$ and $k(v_{\max})$. The expression between brackets in (F.59) is an oscillating function of time, with period of the order of $2\pi \tau_c / (1 - \beta)$, where $\beta = (\partial_p k(p))\Delta q(t, t') / (t' k(p))$ is of the order of unity (and is random as $\Delta q(t, t')$ is). Then

$$
\left| \int_0^t \Xi(t, t') \, dt' \right| \simeq c \frac{\tau_c \gamma(p)^3}{k(p)(1 - \beta)} \tag{F.60}
$$

where c is an O(1) numerical constant. Hence, for $\tau_c \ll \min(t_1, t_2)$,

$$
|\Theta_+(t_1, t_2)| \simeq c^2 \tau_c^2 \gamma_D^6 k^{-2} \tag{F.61}
$$

and we estimate

$$
\Delta_+ = \frac{1}{2} \int_0^t \int_0^t \Theta_+(t_1, t_2) \, dt_1 \, dt_2 \sim c_2 \gamma_D^6 k^{-2} \tau_c^2 t^2 \tag{F.62}
$$

where c_2 is an O(1) numerical constant.

This ends our estimate for Δ_+ but, according to (6.81), the estimate for Θ_- is the double sum restricted to $m_1 \neq m_2$. Therefore, we must subtract from (F.61) the diagonal contribution $(m_1 = m_2 = m)$, with $\delta q(t) = \delta q_1(t) = \delta q_2(t)$, which reads, assuming $0 \leq t_1 \leq t_2$,

$$
\Theta_0(t_1, t_2) = \Theta_0(t_2, t_1)
$$

$$
\equiv \sum_{m=1}^{M} \langle k_m^4 A_m^2 \delta q(t_1) \delta q(t_2) \cos \Psi_-(t_1, t_2) \rangle
$$

$$
\simeq \sum_{m=1}^{M} \kappa_m \int_0^{t_1} \int_0^{t_1'} \int_0^{t_2} \int_0^{t_2'} \langle \cos(k_m(q_{\dot{m}}(t_1) - q_{\dot{m}}(t_2)
$$

$$
- q_{\dot{m}}(t_1'') + q_{\dot{m}}(t_2'')) - \omega_m(t_1 - t_1'' - t_2 + t_2''))\rangle \, dt_2'' \, dt_2' \, dt_1'' \, dt_1'
$$

$$
\simeq \sum_{m=1}^{M} \kappa_m \int_0^{t_1} \int_{t_1''}^{t_1} \int_0^{t_2} \int_{t_2''}^{t_2} \langle \cos(k_m(\Delta q(t_1) - \Delta q(t_2))
$$

$$- \Delta q(t_1'') + \Delta q(t_2'')) + \omega_m (t_1 - t_1'' - t_2 + t_2''))) \, dt_2' \, dt_2'' \, dt_1' \, dt_1''$$

$$(F.63)$$

where $\kappa_m \equiv k_m^6 A_m^4 / 4$. We are led to expressions like (F.55). A similar approach yields

$$|\Theta_0(t_1, t_2)| \sim k^{-1} |\Delta v| \gamma_D^6 t_1^3 \qquad (F.64)$$

and

$$\Delta_- = \frac{1}{2} \int_0^t \int_0^t (\Theta_-(t_1, t_2) - \Theta_0(t_1, t_2)) \, dt_1 \, dt_2$$

$$\sim c_2 \gamma_D^6 k^{-2} \tau_c^2 t^2 + c_5 k^{-1} |\Delta v| \gamma_D^6 t^5 \qquad (F.65)$$

for some c_2, c_5 of the order of 1. The correction due to Θ_0 in Δ_- is of order Δ_0 times $k |\Delta v| \gamma_D^3 t^4 = \mathcal{B}^{-3} (t k |\Delta v|)^4 \ll 1$ for E large and $t \leq \tau_{QL}$, where τ_{QL} is given by (6.92). This correction is, thus, negligible in the time interval of interest here.

The estimate (F.62) for Δ_+ and the first term in the estimate (F.65) for Δ_- are identical and of order Δ_0 times $\gamma_D^3 \tau_c^2 t$. For this ratio to be small, τ_c must be small (i.e. $\tau_c \ll \gamma_D^{-1} |\ln \mathcal{B}|^{-1/2}$), which can be imposed by an appropriate choice of the spectrum of waves. Physically, this requires τ_c to be small enough for the chaotic motion to stay in the velocity domain of resonances of Hamiltonian (6.48). Then $\Delta_\pm$ is negligible with respect to Δ_0 over time τ_{QL}.

Appendix G

Vlasovian formulation of wave–plasma dynamics

In this appendix we outline the connection between our finite-N approach and the more traditional kinetic approach. Section G.1 formulates the kinetic theory of the self-consistent Hamiltonian in the limit $N \to \infty$. Section G.2 sketches the analogue of chapter 3, and section G.3 formulates the nonlinear Bernstein–Greene–Kruskal modes solving the kinetic model in the simplest case, the single-wave model. Section G.4 finally comments on the connections between the Vlasov-wave model and the more traditional Vlasov–Poisson description of plasmas.

The limit $N \to \infty$ in the particle dynamics leads to kinetic theory. While in finite-N dynamics the system state is represented by a single point $(\boldsymbol{p}, \boldsymbol{x})$ in phase space $\Lambda_N = \mathbb{R}^N \times S_L^N$, in kinetic theory the system state is described by the hierarchy of probability distribution functions $f_1(p_1, x_1)$ for one particle in $\Lambda_1 = \mathbb{R} \times S_L$, $f_2(p_1, p_2, x_1, x_2)$ jointly for two particles in $\Lambda_2 = \mathbb{R}^2 \times S_L^2, \ldots$. The kinetic approach also formally covers the finite-N models, if one takes a singular density f in (x, v)-space, corresponding to the so-called empirical distribution

$$f^N(x, v, t) = N^{-1} \sum_{r=1}^{N} \delta(x - x_r(t))\delta(v - v_r(t)) \tag{G.1}$$

where (x_r, v_r) are the position and velocity of particle r and δ is the Dirac distribution (B.12).

The motion of the particles under their interactions and the effect of fields implies the evolution of the distribution functions f_r ($r \geq 1$). In the Coulomb plasma case and in the wave–particle case, the interaction is of the mean-field type, i.e. the coupling constant goes to 0 as $N \to \infty$, so that the kinetic equation is of the simplest type, namely Vlasov.

Kinetic theory can be introduced from various physical viewpoints and its understanding may raise subtle questions. In particular, in kinetic theory, one distinguishes between the 'normal' particles, contributing to the plasma evolution, and 'test' particles, which sample the plasma but do not contribute to the fields: the finite-N approach bypasses this question, as all particles play equivalent roles, and shows why the test particle picture is valid. A second difficulty with kinetic theory is the interpretation of the distribution function $f(x, v, t)$ for a single experimental realization of the plasma: we use it to pick at random one particle out of N. In a different interpretation one considers many plasmas with the same macroscopic properties and the distribution function is used to perform an average among all these many-body realizations; but then understanding the unique electric field interacting with the many possible plasmas is quite uneasy and disturbing. See e.g. Spohn (1991) for a clear discussion of these concepts.

G.1 Vlasov-wave partial differential system

G.1.1 Formulation

Taking the limit $N \to \infty$ in the self-consistent dynamics (2.111), (2.112), (2.114) yields the evolution equation for f_1:

$$\partial_t f_1 + p \partial_x f_1 + \varepsilon \Re \left(\sum_{j=1}^{M} i\beta_j Z_j e^{ik_j x} \right) \partial_p f_1 = 0 \qquad (G.2)$$

coupled to M wave evolution equations

$$\dot{Z}_j = -i\omega_{j0} Z_j + i\varepsilon \beta_j k_j^{-1} N \int_{\Lambda_1} e^{-ik_j x} f_1(p, x, t) \, dp \, dx. \qquad (G.3)$$

Thanks to the mean-field nature of the interaction, these equations involve neither the two-particle function f_2 nor higher joint distributions f_r. This is a decisive advantage (in terms of simplicity) for the Vlasov equation over the Boltzmann equation of gas kinetic theory, where the short-range nature of the interparticle interactions leads to a hierarchy of equations, with $\partial_t f_r$ explicitly involving f_{r+1}.

From now on, we omit the subscript 1 to f. With the scaled coupling constants

$$\varepsilon'_j = \varepsilon \beta_j k_j^{-1} N^{1/2} \qquad (G.4)$$

and the rescaled wave variable

$$\zeta_j = N^{-1/2} Z_j \qquad (G.5)$$

this system reads:

$$\partial_t f + p \partial_x f + \Re \left(\sum_{j=1}^{M} i\varepsilon'_j k_j \zeta_j e^{ik_j x} \right) \partial_p f = 0 \qquad (G.6)$$

$$\dot{\zeta}_j = -i\omega_{j0}\zeta_j + i\varepsilon'_j \int_{\Lambda_1} e^{-ik_j x} f(p, x, t) \, dp \, dx. \qquad (G.7)$$

It is also convenient to introduce the rescaled intensity

$$\psi_j = I_j/N = \zeta_j \zeta_j^*/2. \tag{G.8}$$

The system (G.2), (G.3) preserves three constants of the motion, namely the number of particles, hence the normalization of f

$$\int f(x, p)\,dx\,dp = 1 \tag{G.9}$$

the rescaled total momentum

$$\int pf(x, p)\,dx\,dp + \sum_{j=1}^{M} k_j \psi_j = \sigma \tag{G.10}$$

and the rescaled total energy

$$\int \left(\frac{p^2}{2} - \Re\left(\sum_{j=1}^{M} \varepsilon'_j k_j \zeta_j e^{ik_j x}\right)\right) f(x, p)\,dx\,dp + \sum_{j=1}^{M} \omega_{j0}\psi_j = h. \tag{G.11}$$

Beside these 'mechanical' constants of the motion, the Vlasov-wave system preserves the usual integral (Casimir) invariants of Vlasov equations like

$$C_u[f] \equiv \int u[f(x, p)]\,dx\,dp \tag{G.12}$$

for any function $u : [0, \infty[\to \mathbb{R}$. The case $u[f] = f$ reduces to (G.9), the case $u[f] = f \ln(f/A)$ (with arbitrary constant A and $x \ln x = 0$ for $x = 0$) yields the classical total entropy, and $u[f] = f^2$ yields the square of the 2-norm of f (which plays an important role in numerical analysis too).

Exercise G.1. Show the conservation of (G.12). Compare your result with exercise E.6.

Exercise G.2. Like the Vlasov–Poisson system, the Vlasov-wave system has some symmetry properties. Consider $f(x, v, t)$ and $Z_j(t)$ solutions to (G.2), (G.3) and show that

  (i) (the space-time translation) $f(x + a, v, t + b)$ and $e^{ik_j a} Z_j(t + b)$ are solutions to this system,
  (ii) (Galileo transformation) $f(x + ut, v + u, t)$ and $Z_j(t)$ are solutions to the system with $\omega_j + k_j u$ substituted to ω_j.

In connection with exercise 2.6, how does the Lagrangian (2.72) transform under the corresponding changes of variables? What transformations correspond to time and space reversal?

The wave equation (G.3) looks simpler in the kinetic approach than in the N-particle approach, because it is linear with respect to its unknowns f and Z_j. The Vlasov equation (G.2) does not look much more complicated: it is linear with respect to f and quadratic with respect to the actual unknowns, since the only nonlinear term is $Z\partial_p f$. In contrast, the original $N + M$ degrees of freedom dynamics had the additional nonlinearities associated with the factors $e^{ik_j x_r}$: these nonlinearities have now disappeared because x and p are treated in the kinetic approach as arguments of the unknown $f(x, p)$. This 'exact linearization' has its price[1] : we no longer describe dynamics by ordinary differential equations in a finite-dimensional space of variables (x_r, p_r) but by a partial differential equation for the distribution f.

G.1.2   Derivation from N-body dynamics: kinetic limit

The Vlasov-wave system (G.2), (G.3) is the starting point for the kinetic analysis of the wave–plasma dynamics. A clear presentation of the Vlasov equation requires a sharp formulation of the limit $N \rightarrow \infty$. The natural mathematical framework of kinetic theory is functional analysis (as the relevant objects are functions f_r and we focus on limits such as $N \rightarrow \infty$), which is beyond the scope of this book. The following theorems (Firpo and Elskens 1998, Elskens and Firpo 1998) should reassure one that the limit $N \rightarrow \infty$ is regular for the plasma–wave model. We denote by $\mathcal{F}$ the space of distributions f_1 of interest and by d_{bL} a relevant distance on $\mathcal{F}$, by $\mathcal{Z} = \mathbb{C}$ the space of values for the complex amplitude ζ for each wave, and $\|\zeta\| = \sum_j |\zeta_j| = \sum_j \sqrt{2\psi_j}$.

Existence of dynamics and divergence of evolutions:
Given initial data $(f_0, \zeta(0))$, $(f_0', \zeta'(0)) \in \mathcal{F} \times \mathcal{Z}^M$, with $h_0 = h(f_0, \zeta(0))$ and $h_0' = h(f_0', \zeta'(0))$, the kinetic evolution system (G.2), (G.3) generates states $(f_t, \zeta(t))$ and $(f_t', \zeta'(t))$, respectively, from these data, for all times $t \geq 0$. Moreover,

$$d_{\mathrm{bL}}(f_t, f_t') + \|\zeta(t) - \zeta'(t)\| \leq e^{Ct}(d_{\mathrm{bL}}(f_0, f_0') + \|\zeta(0) - \zeta'(0)\|)$$
(G.13)

for some $C = C(\max(h_0, h_0')) < \infty$.

The constant C is bounded by an upper estimate for the largest Lyapunov exponent of the dynamics with finite N. This theorem implies the corollary:

Finite-time evolution commutes with kinetic limit:
Given

(i)   a distribution $f_0^\infty \in \mathcal{F}$ and a sequence of finite-N empirical distributions $f_0^N \in \mathcal{F}$ for particle initial data, such that

$$\lim_{N\to\infty} d_{\mathrm{bL}}(f_0^N, f_0^\infty) = 0$$
(G.14)

---

[1]   The same remark applies to the Liouville formulation of classical N-body dynamics.

(ii) initial waves $\zeta(0) \in \mathcal{Z}^M$ and
(iii) any time $T > 0$,

consider for all $0 \le t \le T$ the resulting distributions f_t^N and waves $\zeta^N(t)$ generated by $H_{\mathrm{sc}}^{N,M}$ and the kinetic solution $(f_t^\infty, \zeta^\infty(t))$. Then $\lim_{N \to \infty} d_{\mathrm{bL}}(f_t^N, f_t^\infty) = 0$ and $\lim_{N \to \infty} \zeta^N(t) = \zeta^\infty(t)$, uniformly on $[0, T]$.

The proof of these theorems closely follows the proof of analogous theorems for the Vlasov–Poisson system and for other self-consistent mean-field dynamics (Spohn, 1991).

Exercise G.3. Let $\Delta g(N, t) \equiv g^{(N)}(t) - g^\infty(t)$ be the discrepancy between the values of an observable g at time t for the kinetic model and for a 'nearby' finite-N evolution. Assume that $|\Delta g(N, t)| = e^{Ct} N^{-a} A$, with $C > 0$, $A > 0$ and with $a = 1/2$ (for a typical 'central limit' estimate) or $a = 1$ (for a 'smart' interpolation of f^∞ by f^N). Let $\varepsilon > 0$. Denote by $N_\varepsilon(T)$ the smallest value of N such that $|\Delta g(N, t)| \le \varepsilon$ for all $0 \le t \le T$, and by $\tau_\varepsilon(N)$ the largest time T such that $|\Delta g(N, t)| \le \varepsilon$ for all $0 \le t \le T$.

(i) Write the functions $N_\varepsilon(T)$ and $\tau_\varepsilon(N)$ explicitly and sketch their graphs for a few choices of ε. Is there a relation between both functions? How much do they change if one replaces ε by $\varepsilon' = \varepsilon/20$?
(ii) Show that $\lim_{t \to \infty} |\Delta g(N, t)| = \infty$ for any $N > 0$.
Show that $\lim_{N \to \infty} |\Delta g(N, t)| = 0$ for any $t > 0$.
(iii) Show that $\lim_{T \to \infty} N_\varepsilon(T) = \infty$ for any $\varepsilon > 0$.
Show that $\lim_{N \to \infty} \tau_\varepsilon(N) = \infty$ for any $\varepsilon > 0$.
(iv) Show that $\lim_{\varepsilon \to 0} N_\varepsilon(T) = \infty$ for any $T > 0$.

Mathematically, these results stress the difference between pointwise convergence and uniform convergence. Comment on them physically.

These theorems enable the following

Interpretation of f as a probability density for test particles:
Consider a solution $(f_t, \zeta(t))$ of the Vlasov-wave system. Consider $n > 0$ particles for which one draws inital data $(x_r(0), p_r(0))$ according to f_0 independently from each other. Consider a sequence of N-particle initial data, with empirical distributions $f_0^N \in \mathcal{F}$ such that $\lim_{N \to \infty} d_{\mathrm{bL}}(f_0^N, f_0) = 0$. Denote by $(x_r^N(t), p_r^N(t))$ the trajectories of the n 'test' particles for the evolutions $(f_t^N, \zeta^N(t))$ generated by the finite-N dynamics. Then the joint probability distribution of $(x_r^N(t), p_r^N(t))$ converges for $N \to \infty$ to the factorized probability density $\prod_{r=1}^n f_t^N$.

As the latter observation generates some confusions, note that, for finite N and $n = 1$, the analogous statement is: 'given N particles, draw from them at random a particle at time 0; then at time t the probability distribution for the test particle is $N^{-1} \sum_{l=1}^{N} \delta(x - x_l(t)) \delta(p - p_l(t))$', i.e. the test particle is at the same place and has the same momentum as one of the N particles. The factorization, which amounts to independence of the corresponding random variables, is usually called the propagation of molecular chaos. It follows from the fact that, in the limit $N \to \infty$, the probability to draw twice any of the N particles is $O(N^{-1})$.

Moreover, test particles do not affect the waves' evolution in the kinetic limit. Indeed, one may show the corollary:

Motion of test particles:
Consider a solution $(f_t, \zeta(t))$ of the Vlasov-wave system. Consider $n > 0$ 'test' particles for which one takes inital data $(x_r(0), p_r(0))$. Consider a sequence of N-particle initial data, with empirical distributions $f_0^N \in \mathcal{F}$ such that $\lim_{N \to \infty} d_{\mathrm{bL}}(f_0^N, f_0) = 0$ and for which the particles $1 \le r \le n$ have the prescribed initial data. Denote by $(x_r^N(t), p_r^N(t))$ the trajectories of the n test particles for the evolutions $(f_t^N, \zeta^N(t))$ generated by the finite-N dynamics, and denote by $(x_r^\infty(t), p_r^\infty(t))$ the trajectories of the test particles in the non-self-consistent dynamics

$$\dot{x}_r^\infty = p_r \qquad (G.15)$$

$$\dot{p}_r^\infty = \Re\left(\sum_{j=1}^{M} i\varepsilon_j' k_j \zeta_j(t) e^{ik_j x_r(t)} \right). \qquad (G.16)$$

Then, for any $T > 0$ and $1 \le r \le n$,

$$\lim_{N \to \infty} (x_r^N(t), p_r^N(t)) = (x_r^\infty(t), p_r^\infty(t)) \qquad (G.17)$$

uniformly on $[0, T]$.

The 'existence and divergence' limit theorem also ensures that test particles do not influence the motion of the other particles in the limit $N \to \infty$ either. All these features result directly from the mean-field nature of the wave–particle interaction.

A major distinction between the Vlasov and Fokker–Planck equations is stressed by exercises E.6 and G.1. The determinism in the Vlasov-wave system appears not only in the conservation of entropy but also in the fact that the motion of a test particle is deterministic in the kinetic limit, whereas the underlying picture for the Fokker–Planck equation involves infinitely many trajectories originating from the same initial data.

Exercise G.4. Extend the statements of this appendix from $[0, T]$ to $[-T, T]$. Does (G.13) hold with e^{Ct} or with $e^{C'|t|}$?

G.1.3   Fourier-transformed Vlasov-wave system

Given the periodic boundary conditions in x, we introduce the spatial Fourier components of f,

$$f_j(p, t) = \int_0^L e^{-ik_j x} f(x, p, t)\, dx \qquad (G.18)$$

so that

$$f(x, p, t) = \frac{1}{L} \sum_{k_j = -\infty}^{+\infty} e^{ik_j x} f_j(p, t) \qquad (G.19)$$

where the sum runs over all $k_j = n_j 2\pi/L$ with $n_j \in \mathbb{Z}$. For notational convenience, we redefine wave indices j so that $j = n_j$ until the end of this appendix. Then $k_{-j} = -k_j$. We let $\varepsilon'_{-j} = \varepsilon'_j$ for all j corresponding to Langmuir wavenumbers, and $\varepsilon'_j = 0$ for all other wavenumbers. As f is real valued, $f_j = f^*_{-j}$.

Exercise G.5. Show that system (G.6), (G.7) is equivalent to the system

$$\partial_t f_i + pik_i f_i + \sum_{j=1}^{\infty} \frac{i}{2} \varepsilon'_j k_j (\zeta_j \partial_p f_{i-j} - \zeta^*_j \partial_p f_{i+j}) = 0 \qquad (G.20)$$

$$\dot{\zeta}_j = -i\omega_{j0}\zeta_j + i\varepsilon'_j \int_{\mathbb{R}} f_j(p, t)\, dp. \qquad (G.21)$$

If one defines $\zeta_i = 0$ for non-Langmuir wavenumbers and $\zeta_{-j} = \zeta^*_j$, the first equation reads even more simply

$$\partial_t f_i + ik_i p f_i + \sum_{j=-\infty}^{\infty} \frac{i}{2} \varepsilon'_j k_j \zeta_j \partial_p f_{i-j} = 0. \qquad (G.22)$$

G.2   Zero-field solutions and their linear stability

The Vlasov-wave system admits many stationary solutions, with a distribution function $f(x, p, t) = f_0(p)$ and vanishing fields $Z_j(t) = 0$.

Exercise G.6. Check that any normalized positive function f_0 satisfies (G.2), (G.3) with $Z_j = 0$.

The small perturbations of a stationary solution $f_0(p)$, whose evolution determines its linear stability, obey the system of integro-differential equations for $j \neq 0$

$$\partial_t f_j(p, t) + ik_j p f_j(p, t) + \frac{i}{2} \varepsilon'_j k_j f'_0(p) \zeta_j = 0 \qquad (G.23)$$

$$\dot{\zeta}_j + i\omega_{j0}\zeta_j - i\varepsilon'_j \int_{\mathbb{R}} f_j(p, t)\, dp = 0. \qquad (G.24)$$

A salient feature of this system is that all Fourier components with different j are uncoupled. Indeed, the coupling term was quadratic with respect to (f_j, Z_i), and our reference state has the property that $f_j = 0$ and $Z_j = 0$ for all $j \neq 0$. This implies that, to first order, the wave with wavenumber k_j interacts only with perturbations of the particle distribution function with the same wavenumber.

A second remarkable feature of (G.23) is the absence of derivative of f_j with respect to p: the dependence of f_j on p occurs only through the function $f_0'(p)$ and algebraic operations. However, the integral in (G.24) couples Z_j to all momentum values p.

To solve system (G.23), (G.24) we follow the method of chapter 3. We first look for special solutions, depending on time as $e^{\alpha t}$, and then obtain a complete solution as a linear combination of such special solutions. See Case (1959) for a detailed similar treatment of the Vlasov–Poisson system.

G.2.1 Wavelike modes

As an ansatz, consider a solution such that

$$\zeta_j = z e^{\alpha t} \tag{G.25}$$
$$f_j(p, t) = f_j(p, 0) e^{\alpha t} \tag{G.26}$$

where α, z and $f_j(p, 0)$ are complex constants.

Then (G.23) is solved by

$$f_j(p, 0) = -\frac{i}{2} \varepsilon_j' k_j \frac{f_0'(p)}{\alpha + i k_j p} z. \tag{G.27}$$

Then (G.24) reduces to

$$\big(\sigma - \chi(\sigma, k_j)\big) z = 0 \tag{G.28}$$

where $\sigma = i\alpha$ and we define

$$\chi(\sigma, k_j) = \omega_{j0} - \frac{1}{2} \varepsilon_j'^2 k_j \int_{\mathbb{R}} \frac{f_0'(p)}{\sigma - k_j p} \, \mathrm{d}p. \tag{G.29}$$

Equation (G.28) admits a non-zero solution if and only if $\sigma = \chi(\sigma, k_j)$. This is the dispersion relation for the wavelike mode in the kinetic approach, and it coincides with the limit $N \to \infty$ of the dispersion relation discussed in chapter 3.

It is worth noting that, if $\alpha = \gamma - i\omega$ with a small $|\gamma|$, the most important perturbation to f_j occurs near the resonant velocity $p = \omega / k_j$ but also that $f_j(p, 0) \neq 0$ for all velocities p at which $f_0'(p) \neq 0$. Conversely, at velocities p such that $f_0'(p)$ vanishes, the wavelike solution does not perturb $f_j(p)$ (and these velocities do not contribute to the integral in (G.29)).

These wavelike solutions are parametrized by the values of $\alpha = -i\sigma$ for which the dispersion relation $\chi(\sigma, k_j) = \sigma$ is satisfied.

G.2.2 Ballistic modes

The finite-N model admitted also linear perturbations for which $\zeta_j(t) = 0$. The kinetic analogues of such perturbations satisfy (G.23) with the solution

$$f_j(p, t) = e^{-ik_j p t} f_j(p, 0) \tag{G.30}$$

and (G.24) with

$$\varepsilon'_j \int_{\mathbb{R}} e^{-ik_j p t} f_j(p, 0)\, \mathrm{d}p = 0 \tag{G.31}$$

for all time t. If $\varepsilon'_j = 0$, this condition is trivially satisfied for any choice of initial perturbation $f_j(p, 0)$. Such perturbations of the distribution function $f(p, x, t)$ correspond to the ballistic modes of the finite-N model of chapter 3.

But if $\varepsilon'_j \neq 0$, since $k_j t$ runs over all $\mathbb{R}$, condition (G.31) can be satisfied if and only if $f_j(p, 0) = 0$.

G.2.3 Complete solution

The general solution to the linearized Vlasov-wave system is a linear superposition of the wavelike and ballistic perturbations. Indeed, a general solution is represented after a Fourier transformation in space by the data (f_j, ζ_j), and each j-component is independent of the others. Thus, for components with $\varepsilon'_j = 0$, the perturbation is a ballistic mode, whereas for components with $\varepsilon'_j \neq 0$ the perturbation must be in the form

$$\zeta_j(t) = \int_\sigma \zeta_{j\sigma} e^{-i\sigma t}\, \mathrm{d}\sigma \tag{G.32}$$

$$f_j(p, t) = \frac{i\varepsilon'_j k_j}{2} \int_\sigma \frac{f'_0(p)}{\sigma - k_j p} \zeta_{j\sigma} e^{-i\sigma t}\, \mathrm{d}\sigma \tag{G.33}$$

where the integral $\int_\sigma$ runs over the solutions of the dispersion relation: the isolated eigenvalues contribute a sum to (G.32), (G.33), while the continuous spectrum of (generalized) roots on the whole real axis may contribute an integral indeed.

The evolution of the wavelike modes is controlled by the coefficients $\zeta_{j\sigma}$, which must be determined from initial data $\zeta_j(0)$ and $f_j(p, 0)$. Mathematically, the relation between $f_j(p, 0)/f'_0(p)$ and the coefficient $\zeta_{j\sigma}$ on the real σ-axis amounts to a Hilbert transform, which is invertible in the form

$$\frac{i}{2}\varepsilon'_j k_j \zeta_{j\sigma} = -\pi^{-2} \mathrm{PP} \int_{\mathbb{R}} \frac{f_j(p, 0)/f'_0(p)}{p - \sigma/k_j}\, \mathrm{d}p \tag{G.34}$$

where the principal part signals that analytic evaluations require us to cut the complex p and σ planes into half-planes (section B.3).

The similarity between (G.34) and the weight functions (3.96) which appear in the finite-N analysis may be used to develop further insight into both approaches to the linear stability analysis. The fact that the Hilbert transform involves complex half-planes rather than full planes deserves further comments. Mathematically, it enables one to construct solutions to the initial value problem by Laplace transforms, which the physicist relates to the causality principle. Analytically, it also relates to the identity (exercise A.2)

$$e^{-|at|} = \frac{a}{\pi} \int_{-\infty}^{\infty} \frac{e^{i\omega t}}{\omega^2 + a^2} \, d\omega \qquad (G.35)$$

for $a > 0$ and $t \in \mathbb{R}$, which implies that a function decaying purely exponentially for $t \geq 0$ is also exactly represented as a linear combination of purely oscillating functions of t. In turn, this implies that a single damping mode, associated with a pole with $\Re\alpha < 0$, i.e. $\Im\sigma < 0$, can be replaced in eigenmode decomposition, for the evolution for $t > 0$, by a continuum of poles on the marginal stability axis $\Re\alpha = 0$, i.e. $\Im\sigma = 0$. However, for finite N, the decomposition (3.99) stresses the importance of such a dual representation. In the kinetic formalism, the time reversal symmetry incorporated in (3.99) is broken, because the time-reversed dynamics involve the opposite complex α or σ half-plane.

As a result, the finite-N sums over many van Kampen-like eigenmodes thus turn into integrals with denominator $k_j p + i\alpha = k_j p - \omega_r + i\gamma_r$, where $\gamma \to 0$ as $\Delta p \to 0$: this leads to the 'causal $i\varepsilon$ prescriptions' ubiquitous in kinetic theoretical calculations.

G.3 Bernstein–Greene–Kruskal modes

Few exact solutions of the nonlinear Vlasov-wave system are available. Solutions with vanishing wave intensity are trivial. Non-trivial solutions can be found in the form of travelling modes, with a finite wave intensity for $M = 1$, such that f_1 is invariant in the frame moving at the mode velocity v. Interpreting the Vlasov equation as a transport equation for the distribution function f_1, these solutions read:

$$f_1(p, x, t) = g(H_1(p - v, x - \theta/k)) \qquad (G.36)$$
$$\zeta = z e^{-i(\theta + kvt)} \qquad (G.37)$$

where H_1 is the one-particle effective Hamiltonian in the comoving frame (exercise 1.3). Evolution equations (G.6), (G.7) then determine the compatibility integral equation linking the constant z and the function g. As only one equation constrains z and g, the choice of g is quite arbitrary.

Special solutions of this type were first introduced by Bernstein *et al* (1957) for the Vlasov–Poisson system in one space dimension. They may be generalized to various more complicated plasma models and they appear as the nonlinear 'continuations' of the linear solutions of section G.2, as discussed by Dorning and coworkers (see section 8.4) for the Vlasov–Poisson system.

G.4 Vlasov–Poisson system

The Vlasov-wave system focuses on the interaction of the near-resonant particles with the Langmuir wave. The traditional approach to plasma kinetic theory starts from the Vlasov–Poisson system coupling the particle distribution function $f(x, p, t)$ with the electric field $E(x, t)$.

Indeed one can also take the continuum limit $N \to \infty$ for the original Coulombian many-body model of chapter 1. There is, however, a special difficulty due to the singularity for $r \to 0$ of the Coulomb force (like $r^{-(d-1)}$) in space dimension $d > 1$, and even for $d = 1$ the force is discontinuous at $r = 0$. We refer to Spohn (1991) and references therein for a discussion of this difficulty. In this respect, our wave–particle model appears as a regularization of the original singular many-body model[2].

In the kinetic limit, we characterize the plasma by the particle density $f(x, v, t)$, normalized by $\int_{\Lambda_1} f \, dx \, dv = 1$. In this continuum limit, let $\mathcal{N}(x, v, \Delta x, \Delta v)$ be the number of particles in a small domain $\Delta x \Delta v$ near (x, v) and N be the total number of particles. Then $f(x, v) \simeq \mathcal{N}(x, v, \Delta x, \Delta v)/(N \Delta x \Delta v)$ and we assume this is a smooth enough function of (x, v) for all later purposes. Along with the limit $N \to \infty$, we let the contribution of each particle to the field go to zero, by letting its electric charge $q_* \to 0$, keeping the total charge in the system $N q_*$ constant. Simultaneously, we let the mass of each particle go to zero, in such a way that the acceleration it suffers from a given field does not change with q_*. Thus we keep m/q_* constant. In other words, in the limit $N \to \infty$, we 'fragment' each particle of mass m and charge q_* into many smaller particles, with masses m_r and charges q_{*r} such that if all these particles are at the same place $x_r = x$ with the same velocity $v_r = v$ they contribute to the field just as a single particle with charge $q_* = \sum_r q_{*r}$ and they undergo the same acceleration $\dot{v}_r = (q_{*r}/m_r)E(x_r) = (q_*/m)E(x) = \dot{v}$. This limit is easy if the pair interaction potential is smooth but the (non-smooth) Coulomb case allows for ill defined time evolutions of the Vlasov–Poisson dynamics (Majda *et al* 1994).

An important concept in plasma physics is that we focus on collective behaviours of the particles, which are not very sensitive to this 'fine-graining' process. In particular, the plasma frequency and the Debye length are invariant in this operation. Since the charge $q_{*r} \to 0$ in this process, the limit $N \to \infty$ is also called a weak-coupling limit. In this limit, one discards the granular nature of the plasma. Though, in nature, a plasma is made of a finite large number of particles, this continuum model is useful because the long-range nature of the Coulomb interaction ensures that a particle is acted upon by many partners, in contrast with neutral fluids where a particle is acted upon mainly by few neighbours.

---

[2] The wave–particle model is also a regularization of the Coulomb many-body system in that it only involves a finite number M of waves, obeying ordinary differential evolution equations. For the original Coulomb problem, the field involves an infinity of Fourier components and obeys an elliptic partial differential equation, namely the Poisson equation.

The kinetic limit associated with (1.10) reads as the conservation law for the number of particles,

$$\partial_t f + \partial_x(vf) + \partial_v\left(\frac{q_*}{m}E(x)f\right) = 0 \tag{G.38}$$

where the first term estimates how f changes in time, the second term estimates the balance of particles at (x, v) due to their convection at velocity v and the third term estimates the balance due to their acceleration by the field E. As $\partial_x(mv) + \partial_v(q_*E) = 0$ (the plasma is a Hamiltonian system), the conservation law (G.38) can also be written as a transport equation

$$\partial_t f + v\partial_x f + \frac{q_*}{m}E(x)\partial_v f = 0. \tag{G.39}$$

The field E in (G.39) is an auxiliary variable, obtained for finite N from the Poisson equation

$$\partial_x E = \sum_r \frac{q_*}{\epsilon_0}\left(\delta_L(x_r - x) - \delta_L\left(x_r - x + \frac{L}{2}\right)\right) \tag{G.40}$$

where the periodic Dirac distribution (see section B.3) takes into account the periodic boundary conditions and the second term in the sum represents the effect of the 'ghost' particle at $x_r + L/2$ associated with the particle at x_r (exercise 1.2)[3]. In the kinetic limit, (G.40) becomes

$$\partial_x E(x, t) = \int_{\mathbb{R}} \frac{q_*}{\epsilon_0}\left(f(x, v, t) - f\left(x + \frac{L}{2}, v, t\right)\right) dv \tag{G.41}$$

with $x + L/2$ understood modulo L. On Fourier expanding f by (G.18) and E similarly, one obtains a system similar to (G.20), (G.21), which admits the time-independent solution $f(x, p) = f_0(p)$, $E = 0$. This is the starting point for a stability analysis similar to the one discussed through this book.

System (G.39), (G.41) is the Vlasov–Poisson integro-differential system, on which much theoretical work in plasma physics relies. The derivation of the Vlasov equation from the many-body dynamics in the kinetic limit is discussed by Spohn (1991) and in the references therein.

We shall not discuss (G.39), (G.41) here, and we only insist that, in kinetic theory, one does not always require f to be a regular function but only a distribution, mainly[4] used to compute physical averages $\int fg\,dx\,dv$ of functions $g(x, v)$ over the particles. For physical reasons, one always requires f to be

---

[3] If the particles (e.g. electrons) are immersed in a uniform neutralizing background (e.g. ions), (G.40) reads as $\partial_x E = \sum_r (q_*/\epsilon_0)[\delta_L(x_r - x) - 1/L]$, and (G.41) becomes $\partial_x E = (q_*N)/(\epsilon_0 L)[\int_{\mathbb{R}} f\,dp - 1]$.

[4] The entropy density $s(x, t) = \int_{\mathbb{R}} f(x, v, t)\ln[f(x, v, t)/A]\,dv$ is a famous exception. More generally, all nonlinear functionals of f must be discussed separately, even for expressions as simple as $\int\int G[f(x, v, t)]\,dx\,dv$ with 'nice' functions G.

positive (i.e. $f(x, v) \geq 0$ for all (x, v) if f is a function) and normalized (i.e. $\int f \, dx \, dv = 1$).

Exercise G.7. Check that, for any positive normalized function $f_0(v)$, the particle distribution $f(x, v, t) = f_0(v)$ and the field $E(x, t) = 0$ satisfy the Vlasov–Poisson equations.

Spatially uniform solutions of the Vlasov–Poisson system have zero field, i.e. they are electrically neutral. They describe a state of the plasma which is preserved in time. The particles themselves move in the plasma, i.e. $f(x, v)$ does not vanish, in general, for $v \neq 0$ but the contribution of each particle (x_r, v_r) to the field is cancelled by the contribution of other particles with the same velocity and other positions.

Historically, the self-consistent model was first derived from the Vlasov–Poisson model as a reduced model with a finite number of relevant particles and waves (section 2.8).

References

(anonymous) National Institute of Standards and Technology *Digital Library of Mathematical Functions* http://dlmf.nist.gov/

Abdullaev S S 2000 Structure of motion near saddle points and chaotic transport in Hamiltonian systems *Phys. Rev.* E **62** 3508–28

Abraham R H and Shaw C D 1992 *Dynamics—The geometry of Behavior* (Redwood City, CA: Addison-Wesley)

Abramowitz M and Stegun I A 1970 *Handbook of Mathematical Functions* (New York: Dover)

Adam J C, Laval G and Pesme D 1979 Reconsideration of quasilinear theory *Phys. Rev. Lett.* **43** 1671–5

Akhiezer A I and Fainberg Ya B 1951 On the high-frequency oscillations of electronic plasma *Zh. Eksp. Teor. Fiz.* **21** 1262–9

Alekseev V M and Yakobson M V 1981 Symbolic dynamics and hyperbolic dynamical systems *Phys. Rep.* **75** 287–325

Alligood K T, Sauer T D and Yorke J A 1996 *Chaos—An Introduction to Dynamical Systems* (New York: Springer)

Antoni M, Elskens Y and Escande D F 1998 Explicit reduction of N-body dynamics to self-consistent particle–wave interaction *Phys. Plasmas* **5** 841–52

Antoni M and Ruffo S 1995 Clustering and relaxation in hamiltonian long-range dynamics *Phys. Rev.* E **52** 2361–74

Arnold V I and Avez A 1968 *Ergodic Problems of Classical Mechanics* (New York: Benjamin)

Arnold V I, Kozlov V V and Neishtadt A I 1988 *Mathematical Aspects of Classical and Celestial Mechanics (Encyclopaedia of Mathematical Sciences 3)* translated by I Iacob (Berlin: Springer)

Aubry S 1978 The new concept of transitions by breaking of analyticity in a crystallographic model *Solitons and Condensed Matter Physics* ed A R Bishop and T Schneider (Berlin: Springer) pp 264–77

Aubry S and Abramovici G 1990 Chaotic trajectories in the standard map: the concept of anti-integrability *Physica* D **43** 199–219

Balescu R 1960 Irreversible processes in ionized gases *Phys. Fluids* **3** 52–63

——1963 *Statistical Mechanics of Charged Particles* (London: Interscience)

——1975 *Equilibrium and Nonequilibrium Statistical Mechanics* (New York: Wiley–Interscience)

——1997 *Statistical Dynamics: Matter out of Equilibrium* (London: Imperial College Press)

Bender C M and Orszag S A 1978 *Advanced Mathematical Methods for Scientists and Engineers* (New York: McGraw-Hill)

Benettin G and Fassò F 1999 From Hamiltonian perturbation theory to symplectic integrators and back *Appl. Numer. Math.* **29** 73–87

Benettin G and Giorgilli A 1994 On the hamiltonian interpolation of near-to-the-identity symplectic mappings with applications to symplectic integration algorithms *J. Stat. Phys.* **74** 1117–43

Benettin G, Galgani L, Giorgilli A and Strelcyn J-M 1984 A proof of Kolmogorov's theorem on invariant tori using canonical transformations defined by the Lie method *Nuovo Cimento* B **79** 201–23

Bénisti D 1995 Validité de l'équation de diffusion en dynamique hamiltonienne *Thèse de Doctorat* Université de Provence (Marseille)

Bénisti D and Escande D F 1997 Origin of diffusion in hamiltonian dynamics *Phys. Plasmas* **4** 1576–81

——1998a Finite range of large perturbations in hamiltonian dynamics *J. Stat. Phys.* **92** 909–72

——1998b Nonstandard diffusion properties of the standard map *Phys. Rev. Lett.* **80** 4871–4

Berndtson J T, Heikkinen J A, Karttunen S J, Pättikangas T J and Salomaa R R E 1994 Analysis of velocity diffusion of electrons with Vlasov–Poisson simulations *Plasma Phys. Control. Fusion* **36** 57–71

Bernstein I B, Greene J M and Kruskal M D 1957 Exact nonlinear plasma oscillations *Phys. Rev.* **108** 546–50

Bertrand P, Ghizzo A, Karttunen S J, Pättikangas T J H, Salomaa R R E and Shoucri M 1992 Simulations of wave–particle interactions in stimulated Raman forward scattering in a magnetized plasma *Phys. Fluids* B **4** 3590–607

Bliokh P, Sinitsin V and Yaroshenko V 1995 *Dusty and Self-Gravitational Plasmas in Space* (Dordrecht: Kluwer)

Bohm D and Gross E P 1949a Theory of plasma oscillations A—Origin of medium-like behavior *Phys. Rev.* **75** 1851–64

——1949b Theory of plasma oscillations B—Excitation and damping of oscillations *Phys. Rev.* **75** 1864–76

Bohm D and Pines D 1951 A collective description of electron interactions I—Magnetic interactions *Phys. Rev.* **82** 625–34

Boutros-Ghali T and Dupree T H 1981 Theory of two-point correlation function in a Vlasov plasma *Phys. Fluids* **24** 1839–58

Bruhwiler D L and Cary J R 1989 Diffusion of particles in a slowly modulated wave *Physica* D **40** 265–82

Brydges D C and Martin Ph A 1999 Coulomb systems at low density: a review *J. Stat. Phys.* **96** 1163–330

Buchanan M and Dorning J 1994 Near-equilibrium multiple-wave plasma states *Phys. Rev.* E **50** 1465–78

Caglioti E and Maffei C 1998 Time asymptotics for solutions of Vlasov–Poisson equation in a circle *J. Stat. Phys.* **92** 301–23

Cary J R and Doxas I 1993 An explicit symplectic integration scheme for plasma simulation *J. Comput. Phys.* **107** 98–104

Cary J R, Doxas I, Escande D F and Verga A D 1992 Enhancement of velocity diffusion in longitudinal plasma turbulence *Phys. Fluids* B **4** 2062–9

Cary J R, Escande D F and Tennyson J L 1986 Adiabatic-invariant change due to separatrix crossing *Phys. Rev.* A **56** 4256–75

Cary J R, Escande D F and Verga A D 1990 Non quasilinear diffusion far from the chaotic threshold *Phys. Rev. Lett.* **65** 3132–5

Cary J R and Skodje R T 1989 Phase change between separatrix crossings *Physica* D **36** 287–316

Case K M 1959 Plasma oscillations *Ann. Phys., NY* **7** 349–64

Celletti A and Chierchia L 1988 Construction of analytic KAM surfaces and effective stability bounds *Commun. Math. Phys.* **118** 119–61

Chandre C and MacKay R S 2000 Approximate renormalization with codimension-one fixed point for the break-up of some three-frequency tori *Phys. Lett.* A **275** 394–400

Chandre C and Jauslin H R 2000 Strange attractor for the renormalization flow for invariant tori of Hamiltonian systems with two generic frequencies *Phys. Rev.* E **61** 1320–8

Channell P J and Scovel C 1990 Symplectic integration of hamiltonian systems *Nonlinearity* **3** 231–59

Chen F F 1984 *Introduction to Plasma Physics and Controlled Fusion* (New York: Plenum)

Chenciner A, Gerver J, Montgomery R and Simó C 2001 Simple choreographic motions of *N*-bodies, a preliminary study http://www.maia.ub.es/dsg

Chernikov A, Natenzon M Ya, Petrovichev B A, Sagdeev R Z and Zaslavsky G M 1987 Some pecularities of stochastic layers and stochastic web formation *Phys. Lett.* A **122** 39–46

——1988 Strong changes of adiabatic invariants, KAM-tori and web-tori *Phys. Lett.* A **129** 377–80

Chirikov B V 1979 A universal instability of many-dimensional oscillator systems *Phys. Rep.* **52** 263–379

Choquard Ph, Kunz H, Martin Ph A and Navet M 1981 One-dimensional Coulomb systems *Physics in One Dimension* ed J Bernasconi and T Schneider (Berlin: Springer) pp 335–50

Conte R (ed) 1999 *The Painlevé Property—One Century Later* (Heidelberg: Springer)

Crawford J D 1994 Universal trapping scaling on the unstable manifold for a collisionless electrostatic mode *Phys. Rev. Lett.* **73** 656–9

Crawford J D and Jayaraman A 1996 Nonlinear saturation of an electrostatic wave: mobile ions modify trapping scaling *Phys. Rev. Lett.* **77** 3549–52

——1999 First principles justification of a 'single wave model' for electrostatic instabilities *Phys. Plasmas* **6** 666–73

Croquette V and Poitou C 1981 Cascade of period doubling bifurcations and large scale stochasticity in the motions of a compass *J. Physique* **42** L537–9

Dawson J M 1960 Plasma oscillations of a large number of electron beams *Phys. Rev.* **118** 381–9

del-Castillo-Negrete D 2000 Self-consistent chaotic transport in fluids and plasmas *Chaos* **10** 75–88

del-Castillo-Negrete D and Firpo M-C 2002 Coherent structures and self-consistent transport in a mean field hamiltonian model *Chaos* **12** 496–507

del-Castillo-Negrete D, Greene J M and Morrison P J 1997 Renormalization and transition to chaos in area preserving nontwist maps *Physica* D **100** 311–29

Doveil F 1981 Stochastic plasma heating by a large-amplitude stationary wave *Phys. Rev. Lett.* **46** 532–4

Doveil F and Escande D F 1982 Fractal diagrams for non integrable hamiltonians *Phys. Lett.* A **90** 226–30

Doveil F, Firpo M-C, Elskens Y, Guyomarc'h D, Poleni M and Bertrand P 2001 Trapping oscillations, discrete particle effects and kinetic theory of collisionless plasma *Phys. Lett.* A **284** 279–85

Doxas I and Cary J R 1997 Numerical observation of turbulence enhanced growth rates *Phys. Plasmas* **4** 2508–18

Drummond W E and Pines D 1962 Nonlinear stability of plasma oscillations *Nucl. Fusion Suppl.* **3** 1049–57

Dupree T H 1966 A perturbation theory for strong plasma turbulence *Phys. Fluids* **9** 1773–82

Earn D J and Tremaine S 1992 Exact numerical studies of hamiltonian maps: iterating without roundoff error *Physica* D **56** 1–22

Elskens Y 2001 Finite-N dynamics admit no traveling-wave solutions for the hamiltonian XY model and single-wave collisionless plasma model *ESAIM Proc.* **10** 211–15 (Coquel F and Cordier S (ed) http://www.emath.fr/Maths/Proc)

Elskens Y and Antoni M 1997 Equilibrium statistical mechanics of one-dimensional hamiltonian systems with long-range force *Phys. Rev.* E **55** 6575–81

Elskens Y and Escande D F 1991 Slowly pulsating separatrices sweep homoclinic tangles where islands must be small: an extension of classical adiabatic theory *Nonlinearity* **4** 615–67

——1993 Infinite resonance overlap: a natural limit of hamiltonian chaos *Physica* D **62** 66–74

Elskens Y and Firpo M-C 1998 Kinetic theory and large-N limit for wave–particle self-consistent interaction *Phys. Scr.* T **75** 169–72

Escande D F 1982 Renormalization for stochastic layers *Physica* D **6** 119–25

——1985 Stochasticity in classical hamiltonian systems: universal aspects *Phys. Rep.* **121** 165–261

——1989 Description of Landau damping and weak Langmuir turbulence through microscopic dynamics *Nonlinear World* vol 2, ed V G Bar'yakhtar *et al* (Singapore: World Scientific) pp 817–36

——1991 Large scale structures in kinetic plasma turbulence *Large Scale Structures in Nonlinear Physics (Lecture Notes in Physics 392)* ed J D Fournier and P L Sulem (Berlin: Springer) pp 73–104

Escande D F and Doveil F 1981a Renormalization method for the onset of stochasticity in a hamiltonian system *Phys. Lett.* A **83** 307–10

——1981b Renormalization method for computing the threshold of large-scale stochastic instability in two degrees of freedom hamiltonian systems *J. Stat. Phys.* **26** 257–84

Escande D F and Elskens Y 2001 Quasilinear diffusion for the chaotic motion of a particle in a set of longitudinal waves *Acta Phys. Pol.* B **33** 1073–84

Escande D F, Kantz H, Livi R and Ruffo S 1994 Self-consistent check of the validity of Gibbs calculus using dynamical variables *J. Stat. Phys.* **76** 605–26

Escande D F, Mohamed-Benkadda M S and Doveil F 1984 Threshold of global stochasticity *Phys. Lett.* A **101** 309–13

Escande D F, Paccagnella R, Cappello S, Marchetto C and D'Angelo F 2000 Chaos healing by separatrix disappearance and quasisingle helicity states of the reversed field pinch *Phys. Rev. Lett.* **85** 3169–72

Escande D F, Zekri S and Elskens Y 1996 Intuitive and rigorous microscopic description of spontaneous emission and Landau damping of Langmuir waves through classical mechanics *Phys. Plasmas* **3** 3534–9

Feller W 1970 *An Introduction to Probability Theory and its Applications* vols I and II (New York: Wiley)

Firpo M-C 1999 Etude dynamique et statistique de l'interaction onde–particule *Thèse de Doctorat* Université de Provence (Marseille)

Firpo M-C and Doveil F 2002 Velocity width of the resonant domain in wave–particle interaction *Phys. Rev.* E **65** 016411 (8 pages)

Firpo M-C, Doveil F, Elskens Y, Bertrand P, Poleni M and Guyomarc'h D 2001 Long-time discrete particle effects versus kinetic theory in the self-consistent single-wave model *Phys. Rev.* E **64** 026407 (10 pages)

Firpo M-C and Elskens Y 1998 Kinetic limit of *N*-body description of wave–particle self-consistent interaction *J. Stat. Phys.* **93** 193–209

——2000 Phase transition in the collisionless damping regime for wave–particle interaction *Phys. Rev. Lett.* **84** 3318–21

Fried B D, Liu C S, Means R W and Sagdeev R Z 1971 Nonlinear evolution of an unstable electrostatic wave *Plasma Physics Group Report* PPG-93, University of California, Los Angeles

Galeev A A, Sagdeev R Z, Shapiro V D and Shevchenko V I 1980 Is renormalization necessary in the quasilinear theory of Langmuir oscillations? *Sov. Phys.–JETP* **52** 1095–9

Gallavotti G 1999 *Statistical Mechanics: A Short Treatise* (New York: Springer)

Gelfreich V G and Lazutkin V F 2001 Splitting of separatrices: perturbation theory and exponential smallness *Usp. Mat. Nauk* **56** 79–142 (Engl. transl. *Russ. Math. Surv.* **56** 499–558)

Goldstein H 1980 *Classical Mechanics* (Reading, MA: Addison-Wesley)

Gosson M de 2001 The symplectic camel and phase space quantization *J. Phys. A: Math. Gen.* **34** 10 085–96

Gould R W, O'Neil T M and Malmberg J H 1967 Plasma wave echo *Phys. Rev. Lett.* **19** 219–22

Greene J M 1979 A method for determining a stochastic transition *J. Math. Phys.* **20** 1183–201

Gromov M 1985 Pseudoholomorphic curves in symplectic manifolds *Invent. Math.* **82** 307–47

Guckenheimer J and Holmes P 1983 *Nonlinear Oscillations, Dynamical Systems, and Bifurcations of Vector Fields* (New York: Springer)

Guo Y and Strauss W 1995 Instability of periodic BGK equilibria *Commun. Pure Appl. Math.* **48** 861–94

Gutzwiller M C 1991 *Chaos in Classical and Quantum Mechanics* (New York: Springer)

Guyomarc'h D, Doveil F, Elskens Y and Fanelli D 1996 Warm beam–plasma instability beyond saturation *Transport, Chaos and Plasma Physics* vol 2, ed S Benkadda, F Doveil and Y Elskens (Singapore: World Scientific) pp 406–10

Harris E G 1969 Classical plasma phenomena from a quantum mechanical viewpoint *Adv. Plasma Phys.* vol 3, ed A Simon and W B Thompson (New York: Wiley–Interscience) pp 157–248

Hartmann D A, Driscoll C F, O'Neil T M and Shapiro V D 1995 Measurements of the weak warm beam instability *Phys. Plasmas* **2** 654–77

Helander P and Kjellberg L 1994 Simulation of nonquasilinear diffusion *Phys. Plasmas* **1** 210–12

Henrard J 1993 The adiabatic invariant in classical mechanics *Dynamics Reported* vol 2, ed C K R T Jones, U Kirchgraber and H O Walther (Berlin: Springer) pp 117–235

Hirsch M W and Smale S 1974 *Differential Equations, Dynamical Systems, and Linear Algebra* (New York: Academic)

Holloway J P and Dorning J J 1989 Undamped longitudinal plasma waves *Phys. Lett.* A **138** 279–84

Iooss G and Joseph D D 1990 *Elementary Stability and Bifurcation Theory* (New York: Springer)

Ishihara O, Xia H and Watanabe S 1993 Long-time diffusion in plasma turbulence with broad uniform spectrum *Phys. Fluids* B **5** 2786–92

Janssen P A E M and Rasmussen J J 1981 Limit cycle behavior of the bump-on-tail instability *Phys. Fluids* **24** 268–73

Kaufman A N 1972 Reformulation of quasilinear theory *J. Plasma Phys.* **8** 1–5

Khinchin A Ya 1964 *Continued Fractions* (Chicago, IL: University of Chicago Press)

Koch H 2002 A renormalization group fixed point associated with the breakup of golden invariant tori, http://www.ma.utexas.edu/mp_arc02-175

Krivoruchko S M, Bashko V A and Bakai A S 1981 Experimental investigations of correlation phenomena in the relaxation of velocity-spread beam in a plasma *Sov. Phys.–JETP* **53** 292–8

Kupersztych J 1985 Electron acceleration in high-frequency longitudinal waves, Doppler-shifted ponderomotive forces, and Landau damping *Phys. Rev. Lett.* **54** 1385–7

Kuramoto Y 1984 *Chemical Oscillations, Waves and Turbulence* (Berlin: Springer)

Lancellotti C and Dorning J J 1999 Time-asymptotic traveling-wave solutions to the nonlinear Vlasov–Poisson–Ampère equations *J. Math. Phys.* **40** 3895–917

——2000 Time-asymptotic wave propagation in collisionless plasmas *Preprint* http://www.rci.rutgers.edu/~carlo/pdfs/pre.pdf

Landau L D 1937 The transport equation in the case of Coulomb interactions *Zh. Eksp. Teor. Fiz.* **7** 203–9 (reprinted 1965 *Collected Papers of Landau* ed D ter Haar (Oxford: Pergamon) pp 163–70)

——1946 On the vibrations of the electronic plasma *Zh. Eksp. Teor. Fiz.* **16** 574–86 (reprinted 1965 *Collected Papers of Landau* ed D ter Haar (Oxford: Pergamon) pp 445–60)

Laskar J 1993 Frequency analysis for multi-dimensional systems: global dynamics and diffusion *Physica* D **67** 257–81

——1999 Introduction to frequency map analysis *Hamiltonian Systems with Three or More Degrees of Freedom (NATO Advanced Study Institute, S'Agaró-Barcelona, June 1995)* ed C Simó (Dordrecht: Kluwer) pp 134–50

Laval G and Pesme D 1983a Breakdown of quasilinear theory for incoherent 1-D Langmuir waves *Phys. Fluids* **26** 52–65

——1983b Inconsistency of quasilinear theory *Phys. Fluids* **26** 66–8

——1984 Self-consistency effects in quasilinear theory: a model for turbulent trapping *Phys. Rev. Lett.* **53** 270–3

——1999 Controversies about quasilinear theory *Plasma Phys. Control. Fusion* **41** A239–46

Lazutkin V 1984 Splitting of separatrices for the Chirikov standard map *Preprint* VINITI 6372/84

Lenard A 1960 On Bogoliubov's kinetic equation for a spatially homogeneous plasma *Ann. Phys.* **3** 390–400

——1961 Exact statistical mechanics of a one-dimensional system with Coulomb forces *J. Math. Phys.* **2** 682–93

——1963 —III. Statistics of the electric field *J. Math. Phys.* **4** 533–43

Levin M B, Lyubarskiĭ, Onishchenko I N, Shapiro V D and Shevchenko V I 1972 Contribution to the nonlinear theory of kinetic instability of an electron beam in plasma *Zh. Eksp. Teor. Fiz.* **62** 1725–32 (Engl. transl. *Sov. Phys.–JETP* **35** 898–901)

Liang Y-M and Diamond P H 1993a Weak turbulence theory of Langmuir waves: a reconsideration of the validity of quasilinear theory *Comment. Plasma Phys. Control. Fusion* **15** 139–49

——1993b Revisiting the validity of quasilinear theory *Phys. Fluids* B **5** 4333–40

Lichtenberg A J and Lieberman M A 1983 *Regular and Stochastic Motion* (New York: Springer)

Lifshitz E M and Pitaevskiĭ L P 1981 *Landau and Lifshitz's Course of Theoretical Physics Vol 10: Physical Kinetics* translated by J B Sykes and R N Franklin (Oxford: Pergamon)

Lochak P and Meunier C 1988 *Multiphase Averaging for Classical Systems* translated by H S Dumas (New York: Springer)

Majda A J, Majda G and Zheng Y 1994 Concentrations in the one-dimensional Vlasov–Poisson equations I: Temporal development and non-unique weak solutions in the single component case *Physica* D **74** 268–300

Malmberg J H and Wharton C B 1964 Collisionless damping of electrostatic plasma waves *Phys. Rev. Lett.* **13** 184–6

Mattis D C (ed) 1993 *The Many-Body Problem* (Singapore: World Scientific)

MacKay R S 1982 Renormalization in area-preserving maps *PhD Thesis* Princeton University (reprinted 1992 (Singapore: World Scientific))

——1983 A renormalisation approach to invariant circles in area-preserving maps *Physica* D **7** 283–300

——1988 Exact results for an approximate renormalisation scheme and some predictions for the breakup of invariant tori *Physica* D **33** 240–65

Manfredi G and Bertrand P 2000 Stability of Bernstein–Greene–Kruskal modes *Phys. Plasmas* **7** 2425–31

Mazitov R K 1965 On the damping of plasma waves *Prikl. Mekh. Tekhn. Fiz.* **1** 27–31

Mehr A and Escande D F 1984 Destruction of KAM tori in hamiltonian systems: link with the destabilization of nearby cycles and calculation of residues *Physica* D **13** 302–38

Meiss J D 1992 Symplectic maps, variational principles, and transport *Rev. Mod. Phys.* **64** 795–848

Melnikov V K 1963 On the stability of the center for time-periodic perturbations *Trans. Mosk. Mat. O-va* **12** 3–52 (Engl. transl. 1965 *Trans. Moscow Math. Soc.* **12** 1–56)

Menyuk C R 1985 Particle motion in the field of a modulated wave *Phys. Rev.* A **31** 3282–90

Montgomery R 2001 A new solution to the three-body problem *Not. Am. Math. Soc.* **48** 471–81

Mynick H E and Kaufman A N 1978 Soluble theory of nonlinear beam–plasma interaction *Phys. Fluids* **21** 653–63

Neishtadt A I, Sidorenko V V and Treschev D V 1997 Stable periodic motion in the problem on passage through a separatrix *Chaos* **7** 1–11

Nicholson D R 1983 *Introduction to Plasma Theory* (New York: Wiley)

O'Neil T M 1965 Collisionless damping of nonlinear plasma oscillations *Phys. Fluids* **8** 2255–62

O'Neil T M and Malmberg J H 1968 Transition of the dispersion roots from beam-type to Landau-type solutions *Phys. Fluids* **11** 1754–60

O'Neil T M and Winfrey J H 1972 Nonlinear interaction of a small cold beam and a plasma. II *Phys. Fluids* **15** 1514–22

O'Neil T M, Winfrey J H and Malmberg J H 1971 Nonlinear interaction of a small cold beam and a plasma *Phys. Fluids* **14** 1204–12

Onishchenko I N, Linetskiĭ A R, Matsiborko N G, Shapiro V D and Shevchenko V I 1970 Contribution to the nonlinear theory of excitation of a monochromatic plasma wave by an electron beam *ZhETF Pis. Red.* **12** 407–11 (Engl. transl. *JETP Lett.* **12** 281–5)

Ottino J M 1989 *The Kinematics of Mixing: Stretching, Chaos, and Transport* (Cambridge: Cambridge University Press)

Penrose O 1960 Electrostatic instability of a non-maxwellian plasma *Phys. Fluids* **3** 258–65

Percival I C 1979 Variational principles for invariant tori and cantori *Nonlinear Dynamics and the Beam–Beam Interaction (Conf. Proc. 57)* ed M Month and J C Herrera (New York: American Institute of Physics) pp 302–10

Percival I and Richards D 1982 *Introduction to Dynamics* (Cambridge: Cambridge University Press)

Pesme D 1994 The Fokker–Planck description in the context of the quasilinear theory *Phys. Scr.* T **50** 7–14

Pesme D and DuBois D F 1982 Nonlinear problems: present and future *Los Alamos Report* LA-UR-81-2234, ed A Bishop (Amsterdam: North-Holland)

Pesme D, Mora P, Verga A D and Laval G 1993 The contribution of statistical theory to the understanding of dynamical systems behavior. Example of deviation from the quasilinear theory. Description of plasma turbulence *Spatio-Temporal Analysis for Resolving Plasma Turbulence (ESA WPP-047 Proceedings)* ed F Lefeuvre, D Le Quéau and A Roux (Paris: European Space Agency) pp 19–27

Peyrard M and Aubry S 1983 Critical behaviour at the transition by breaking of analyticity in the discrete Frenkel–Kontorova model *J. Phys. C: Solid State Phys.* **16** 1593–608

Pines D and Bohm D 1952 A collective description of electron interactions II—Collective *vs* individual aspects of the interaction *Phys. Rev.* **85** 338–53

Prager S 1962 The one-dimensional plasma *Advances in Chemical Physics* vol 4, ed I Prigogine (New York: Wiley–Interscience) pp 201–24

Ragot B R 1998 Nonlinear particle dynamics in a broadband turbulence wave spectrum *J. Plasma Phys.* **60** 299–329

Rechester A B, Rosenbluth M N and White R B 1979 Calculation of the Kolmogorov entropy for motion along a stochastic magnetic field *Phys. Rev. Lett.* **42** 1247–50 (erratum *Phys. Rev. Lett.* **45** 851)

Rechester A B and White R B 1980 Calculation of turbulent diffusion for the Chirikov–Taylor model *Phys. Rev. Lett.* **44** 1586–9

Roberson C and Gentle K W 1971 Experimental test of the quasilinear theory of the gentle bump instability *Phys. Fluids* **14** 2462–9

Rom-Kedar V 1994 The topological approximation method *Transport, Chaos and Plasma Physics* vol 2, ed S Benkadda, F Doveil and Y Elskens (Singapore: World Scientific) pp 39–57

Romanov Yu A and Filippov G F 1961 The interaction of fast electron beams with longitudinal plasma waves *Zh. Eksp. Theor. Phys.* **40** 123–32 (Engl. transl. *Sov. Phys.–JETP* **13** 87–92)

Rouet J L and Feix M R 1991 Relaxation for a one-dimensional plasma: test particles versus global distribution behaviour *Phys. Fluids* B **3** 1830–4

——1996 Computer experiments on dynamical cloud and space time fluctuations in one-dimensional meta-equilibrium plasmas *Phys. Plasmas* **3** 2538–45

Ruelle D 1989 *Chaotic Evolution and Strange Attractors: The Statistical Analysis of Time Series for Deterministic Nonlinear Systems (Lezioni Lincee)* (Cambridge: Cambridge University Press)

Ryutov D D 1975 Analog of Landau damping the problem of sound-wave propagation in a liquid with gas bubbles *JETP Lett.* **22** 215–17

——1999 Landau damping: half a century with the great discovery (Proc. EPS Conf. Control. Fusion Plasma Phys. and Int. Cong. Plasma Phys. Prague 1998) *Plasma Phys. Control. Fusion* **41** (Suppl. 3A) A1–12

Schmidt G 1980 Stochasticity and fixed-point transitions *Phys. Rev.* A **22** 2849–54

Schmidt G and Bialek J 1982 Fractal diagrams for hamiltonian stochasticity *Physica* D **5** 397–404

Schutz B 1980 *Geometrical Methods of Mathematical Physics* (Cambridge: Cambridge University Press)

Self S A, Shoucri M M and Crawford F W 1971 Growth rates and stability limits for beam–plasma interaction *J. Appl. Phys.* **42** 704–13

Shapiro V D and Sagdeev R Z 1997 Nonlinear wave–particle interaction and conditions for the applicability of quasilinear theory *Phys. Rep.* **283** 49–71

Simon A and Rosenbluth M N 1976 Single-mode saturation of the bump-on-tail instability: immobile ions *Phys. Fluids* **19** 1567–80

Sonnendrücker E, Roche J, Bertrand P and Ghizzo A 1999 The semi-lagrangian method for the numerical resolution of Vlasov equations *J. Comput. Phys.* **149** 201–20

Spohn H 1991 *Large Scale Dynamics of Interacting Particles* (Berlin: Springer)

Stark J 1988 Unstable manifolds for the MacKay approximate renormalisation *Physica* D **34** 208–22

Stubbe P and Sukhorukov A I 1999 On the physics of Landau damping *Phys. Plasmas* **6** 2976–88

Tennyson J L, Cary J R and Escande D F 1986 Change of the adiabatic invariant due to separatrix crossing *Phys. Rev. Lett.* **56** 2117–20

Tennyson J L, Meiss J D and Morrison P J 1994 Self-consistent chaos in the beam–plasma instability *Physica* D **71** 1–17

Theilhaber K, Laval G and Pesme D 1987 Numerical simulations of turbulent trapping in the weak beam–plasma instability *Phys. Fluids* **30** 3129–49

Treschev D 1998 Width of stochastic layers in near-integrable two-dimensional symplectic maps *Physica* D **116** 21–43

Tsunoda S I, Doveil F and Malmberg J H 1991 Experimental test of quasilinear theory *Phys. Fluids* B **3** 2747–57

van Kampen N G 1955 On the theory of stationary waves in plasmas *Physica* **21** 949–63

——1957 The dispersion equation for plasma waves *Physica* **23** 641–50

——1992 *Stochastic Processes in Physics and Chemistry* (Amsterdam: Elsevier)

van Kampen N G and Felderhof B U 1967 *Theoretical Methods in Plasma Physics* (Amsterdam: North-Holland)

Vedenov A A, Velikhov E P and Sagdeev R Z 1962 Quasilinear theory of plasma oscillations *Nucl. Fusion Suppl.* **2** 465–75

Whittaker E T 1964 *A Treatise on the Analytical Dynamics of Particles and Rigid Bodies* (Cambridge: Cambridge University Press)

Whittaker E T and Watson G N 1980 *A Course of Modern Analysis* (Cambridge: Cambridge University Press)

Wiggins S 1992 *Chaotic Transport in Dynamical Systems* (New York: Springer)

Index